Communications and Control Engineering

Published titles include:

Stability and Stabilization of Infinite Dimensional Systems with Applications
Zheng-Hua Luo, Bao-Zhu Guo and Omer Morgul

Nonsmooth Mechanics (Second edition)
Bernard Brogliato

Nonlinear Control Systems II
Alberto Isidori

L_2-Gain and Passivity Techniques in Nonlinear Control
Arjan van der Schaft

Control of Linear Systems with Regulation and Input Constraints
Ali Saberi, Anton A. Stoorvogel and Peddapullaiah Sannuti

Robust and H∞ Control
Ben M. Chen

Computer Controlled Systems
Efim N. Rosenwasser and Bernhard P. Lampe

Dissipative Systems Analysis and Control
Rogelio Lozano, Bernard Brogliato, Olav Egeland and Bernhard Maschke

Control of Complex and Uncertain Systems
Stanislav V. Emelyanov and Sergey K. Korovin

Robust Control Design Using H∞ Methods
Ian R. Petersen, Valery A. Ugrinovski and Andrey V. Savkin

Model Reduction for Control System Design
Goro Obinata and Brian D.O. Anderson

Control Theory for Linear Systems
Harry L. Trentelman, Anton Stoorvogel and Malo Hautus

Functional Adaptive Control
Simon G. Fabri and Visakan Kadirkamanathan

Positive 1D and 2D Systems
Tadeusz Kaczorek

Identification and Control Using Volterra Models
F.J. Doyle III, R.K. Pearson and B.A. Ogunnaike

Non-linear Control for Underactuated Mechanical Systems
Isabelle Fantoni and Rogelio Lozano

Robust Control (Second edition)
Jürgen Ackermann

Flow Control by Feedback
Ole Morten Aamo and Miroslav Krstić

Learning and Generalization (Second edition)
Mathukumalli Vidyasagar

Constrained Control and Estimation
Graham C. Goodwin, María M. Seron and José A. De Doná

Randomized Algorithms for Analysis and Control of Uncertain Systems
Roberto Tempo, Giuseppe Calafiore and Fabrizio Dabbene

Zhendong Sun and Shuzhi S. Ge

Switched Linear Systems

Control and Design

With 52 Figures

 Springer

Zhendong Sun, PhD
Hamilton Institute, National University of Ireland, Maynooth, County Kildare, Ireland
Shuzhi Sam Ge, PhD, DIC
Department of Electrical and Computer Engineering, National University of Singapore,
Singapore 117576

Series Editors
E.D. Sontag • M. Thoma • A. Isidori • J.H. van Schuppen

British Library Cataloguing in Publication Data
Sun, Zhendong, 1968–
    Switched linear systems : control and design. -
    (Communications and control engineering)
    1. Linear control systems   2. Switching theory
    I. Title   II. Ge, S. S. (Shuzhi S.)
    629.8′32
    ISBN 1852338938

Library of Congress Cataloging-in-Publication Data
Ge, S. S. (Shuzhi S.)
    Switched linear systems : control and design / Zhendong Sun and S.S. Ge.
      p. cm. — (Communications and control engineering series)
    Includes bibliographical references and index.
    ISBN 1-85233-893-8 (alk. paper)
      1. Linear control systems.   2. Switching theory.   I. Sun, Zhendong, 1968–   II. Title.   III.
    Series.
    TJ220.G4   2004
    629.8′32—dc22   2004056549

Communications and Control Engineering Series ISSN 0178-5354
ISBN 1-85233-893-8 Springer Science+Business Media
springeronline.com

Typesetting: Camera-ready by authors
Printed in the United States of America
69/3830-543210   Printed on acid-free paper   SPIN 10990476

For our parents with love and gratitude.
For Hui and Jinlan, and our children with love and pride.

对父母，我们献上无限的爱和感激。

对妻子和孩子，我们有深深的眷恋和莫名的骄傲。

I keep the subject constantly before me and wait till the first dawnings open little by little into the full light.

— Sir Isaac Newton

Preface

A switched linear system is a hybrid system which consists of several linear subsystems and a rule that orchestrates the switching among them. Switched linear systems provide a framework which bridges the linear systems and the complex and/or uncertain systems. On one hand, switching among linear systems may produce complex system behaviors such as chaos and multiple limit cycles. On the other hand, switched linear systems are relatively easy to handle as many powerful tools from linear and multilinear analysis are available to cope with these systems. Moreover, the study of switched linear systems provides additional insights into some long-standing and sophisticated problems, such as intelligent control, adaptive control, and robust analysis and control.

Switched linear systems have been investigated for a long time in the control literature and have attracted increasingly more attention since the 1990s. The literature grew exponentially and quite a number of fundamental concepts and powerful tools have been developed from various disciplines. Despite the rapid progress made so far, many fundamental problems are still either unexplored or less well understood. In particular, there still lacks a unified framework that can cope with the core issues in a systematic way. This motivated us to write the current monograph.

The book presents theoretical explorations on several fundamental problems for switched linear systems. By integrating fresh concepts and state-of-the-art results to form a systematic approach for the switching design and feedback control, a basic theoretical framework is formed towards a switched system theory which not only extends the theory of linear systems, but also applies to more realistic problems.

The book is primarily intended for researchers and engineers in the system and control community. It can also serve as complementary reading for linear/nonlinear system theory at the post-graduate level.

The book contains seven chapters which exploit several independent yet related topics in detail.

Chapter 1 introduces the system description, background and motivation of the study, and presents several general concepts and fundamental observations which provide a sound base for the book.

Chapter 2 concisely reviews some necessary facts from Wonham's geometric method, linear algebra and stability theory, as well as other useful tools such as differential inclusion theory, Lie algebra, and automata theory.

Chapter 3 deals with stabilizing switching design for switched autonomous systems. Besides stability, the switching frequency and robustness are also considered. First, general results are presented for pointwise/consistent stabilization. Second, based on the average method, periodic switching laws are designed to steer the switched systems stable. Third, a state-feedback switching law is presented based on an appropriate partition of the state space. To avoid possible chattering induced by the perturbations, a modified strategy is proposed by introducing a positive level set. Fourth, to further reduce the switching frequency, a combined switching law is developed by integrating the time-driven switching mechanism with the state-feedback switching scheme. An observer-based switching law is also formulated for the case when the state information is not available. Finally, the discrete-time counterparts are briefly discussed.

In Chapter 4, we address the controllability, observability, feedback equivalence and canonical forms for switched control systems. For continuous-time systems, we prove that both the controllable set and the unobservable set are subspaces of the total space. Verifiable geometric characterizations are presented for the controllable/unobservable subspaces. Based on these, a switched control system can be brought into the canonical decomposition via suitable coordinate and feedback transformations. Parallel results are presented for discrete-time switched systems. The chapter also investigates the problem of sampling without loss of controllability. Sampling criteria are obtained and various regular switching and digital control schemes are discussed in detail. In addition, we discuss in depth several other issues including the controllability under constrained switching/input, local controllability, and decidability of controllability/observability.

The problem of feedback stabilization is investigated in Chapter 5. Stabilizing design for switched control systems is challenging since both the control input and the switching law are design variables, and their interaction must be fully understood. Based on the canonical decomposition, we are able to design stabilizing switching/input laws for several classes of switched control systems. In particular, for a linear system controlled by multiple controllers and measured by multiple sensors, an elegant and complete treatment is presented for the problem of dynamic output feedback stabilization. We also solve the stabilization problem for the switched linear system where the summation of the controllability subspaces of the individual subsystems is the total state space. For the case that the system is completely controllable, several sufficient conditions are presented for quadratic stabilizability and non-quadratic stabilizability as well. For the most general setting where both controllable part and uncontrollable part exist, we discuss the main obstacles involved and propose possible ways for handling it.

Chapter 6 addresses two optimization problems for switched linear systems. The first is the optimal switching problem where the switching signal is the only design variable to achieve the optimality. In this scheme, we compute the optimal convergence rate for systems which are simultaneously triangularizable, and present several nice properties of the optimal cost for a general infinite time horizon cost function. Next, we present a two-state optimization methodology for solving the mixed optimal switching/control problem of the switched linear system where both the switching signal and the control input are design variables. In this scheme, we discuss how to find the optimal switching signal and the optimal control input for solving the optimality and the sub-optimality problems.

Finally, Chapter 7 concludes the book by briefly summarizing the main results presented in the book and presenting related open problems for further investigation.

To summarize, the monograph presents the most recent theoretical development on switching design and feedback control for switched linear dynamical systems. By integrating novel ideas, fresh insights, and rigorous results in a systematic way, the book provides a sound base for further theoretical research as well as a design guide for engineering applications.

The logic connections among the chapters are as follows:

$$
\left\{ \begin{array}{c} \text{Chap. 1} \\ \text{Chap. 2} \end{array} \right\} \implies \left\{ \begin{array}{c} \text{Chap. 3} \\ \text{Chap. 4} \end{array} \right\} \implies \left\{ \begin{array}{c} \text{Chap. 5} \\ \text{Chap. 6} \end{array} \right\} \implies \text{Chap. 7.}
$$

Acknowledgements

For the creation of the book, we are very fortunate to have many helpful suggestions from our colleagues, friends and co-workers, through many stimulating and fruitful discussions. First of all, we would like to express our sincere gratitude to the careful scrutiny, numerous valuable suggestions and comments from D. Z. Cheng of Chinese Academy of Science, K. Narendra of Yale University, L. Noakes of The University of Western Australia, Y. Z. Wang of Shandong University, C. Xiang of National University of Singapore, and J. Zhao of Northeastern University. We would also like to express our sincere appreciation to our co-workers who have contributed to the collaborative studies of switched systems, which form part of the monograph. They are D. Z. Zheng of Tsinghua University, T. H. Lee of National University of Singapore, and R. N. Shorten of National University of Ireland, Maynooth. Special thanks go to O. Jackson, Assistant Editor of Springer-Verlag London, and A. J. Abrams, Senior Production Editor of Springer-Verlag New York, for their consistent and reliable support and comments in the process of publishing the book.

Z. Sun thanks his colleagues in Hamilton Institute, NUI Maynooth for many helpful discussions and assistance provided during the preparation of the book. S. S. Ge would like to express his sincere gratitude to M. Grimble for bringing the work to the attention of Springer-Verlag. S. S. Ge also thanks his current and former graduate students, C. H. Fua, T. T. Han, X. C. Lai, K. P. Tee, and Z. P. Wang for their help provided in formatting, graphing, and many detailed discussions.

We are greatly indebted to our parents and families for their love, support and sacrifice over the years. It is our pleasure to dedicate the book to them for their invisible yet invaluable contribution.

Maynooth, Ireland,   *Zhendong Sun*
Kent Ridge Crescent, Singapore,   *Shuzhi S. Ge*

January, 2005

Contents

List of Symbols

Throughout this book, the following conventions and notations are adopted:

$a \stackrel{def}{=} b$	defines	a to be b
$s.t.$	means	such that
$w.r.t.$	means	with respect to

$\mathbf{R}$	$\stackrel{def}{=}$	the field of real numbers
$\mathbf{C}$	$\stackrel{def}{=}$	the field of complex numbers
$\mathbf{N}$	$\stackrel{def}{=}$	the set of integers
$\mathbf{R}^+$	$\stackrel{def}{=}$	the set of positive real numbers
$\mathbf{R}_+$	$\stackrel{def}{=}$	the set of nonnegative real numbers
$\mathbf{N}^+$	$\stackrel{def}{=}$	the set of natural numbers
$\mathbf{N}_+$	$\stackrel{def}{=}$	the set of nonnegative integers
$\mathbf{R}^n$	$\stackrel{def}{=}$	the set of n-dimensional real vectors
$\mathbf{R}^{n \times m}$	$\stackrel{def}{=}$	the set of $n \times m$-dimensional real matrices
$\sqrt{-1}$	$\stackrel{def}{=}$	the imaginary unit
$\Re s$	$\stackrel{def}{=}$	the real part of complex number s
$I\ (I_n)$	$\stackrel{def}{=}$	the identity matrix (of dimension $n \times n$)
e_i	$\stackrel{def}{=}$	the i column of I
x^T or A^T	$\stackrel{def}{=}$	the transpose of vector x or matrix A
A^{-1}	$\stackrel{def}{=}$	the inverse of matrix A
A^+	$\stackrel{def}{=}$	the Moore-Penrose pseudo-inverse of matrix A
$P > 0\ (P \geq 0)$	$\stackrel{def}{=}$	matrix P is real symmetric and (semi-)positive definite
$P < 0\ (P \leq 0)$	$\stackrel{def}{=}$	matrix P is real symmetric and (semi-)negative definite

$\det A$ $\stackrel{def}{=}$ the determinant of matrix A

$\operatorname{sr} A$ $\stackrel{def}{=}$ the spectral radius of matrix A

$\operatorname{sv} A$ $\stackrel{def}{=}$ the set of singular values of matrix A

$\lambda(A)$ $\stackrel{def}{=}$ the set of eigenvalues of A

$\lambda_{max}(A)$ $\stackrel{def}{=}$ the maximum eigenvalue of real symmetric matrix A

$\lambda_{min}(A)$ $\stackrel{def}{=}$ the minimum eigenvalue of real symmetric matrix A

$x_i, x(j)$ $\stackrel{def}{=}$ the i-th element of vector x

$A(i,j), A_{ij}$ $\stackrel{def}{=}$ the ij-th element of matrix A

$|a|$ $\stackrel{def}{=}$ the absolute value of number a

$|w|$ $\stackrel{def}{=}$ the length of string w

$\|x\|$ $\stackrel{def}{=}$ the Euclidean norm of vector x

$\|A\|$ $\stackrel{def}{=}$ the induced Euclidean norm of matrix A

$\|x\|_p$ $\stackrel{def}{=}$ the l_p norm of vector x

$\|A\|_p$ $\stackrel{def}{=}$ the induced l_p-norm of matrix A

$<x,y>$ $\stackrel{def}{=}$ the inner product of x and y in $\mathbf{R}^n$

$\operatorname{Im} A$ $\stackrel{def}{=}$ the image of operator/matrix A

$\operatorname{Ker} A$ $\stackrel{def}{=}$ the kennel of operator/matrix A

$\mathcal{W}^\perp$ $\stackrel{def}{=}$ the annihilator of $\mathcal{W}$

$\mathcal{X}/\mathcal{Y}$ or $\frac{\mathcal{X}}{\mathcal{Y}}$ $\stackrel{def}{=}$ the factor (quotient) space

$\Gamma_A \mathcal{Y}$ $\stackrel{def}{=}$ the smallest A-invariant subspace containing subspace $\mathcal{Y}$

$\Gamma_{\{A_1,\cdots,A_m\}}\mathcal{Y}$ $\stackrel{def}{=}$ the smallest multiple A_i-invariant subspace containing subspace $\mathcal{Y}$

$\{x, y, \cdots\}$ $\stackrel{def}{=}$ a set of quantities x, y, etc.

$\max S$ $\stackrel{def}{=}$ the maximum element of set S

$\min S$ $\stackrel{def}{=}$ the minimum element of set S

$\sup S$ $\stackrel{def}{=}$ the smallest number that is larger than or equal to each element of set S

$\inf S$ $\stackrel{def}{=}$ the largest number that is smaller than or equal to each element of set S

$\arg\max S$ $\stackrel{def}{=}$ the index of maximum element of ordered set S

$\arg\min S$ $\stackrel{def}{=}$ the index of minimum element of ordered set S

$\emptyset$ $\stackrel{def}{=}$ the empty set

$\operatorname{meas}\Omega$ $\stackrel{def}{=}$ the Lebesgue measure of set Ω in $\mathbf{R}^n$

$\operatorname{int}\Omega$ $\stackrel{def}{=}$ the interior of set Ω

$\operatorname{co}\Omega$ $\stackrel{def}{=}$ the convex hull of set Ω

$\mathbf{B}_r$ $\stackrel{def}{=}$ the ball centered at the origin with radius r

$\mathbf{S}_r$ $\stackrel{def}{=}$ the sphere centered at the origin with radius r

$\bar{k}$ $\stackrel{def}{=}$ set $\{1, \cdots, k\}$

$\underline{k}$ $\stackrel{def}{=}$ set $\{0, 1, \cdots, k-1\}$

$k!$ $\stackrel{def}{=}$ the factorial of k

$\mathrm{mod}(a, b)$ $\stackrel{def}{=}$ the remainder of a divided by b

$\mathrm{sgn}(\cdot)$ $\stackrel{def}{=}$ the signum function

$\mathrm{sat}(\cdot)$ $\stackrel{def}{=}$ the saturation function with unit limits

$\lim_{s \uparrow t} f(s)$ $\stackrel{def}{=}$ the limit from the left of function $f(\cdot)$ at t

$\lim_{s \downarrow t} f(s)$ $\stackrel{def}{=}$ the limit from the right of function $f(\cdot)$ at t

M $\stackrel{def}{=}$ the fixed index set $\{1, 2, \cdots, m\}$

$\{a_i\}_M$ or $\{a_i\}_{i=1}^m$ $\stackrel{def}{=}$ the set $\{a_1, a_2, \cdots, a_m\}$

$[a, b)$ $\stackrel{def}{=}$ the real number set $\{t \in \mathbf{R} \colon a \le t < b\}$ or the integer set $\{t \in \mathbf{N} \colon a \le t < b\}$

$[a, b]$ $\stackrel{def}{=}$ the real number set $\{t \in \mathbf{R} \colon a \le t \le b\}$ or the integer set $\{t \in \mathbf{N} \colon a \le t \le b\}$

$\mathcal{S}_{[a,b)}$ $\stackrel{def}{=}$ the set of well-defined switching paths over $[a, b)$

$\mathcal{S}_{[t_0,\infty)}$ or $\mathcal{S}$ $\stackrel{def}{=}$ the set of well-defined switching paths over $[t_0, \infty)$

$\Sigma(C_i, A_i, B_i)_M$ $\stackrel{def}{=}$ the switched linear system with subsystems (C_i, A_i, B_i), $i \in M$

$\Sigma(A_i, B_i)_M$ $\stackrel{def}{=}$ the switched linear system with subsystems (A_i, B_i), $i \in M$

$\Sigma(C_i, A_i)_M$ $\stackrel{def}{=}$ the unforced switched linear system with subsystems (C_i, A_i), $i \in M$

$\Sigma(A_i)_M$ $\stackrel{def}{=}$ the switched autonomous system with subsystems A_i, $i \in M$

σ $\stackrel{def}{=}$ the switching signal of the switched system

$(\sigma)^{\to h}$ $\stackrel{def}{=}$ time-transition of σ by h

$\phi(t; t_0, x_0, u, \sigma)$ $\stackrel{def}{=}$ the solution of the switched system

$\Phi(t_1, t_2, u, \sigma)$ $\stackrel{def}{=}$ the state transition matrix of the switched system

$\mathcal{C}(A_i, B_i)_M$ $\stackrel{def}{=}$ the controllable set of switched linear system $\Sigma(A_i, B_i)_M$

$\mathcal{R}(A_i, B_i)_M$ $\stackrel{def}{=}$ the reachable set of switched linear system $\Sigma(A_i, B_i)_M$

$\mathcal{UO}(C_i, A_i)_M$ $\stackrel{def}{=}$ the unobservable set of switched system $\Sigma(C_i, A_i)_M$

$\mathcal{UR}(C_i, A_i)_M$ $\stackrel{def}{=}$ the unreconstructible set of switched system $\Sigma(C_i, A_i)_M$

1

Introduction

1.1 Switched Dynamical Systems

In general, a *switched (nonlinear) system* is composed of a family of subsystems and a rule that governs the switching among them, and is mathematically described by

$$\delta x(t) = f_\sigma(x(t), u(t), d(t)) \quad x(t_0) = x_0$$
$$y(t) = g_\sigma(x(t), w(t)) \tag{1.1}$$

where $x(t)$ is the state, $u(t)$ is the controlled input, $y(t)$ is the measured output, $d(t)$ and $w(t)$ stand for external signals such as perturbations, σ is the piecewise constant signal taking value from an index set $M \stackrel{def}{=} \{1, \cdots, m\}$, f_k, $k \in M$ are vector fields, and g_k, $k \in M$ are vector functions, while the symbol δ denotes the derivative operator in continuous time (*i.e.*, $\delta x(t) = \frac{d}{dt}x(t)$) and the shift forward operator in discrete time (*i.e.*, $\delta x(t) = x(t+1)$).

Figure 1.1 presents an illustrating diagram for the architecture of a switched system. It is clear that a switched system is multi-model in nature. Each individual component model

$$\delta x(t) = f_k(x(t), u(t), d(t))$$
$$y(t) = g_k(x(t), w(t)) \tag{1.2}$$

for $k \in M$ is said to be a *subsystem* or *mode* of the switched system. Besides the subsystems, the switched system also consists a switching device usually called the *supervisor*. The supervisor produces the switching rule σ, denoting the *switching signal* or *switching law*, which orchestrates the switching among the subsystems.

Generally speaking, the subsystems represent the low-level 'local' dynamics governed by conventional differential and/or difference equations, while the supervisor is the high-level coordinator producing the switches among

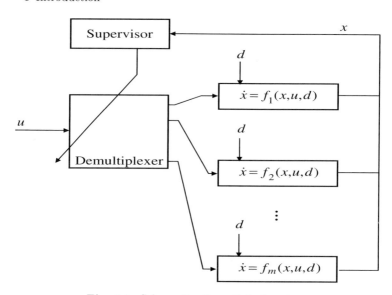

Fig. 1.1. Schematic of a switched system

the local dynamics. The dynamics of the system is determined by both the subsystems and the switching signal.

In general, a switching signal may depend on the time, its own past value, the state/output, and/or possibly an external signal as well

$$\sigma(t+) = \varphi(t, \sigma(t), x(t)/y(t), z(t)) \quad \forall\, t \tag{1.3}$$

where $z(t)$ is an external signal produced by other devices, $\sigma(t+) = \lim_{s \downarrow t} \sigma(s)$ in continuous time and $\sigma(t+) = \sigma(t+1)$ in discrete time.

If $\sigma(t) = i$, then we say that the ith subsystem is *active at time t*. It is clear that at any instant there is one (and only one) active subsystem.

Given a function pair $(x(\cdot), \theta(\cdot))$ over $[t_0, t_1)$, where $x\colon [t_0, t_1) \mapsto \mathbf{R}^n$ is absolutely continuous, and $\theta\colon [t_0, t_1) \mapsto M$ is piecewise constant. Function $x(\cdot)$ is said to be a *solution (or state trajectory) of (1.1) via switching signal (1.3) at x_0 over $[t_0, t_1)$*, if $x(t_0) = x_0$, and for almost all $t \in [t_0, t_1)$, we have

$$\delta x(t) = f_{\theta(t)}(x(t), u(t), d(t))$$
$$y(t) = g_{\theta(t)}(x(t), w(t))$$
$$\theta(t+) = \varphi(t, \theta(t), x(t)/y(t), z(t)).$$

The term "for almost all $t \in [t_0, t_1)$" means that "for all $t \in [t_0, t_1)$ except for possibly a set of isolated instants" in continuous time and "for all integers in $[t_0, t_1)$" in discrete time. The corresponding function $\theta(\cdot)$ is said to be *generated by the switching signal (1.3) along $x(\cdot)$ at x_0 over $[t_0, t_1)$*. The solution over other types of time intervals, such as $[t_0, t_1]$ and $[t_0, \infty)$, can be understood in the same way.

According to the above definition, we had, in fact, excluded any impulse in the state and input variables.

In this book, we focus on a special but very important class of switched systems where all the subsystems are linear time-invariant and the switching signals are governed by deterministic processes. These systems are termed as *switched linear systems* and are described by

$$\delta x(t) = A_\sigma x(t) + B_\sigma u(t) + E_\sigma d(t) \quad x(t_0) = x_0$$
$$y(t) = C_\sigma x(t) + G_\sigma w(t) \tag{1.4}$$

where A_k, B_k, C_k, E_k and G_k are linear mappings (matrices) in appropriate spaces. The *nominal system* is the system free of disturbances, that is

$$\delta x(t) = A_\sigma x(t) + B_\sigma u(t)$$
$$y(t) = C_\sigma x(t). \tag{1.5}$$

If no control input is imposed on the system, then, the system is said to be a *switched autonomous system*, or an *unforced switched system*. The unforced switched linear system is described by

$$\delta x(t) = A_\sigma x(t)$$
$$y(t) = C_\sigma x(t). \tag{1.6}$$

In this book, we denote system (1.5) by $\Sigma(C_i, A_i, B_i)_M$. Similarly, we denote by $\Sigma(A_i, B_i)_M$, $\Sigma(C_i, A_i)_M$, and $\Sigma(A_i)_M$ the switched system without output and/or input, respectively. In the case that we need to distinguish between continuous time and discrete time, we simply denote $\sum_c(C_i, A_i, B_i)_M$ for continuous-time systems and $\sum_d(C_i, A_i, B_i)_M$ for discrete-time systems.

1.2 Background and Examples

1.2.1 Background and Motivations

In this subsection, we briefly discuss the background and motivation in the study of switched systems.

Switched systems deserve investigation for theoretical reasons as well as for practical reasons.

First, switched systems can be used to model systems that are subject to known or unknown abrupt parameter variations such as synchronously switched linear systems [76], networks with periodically varying switchings [19], and sudden change of system structures due to various reasons [149, 24]. For example, the failure of a component or subsystem may have taken place in so short a time interval as to be considered an instantaneous event by comparison with the nominal time constants of the plant model. Hence,

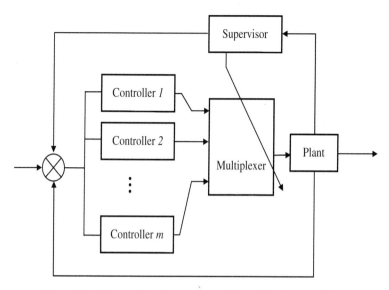

Fig. 1.2. Schematic of a system with multiple controllers

switching among different system structures is an essential feature of many engineering and practical real world systems.

Second, when we try to control a single process by means of multi-controller switching, the overall system can be described by a switched system. Indeed, suppose that for a single process described by

$$\delta x(t) = f(x(t), u(t), d(t))$$

there are several candidate controllers for the system

$$u(t) = \xi_i(x(t), w_i(t)) \quad i = 1, \cdots, m$$

where w_i stands for the external signals associated with the ith control device. Then, the overall system can be described by the switched system

$$\delta x(t) = f_\sigma(x(t), w(t), d(t))$$

where $f_i = f(x(t), \xi_i(x(t), w_i(t)), d(t))$. An illustrative diagram for this architecture is shown in Figure 1.2. In the literature, this multi-controller switching scheme is also known as the *hybrid control architecture*.

The multi-controller switching scheme provides an effective and powerful mechanism to cope with highly complex systems and/or systems with large uncertainties [83, 109, 88, 175]. One such example is hybrid control for nonholonomic systems which are not stabilizable by means of any individual continuous state feedback controller [82, 62, 54]. For these systems, multi-controller switching among smooth controllers provides a good conceptual framework to solve the problem. Even for simple linear time-invariant (LTI) systems, the performance (*e.g.*, transient response) can be improved through controllers/compensators switching [45, 98, 73, 87]. For example, as a common practice in stabilizing an LTI system, we use a hybrid controller that involves switching between a time-optimal or near time-optimal controller when the state is far from the equilibrium, and a linear controller near the equilibrium. This strategy can steer the system to the equilibrium quickly without exciting high-frequency dynamics or exceeding realistic actuator bandwidth.

As a special but very important class of switched systems, switched linear systems provide an attractive framework which bridges the gap between linear systems and the highly complex and/or uncertain systems. On one hand, switched linear systems are relatively easy to handle as many powerful tools from linear and multilinear analysis are applicable to cope with these systems. On the other hand, these systems are accurate enough to represent many practical engineering systems with complex dynamics. In addition, the study of switched linear systems provides additional insights to some long-standing and sophisticated problems, such as

- robust analysis and control [14, 32, 122, 131];
- adaptive control [48, 51, 63];
- intelligent control [151, 115, 43];
- gain scheduling [120, 15]; and
- multi-rate digital control [126, 26].

Practically, the study of switched systems has benefited, or will potentially benefit, many real world systems. These include

- power systems and power electronics;
- automotive control;
- aircraft and air traffic control;
- computer disks; and
- network and congestion control.

1.2.2 Examples

In this subsection, we present several examples to highlight some essential features and implications of switched systems.

Example 1.1. Suppose that we have two force-free linear systems

$$\dot{x} = A_1 x = \begin{bmatrix} 1 & -2 \\ 2 & 1 \end{bmatrix} x \tag{1.7}$$

and

$$\dot{x} = A_2 x = \begin{bmatrix} 3 & -2 \\ 1 & -1 \end{bmatrix} x. \tag{1.8}$$

Simple computation shows that the spectrum of A_1 is $\{1 \pm 2\sqrt{-1}\}$ and the spectrum of A_2 is $\{1 \pm \sqrt{2}\}$. Accordingly, neither of the systems is stable.

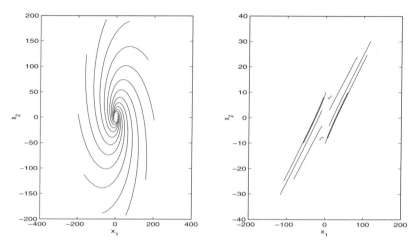

Fig. 1.3. Phase portraits of the two LTI systems

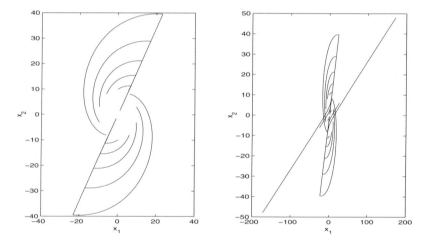

Fig. 1.4. Phase portraits within shorter (left) and longer horizons (right)

The first system has an unstable focus, hence it could rotate from any given non-origin state to any given direction of the phase plane. The second system, on the other hand, has a stable mode and an unstable mode. The stable mode is the subspace

$$\mathcal{W}_s = \text{span}\left\{ \begin{bmatrix} 2 - \sqrt{2} \\ 1 \end{bmatrix} \right\}.$$

Figure 1.3 shows their phase portraits. From these, a switching strategy can be proposed as follows to produce globally stable behaviors.

Given any initial condition $x(t_0) = x_0$, we let the second system be active if the state is in $\mathcal{W}_s$, the stable mode of A_2. Otherwise, let the first system

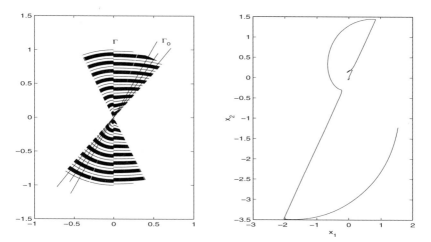

Fig. 1.5. The partition sectors (left) and a sample trajectory (right) in the phase plane

be active. As $\mathcal{W}_s$ is an invariant subspace of A_2, this strategy always steers any state to approach the origin via, at most, one switch. In this scheme, the stable mode serves as a slide mode, just as in the variable structure control scheme. However, this strategy requires the slide mode to be accessed exactly. Any derivation will cause the state trajectory to diverge to infinity. In this sense, the above strategy is only theoretically feasible. Figure 1.4 illustrates the phase portraits of the state trajectories under this strategy for a shorter time horizon ($t \leq 5$) and for a longer time ($t \leq 7.88$), respectively. It is clear that the the state converges along the 'slide mode' at the beginning but finally diverges to infinity.

To overcome the sensitivity problem, we modify the strategy as follows. Let $\xi \in \mathcal{W}_s$ be an eigenvector of A_2 corresponding to the stable eigenvalue $\lambda = 1 - \sqrt{2}$. It can be seen that $\xi^T (A_2 + A_2^T)\xi = 2\lambda\xi^T\xi < 0$. As a result, there is a sector Γ of $\mathbf{R}^2$:

$$\Gamma = \{(x, y) \in \mathbf{R}^2 : k_1 x^2 \leq xy \leq k_2 x^2\}$$

which includes ξ as an interior point such that

$$x^T (A_2 + A_2^T)x < 0 \quad \forall\, x \in \Gamma.$$

In addition, we fix a small neighborhood of the stable mode

$$\Gamma_0 = \{(x, y) \in \mathbf{R}^2 : k_3 x^2 \leq xy \leq k_4 x^2\}$$

such that Γ_0 is a strict subset of Γ. Let us assign a switching signal using the following rules:

(i) if the state is in Γ_0, then let the second subsystem be active;

(ii) if the state is outside Γ, then let the first subsystem be active;
(iii) otherwise, keep the active system unchanged.

It can be seen that this switching strategy always produces stable trajectories. The advantage of this strategy is that it is insensitive to small derivations of the state. Figure 1.5 illustrates the sectors Γ and Γ_0 as well as a sampling phase portrait in the time interval $[0, 50]$. It clearly shows that, though the trajectory converges globally, it locally diverges twice when deviations from the 'slide mode' occur. Another feature is that the switching signal is initial-state dependent, *i.e.*, different initial states correspond to different switching signals. Figure 1.6 illustrates this by depicting the switching signals corresponding to two different initial states.

From this example, we have the following observations:

(a) Appropriate switching among the unstable systems produces stable global system behavior.
(b) The stabilizing switching strategies may involve either a finite or infinite number of switches, while the infinite switching strategy performs better than the finite one in practice.
(c) Both the stabilizing switching strategies are initial-state dependent, *i.e.*, different initial states correspond to different switching time and index sequences.

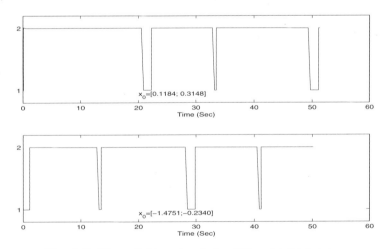

Fig. 1.6. The switching signals for different initial states

In addition, for this example, there is no initial-state independent switching strategy that produces stable state trajectories (*c.f.* Theorem 3.4). That is, for any switching signal of the form $\sigma : [t_0, \infty) \mapsto \{1, 2\}$, the resultant time-varying system

$$\dot{x}(t) = A_{\sigma(t)}x(t)$$

is unstable.

Example 1.2. (PWM-Driven Boost Converter) [33]

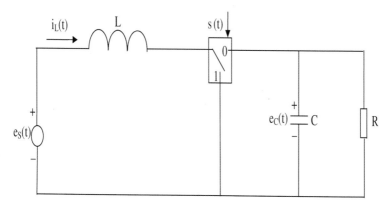

Fig. 1.7. The Boost converter

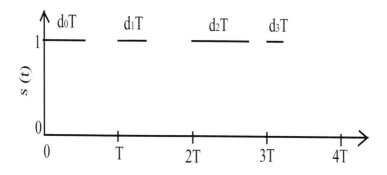

Fig. 1.8. Pulse-width modulation

In this example, we illustrate the modelling of PWM-driven Boost converter. Figure 1.7 shows the schematic of a Boost converter. Here, L is the inductance, C the capacitance, R the load resistance, and $e_s(t)$ the source voltage. With this converter, it is possible to transform the source voltage $e_s(t)$ into a higher voltage $e_C(t)$ over the load R.

The switch $s(t)$ is controlled by a PWM device. Suppose that the switch can have two states, namely, 0 and 1. Then, we have $s(t) \in \{0,1\}$. For simplicity, assume that $s(t)$ can switch at most once in each period. In Figure 1.8, we depict the periods as $0, 1, 2, 3$ and the relative pulse width as $d_0, \cdots, d_3$, where $d_i \in [0, 1]$.

The schematic of the PWM-driven Boost converter is shown in Figure 1.9.

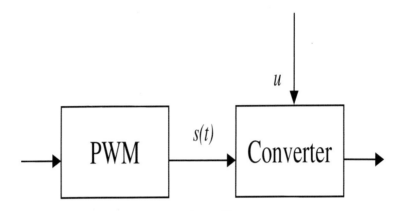

Fig. 1.9. PWM driven Boost converter

By introducing the normalized variables $\tau = t/T$, $L_1 = L/T$, and $C_1 = C/T$, the differential equations for the Boost converter are as follows:

$$\dot{e}_C(\tau) = -\frac{1}{RC_1}e_C(\tau) + (1 - s(\tau))\frac{1}{C_1}i_L(\tau)$$

$$\dot{i}_L(\tau) = -(1 - s(\tau))\frac{1}{L_1}e_C(\tau) + s(\tau)\frac{1}{L_1}e_S(\tau). \qquad (1.9)$$

Let $x_1 = e_C$, $x_2 = i_L$, $u = e_S$, $\sigma = s + 1$, and

$$A_1 = \begin{bmatrix} -\frac{1}{RC_1} & \frac{1}{C_1} \\ -\frac{1}{L_1} & 0 \end{bmatrix} \text{ and } B_1 = \begin{bmatrix} 0 \\ 0 \end{bmatrix}$$

$$A_2 = \begin{bmatrix} -\frac{1}{RC_1} & 0 \\ 0 & 0 \end{bmatrix} \text{ and } B_2 = \begin{bmatrix} 0 \\ \frac{1}{L_1} \end{bmatrix}.$$

Then, equations in (1.9) can be described by

$$\dot{x} = A_\sigma x + B_\sigma u \quad \sigma \in \{1, 2\}$$

which is exactly the switched linear system with two subsystems.

Note that, in the above model, both the input u and the switching signal σ are design variables. A constraint imposed on the switching signal is that at most one switch occurs in any unit time interval.

Example 1.3. (Hybrid Stabilization of a Wheeled Mobile Robot) [4]

The next example addresses the problem of parking the wheeled mobile robot of the unicycle type, shown in Figure 1.10, where x_1 and x_2 are the coordinates

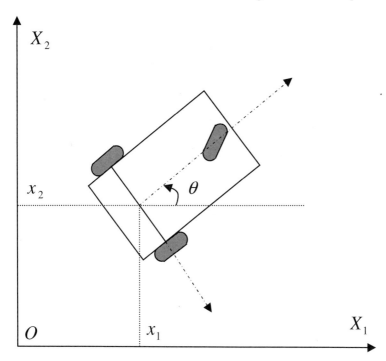

Fig. 1.10. Wheeled mobile robot of unicycle type

of the point in the middle of the rear axle, and θ denotes the angle that the vehicle makes with the x_1-axis. The kinematics of the robot can be modelled by the equations

$$
\begin{aligned}
\dot{x}_1 &= u_1 \cos \theta \\
\dot{x}_2 &= u_1 \sin \theta \\
\dot{\theta} &= u_2
\end{aligned}
\tag{1.10}
$$

where u_1 and u_2 are the control inputs (the forward and the angular velocity, respectively). By parking the vehicle we mean making x_1, x_2 and θ tend to zero by applying state feedback. What makes this problem especially interesting is that the corresponding system is nonholonomic and hence cannot be asymptotically stabilized by any time-invariant continuous state feedback law. As a result, the classical smooth theory and design mechanism of nonlinear control systems cannot be applied. However, if we address the problem using the hybrid control scheme, the obstruction disappears.

To see this, we first introduce some intermediate variables. Let

$$
\begin{aligned}
y_1 &= \theta \\
y_2 &= x_1 \cos \theta + x_2 \sin \theta \\
y_3 &= x_1 \sin \theta - x_2 \cos \theta
\end{aligned}
$$

and

$$D_1 = \{x \in \mathbf{R}^3 : |x_3| > \frac{\|x\|}{2}\}$$
$$D_2 = \{x \in \mathbf{R}^3 : x \notin D_1\}.$$

Next, let us define a set of candidate controllers

$$u^1 = \begin{bmatrix} u_1^1 \\ u_2^1 \end{bmatrix} = \begin{bmatrix} -4y_2 - 6\frac{y_3}{y_1} - y_3 y_1 \\ -y_1 \end{bmatrix}$$

$$u^2 = \begin{bmatrix} u_1^2 \\ u_2^2 \end{bmatrix} = \begin{bmatrix} -y_2 - \operatorname{sgn}(y_2 y_3)y_3 \\ -\operatorname{sgn}(y_2 y_3) \end{bmatrix}$$

where $\operatorname{sgn}(\cdot)$ is the signum function

$$\operatorname{sgn}(t) = \begin{cases} 1 & \text{if } t \geq 0 \\ -1 & \text{otherwise.} \end{cases}$$

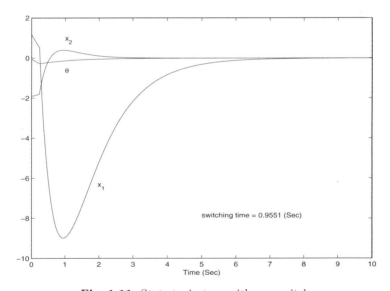

Fig. 1.11. State trajectory with one switch

With these controllers, system (1.10) can be represented as an unforced switched nonlinear system

$$\dot{x}(t) = f_\sigma(x(t))$$

where $x = [x_1, x_2, \theta]^T$, $\sigma \in \{1, 2\}$, and

$$f_i(x) = \begin{bmatrix} u_1^i \cos\theta \\ u_1^i \sin\theta \\ u_2^i \end{bmatrix} \quad i = 1,2.$$

Finally, define the switching law by

$$\sigma(t) = \begin{cases} 1 & \text{if } x(t) \in D_1 \\ 2 & \text{if } x(t) \in D_2. \end{cases} \tag{1.11}$$

Then, it can be proven that the switched system is exponentially stabilizable.

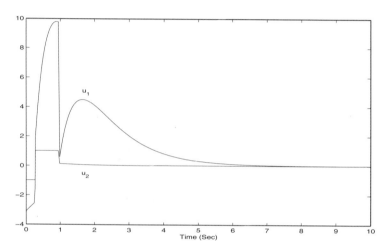

Fig. 1.12. Input trajectory with one switch

Roughly speaking, the switching strategy is to first adjust the angle (away from zero) of the robot if necessary, and then drive the robot smoothly to the origin. Figures 1.11 and 1.12 show the state and input trajectories of the switched system initialized at

$$x(0) = [1.1909, -1.8916, -0.0376]^T$$

respectively. It can be seen that both trajectories converge exponentially to the origin. Figure 1.13 simulates the parking process which clearly shows the two-stage feature of the parking strategy (first backward, then forward).

When the original angle of the robot is far from zero, then, the parking can be made smoothly without any switch. Figures 1.14 and 1.15 show the state/input trajectories as well as the parking simulation, respectively. The switched system is initialized at

$$x(0) = [-1.3362, 0.7143, 1.6236]^T.$$

It can be seen that, though the state and input are smooth vectors, the parking process still exhibits the two-stage feature.

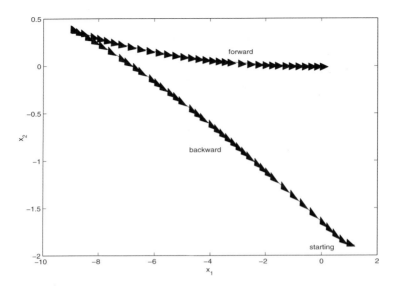

Fig. 1.13. Parking animation with one switch

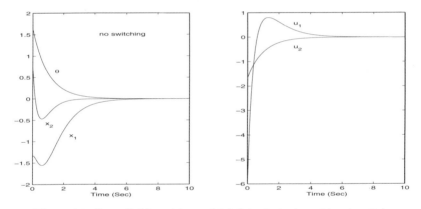

Fig. 1.14. State (left) and input (right) trajectories without switch

This example clearly exhibits the power of the hybrid control methodology in addressing complex dynamical systems.

1.3 Elementary Analysis

1.3.1 Classification of Switching Signals

For the switched system

$$\delta x(t) = f_\sigma(x(t), u(t))$$
$$y(t) = h_\sigma(x(t)) \tag{1.12}$$

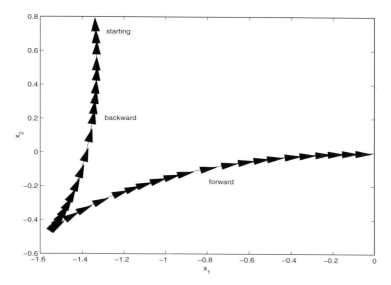

Fig. 1.15. Parking animation without switch

suppose that the subsystem dynamics f_i, h_i, $i \in M$ are given and fixed. In addition, assume that the input u is also given in advance.

For the above system, the performance of the overall system totally relies on the switching signal. As we have mentioned, different switching signals may produce totally different system behaviors. In this subsection, we take a close look at the possible expressions of the switching signals and present a classification for different types of switching signals.

In general, the switching signal is a piecewise constant function of time, its own past values, the state/output, and possibly the external signal. A general representation can thus be given by

$$\sigma(t) = \psi\left([t_0\ \infty), \sigma([t_0\ \infty)), x([t_0, \infty))/y([t_0, \infty)), z([t_0, \infty))\right) \quad t \geq t_0$$

where t_0 is the initial time, $z \colon [t_0, \infty) \mapsto \mathbf{R}^l$ is an external signal produced by an external device (*e.g.*, an observer). Note that the expression does not exclude possible non-causal relationships, that is, the switching signal at a time may depend on the future measurements from its own or other variables. In reality, most relationships are causal. A causal description of the switching signal is given by

$$\sigma(t+) = \psi\left([t_0, t], \sigma([t_0, t]), x([t_0, t])/y([t_0, t]), z([t_0, t])\right) \quad t \geq t_0$$

where $\sigma(t+) = \lim_{s \downarrow t} \sigma(s)$ in continuous time and $\sigma(t+) = \sigma(t+1)$ in discrete time. It can be seen that the switching signal at a time may rely on the past measurements of time, state/output, *etc.* For example, the discrete-time switching signal with initial condition

$$\sigma(t_0) = \arg\min\{x^T(t_0)A_i x(t_0)\}_{i \in M}$$
$$\sigma(t_0 + 1) = \sigma(t_0)$$

and recursion

$$\sigma(t + 1) = \max\{\arg\min\{x^T(t)A_i x(t)\}_{i \in M}, \sigma(t - 1)\} \quad t = 1, 2 \cdots$$

includes a one-step lag delay. In this book, we further assume that the switching signal only depends on the instantly past (on-line) measurements, $i.e.$,

$$\sigma(t+) = \psi(t, \sigma(t), x(t)/y(t), z(t)) \quad t \geq t_0. \tag{1.13}$$

In the following, we briefly classify the switching signals into several types, which we will frequently use in the book.

1) Switching Path

   A switching signal is said to be a *switching path* if it is a function of time. Given an initial time t_0, a switching path is defined on a time interval $[t_0, t_1)$ with $t_0 < t_1 \leq \infty$

$$\sigma \colon [t_0, t_1) \mapsto M.$$

   Note that the switching path is independent of the initial state. That is, such a switching signal is consistent with any initial state.
   In the continuous-time case, two switching paths θ_1 and θ_2 over $[t_0, t_1)$ are said to be *indistinguishable*, if they coincide almost everywhere, $i.e.$, the time set

$$\{t \in [t_0, t_1)\colon \theta_1(t) \neq \theta_2(t)\}$$

   is a set of isolated real numbers. Two indistinguishable paths are seen to be one path as they produce the same system dynamics. For simplicity, we let $\theta_{[t_0,t_1)}$ denote a switching path over $[t_0, t_1)$.

2) Time-driven Switching Law

   If the switching signal only relies on the time and its past values, then we say that the switching signal is a *time-driven switching law*. A time-driven switching law can be described by

$$\sigma(t+) = \psi(t, \sigma(t)) \quad t \geq t_0. \tag{1.14}$$

   A typical feature is that it is independent of the state/output variables. A switching path $\theta_{[t_0,t_1)}$ is said to be *generated by the time-driven switching law*, if for almost all $t \in [t_0, t_1)$, we have

$$\theta(t+) = \psi(t, \theta(t)).$$

   As a special case, a switching path always generates itself.

As an example, we define recursively the time-driven switching law

$$\sigma(t_0) = \arg\max\{0.5, t_0^2\}$$

$$\sigma(t+) = \arg\min\{-(\sigma(t))^2, -\frac{27}{8}(t+1)(t-1)\sigma(t) - 4\}. \quad (1.15)$$

In continuous time, the switching law generates the switching path θ with

$$\theta(t) = \begin{cases} 2 & \text{over } [t_0, -1) & \text{if } t_0 < -1 \\ 1 & \text{over } [t_0, \frac{1}{3}) & \text{if } t_0 \in [-\frac{1}{3}, \frac{1}{3}) \\ 2 & \text{over } [t_0, \infty) & \text{if } t_0 > 1 \end{cases}$$

but generates no switching path if $t_0 \in (-1, -\frac{1}{3})$ or $t_0 \in (\frac{1}{3}, 1)$ as there is a deadlock in this situation. Indeed, by the recursive formula, for any $t \in (-1, -\frac{1}{3}) \cup (\frac{1}{3}, 1)$, it should switch to the other subsystem at the 'next time of t'. As the time is continuous, this clearly leads to a deadlock. In discrete time, however, the switching law generates the switching path

$$\sigma = \begin{cases} 1 & \text{if } t = 0 \\ 2 & \text{otherwise} \end{cases}$$

which is defined everywhere.

3) Event-driven Switching Law

A switching signal is said to be an *event-driven switching law*, if the time does not explicitly appear in the expression (1.13), *i.e.*,

$$\sigma(t+) = \psi\left(\sigma(t), x(t)/y(t), z(t)\right). \quad (1.16)$$

In particular, a switching signal is said to be a *state-feedback switching law*, if it only depends on its past value and the state variables, *i.e.*,

$$\sigma(t+) = \psi\left(\sigma(t), x(t)\right). \quad (1.17)$$

Similarly, a switching signal is said to be an *output-feedback switching law*, if it only depends on its past value and the output variables, *i.e.*,

$$\sigma(t+) = \psi\left(\sigma(t), y(t)\right) \quad (1.18)$$

and to be a *dynamic-output-feedback switching law*, if it depends on its past value, the output variables, and possibly an external signal z, *i.e.*,

$$\sigma(t+) = \psi\left(\sigma(t), y(t), z(t)\right). \quad (1.19)$$

Given an initial state x_0 and a time interval $[t_0, t_1)$ with $t_1 > t_0$, a switching path $\theta_{[t_0,t_1)}$ is said to be *generated by the event-driven switching law (1.16) at x_0 over $[t_0, t_1)$* w.r.t. the switched system, if there is a state trajectory $x \colon [t_0, t_1) \mapsto \mathbf{R}^n$ with $x(t_0) = x_0$, such that for almost any $t \in [t_0, t_1)$, we have

$$\delta x(t) = f_{\theta(t)}(x(t), u(t))$$
$$y(t) = h_{\theta(t)}(x(t))$$
$$\theta(t+) = \psi\left(\theta(t), x(t)/y(t), z(t)\right).$$

Note that a switching law may generate different switching paths *w.r.t.* different switched systems.

4) Pure-state/output-feedback Switching Law

A switching signal is said to be a *pure-state-feedback switching law*, if it only depends on the state variables, *i.e.*,

$$\sigma(t+) = \psi\left(x(t)\right). \tag{1.20}$$

Similarly, switching signal is said to be a *pure-output-feedback switching law*, if it only depends on the output variables, *i.e.*,

$$\sigma(t+) = \psi\left(y(t)\right). \tag{1.21}$$

A typical pure-state-feedback switching law is the quadratic form

$$\sigma(t+) = \arg\min\{x^T(t)P_1x(t), \cdots, x^T(t)P_mx(t)\}$$

where P_i, $i \in M$ are real matrices.

Roughly speaking, by a switching law we mean that the switching signal is initial-condition-dependent. In contrast, a switching path is simply a function of time and hence is independent of the system state/output. A switching signal is said to be *determinant at x_0 over* $[t_0, t_1)$, if it generates a unique switching path at x_0 over $[t_0, t_1)$. It is said to be *completely determinant* over $[t_0, t_1)$, if for any $x_0 \in \mathbf{R}^n$ and $t_0' \in [t_0, t_1)$, the switching law generates a unique switching path at x_0 over $[t_0', t_1)$.

It should be noticed that, as a special case, a switching path defined on $[t_0, t_1)$ generates itself, and hence is completely determinant over $[t_0, t_1)$.

1.3.2 Operations on Switching Signals

Given a switched system in advance, suppose that σ_1 and σ_2 are two switching signals defined by

$$\sigma_i(t+) = \psi_i(t, \sigma_i(t), x(t)/y(t), z(t)) \quad i = 1, 2.$$

σ_1 is said to be the *time-transition of σ_2 by h (w.r.t. the switched system)*, denoted by $\sigma_1 = (\sigma_2)^{\rightarrow h}$, if

$$\psi_1(t, \sigma_1(t), x(t)/y(t), z(t)) = \psi_2(t+h, \sigma_2(t+h), x(t+h)/y(t+h), z(t+h)).$$

Note that, if σ_1 is the time-transition of σ_2 by h, then σ_2 is the time-transition of σ_1 by $-h$. A switching signal σ is said to be *time-invariant at x_0 over $[t_0, t_1)$*, provided that, for any h, and switching path $\theta_{[t_0,t_1)}$ generated by σ at x_0 over $[t_0, t_1)$, the path θ_h defined by

$$\theta_h(t) = \theta(t - h) \quad \forall\, t \in [t_0 + h, t_1 + h)$$

is a switching path generated by $\sigma^{\to h}$ at x_0 over $[t_0+h, t_1+h)$. In addition, the switching signal is said to be *time-invariant over* $[t_0, t_1)$ if it is time-invariant at each state in $\mathbf{R}^n$ over $[t_0, t_1)$. The switching signal is said to be *(completely) time-invariant* if it is time-invariant over any time interval. It can be seen that any switching path is time-invariant over its time domain, so is any switching signal in state/output-feedback form. However, a time-driven switching signal may not be time-invariant. As an example, it can be verified that the time-driven switching signal defined by (1.15) in the continuous-time case is not time-invariant.

Similarly, for two switching signals defined by

$$\sigma_i(t+) = \psi_i(t, \sigma_i(t), x(t)/y(t), z(t)) \quad i = 1, 2$$

σ_1 is said to be the *radial-transition of* σ_2 *by* $\lambda \in \mathbf{R}$ *(w.r.t. the switched system)*, if

$$\psi_1(t, \sigma_1(t), x(t)/y(t), z(t)) = \psi_2(t, \sigma_2(t), \lambda x(t)/\lambda y(t), \lambda z(t)).$$

Note that, if σ_1 is the radial-transition of σ_2 by $\lambda \neq 0$, then σ_2 is the radial-transition of σ_1 by $\frac{1}{\lambda}$. A switching signal σ is said to be *radially invariant*, if for any $\lambda \in \mathbf{R}$, the radial-transition of the switching signal by λ coincides with the switching signal itself, *i.e.*,

$$\psi(t, \sigma(t), x(t)/y(t), z(t)) = \psi(t, \sigma(t), \lambda x(t)/\lambda y(t), \lambda z(t)).$$

It can be seen that any switching path is radial-invariant, so is any time-driven switching law. However, an event-driven switching signal may not be radial-invariant. For example, the pure-state-feedback switching law

$$\sigma(t+) = \arg\min\{\|x(t)\|, x^T(t)x(t)\}$$

is in general not radial-invariant.

A switching signal is said to be *transition-invariant*, if it is both time-invariant and radial-invariant. In the book, when we design switching signals, we will always focus on the transition-invariant switching signals.

Given two switching paths $\theta_i : [t_i, s_i) \mapsto M$, $i = 1, 2$, θ_2 is said to be a *sub-path of* θ_1 *on* $[t_2, s_2)$, denoted by $\theta_2 = \theta_{1[t_2,s_2)}$, if $[t_2, s_2) \subseteq [t_1, s_1)$, and for almost any $t \in [t_2, s_2)$

$$\theta_1(t) = \theta_2(t).$$

Another operation on switching signals is the concatenation of switching signals. Given two switching paths $\theta_{1[t_1,s_1)}$ and $\theta_{2[t_2,s_2)}$, the *concatenation of*

θ_1 *with* θ_2, denoted by $\theta_1 \wedge \theta_2$, is a new switching path defined on $[t_1, s_1 + s_2 - t_2)$ with

$$(\theta_1 \wedge \theta_2)(t) = \begin{cases} \theta_1(t) & \text{if } t \in [t_1, s_1) \\ \theta_2(t - s_1 + t_2) & \text{if } t \in [s_1, s_1 + s_2 - t_2). \end{cases}$$

The concatenation of more than two switching paths can be defined in the same way by recursion.

The operation of concatenation can be extended to general switching signals as follows. Suppose that a switching signal σ_1 is determinant at x_0 over $[t_1, s_1)$ and hence it generates a unique switching path $\theta_{1[t_1, s_1)}$. Let $x_1 = \lim_{t \uparrow s_1} x(t)$, where $x(\cdot)$ is a solution of the switched system via σ_1 at x_0 over $[t_1, s_1)$. Suppose that a switching signal σ_2 is determinant at x_1 over $[t_2, s_2)$, and hence it generates a unique switching path $\theta_{2[t_2, s_2)}$. Then, a switching signal σ is said to be a *concatenation of* σ_1 *with* σ_2 *at* $(x_0, [t_1, s_1), [t_2, s_2))$, if it generates uniquely the switching path $\theta_{1[t_1, s_1)} \wedge \theta_{2[t_2, s_2)}$ at x_0 over $[t_1, s_1 + s_2 - t_2)$.

A more useful concept is the concatenation of two switching signals via a given region. Suppose that σ_1 and σ_2 are two switching signals, Ω is a closed region in the state space. Then, a switching signal σ is said to be a *concatenation of* σ_1 *with* σ_2 *at* $(x_0, t_1, [t_2, s_2))$ *via* Ω, if it generates a unique switching path $\theta_{[t_1, s_1 + s_2 - t_2)}$ with

$$s_1 = \min\{t \geq t_1 : x(t) \in \Omega\}$$

such that σ is the concatenation of σ_1 with σ_2 at $(x_0, [t_1, s_1), [t_2, s_2))$.

Example 1.4. For continuous-time switched system $\Sigma(A_i)_{\{1,2\}}$ with

$$A_1 = \begin{bmatrix} 1 & 2 \\ -2 & 1 \end{bmatrix} \text{ and } A_2 = \begin{bmatrix} -1 & 0 \\ 0 & 10 \end{bmatrix}$$

let $\Omega = \{x \in \mathbf{R}^2 : x_2 = 0\}$, and $\sigma_1 \equiv 1$ and $\sigma_2 \equiv 2$. It can be seen that the concatenation of σ_1 with σ_2 via Ω is the following pure-state-feedback switching law

$$\sigma(t) = \begin{cases} 1 & \text{if } x(t) \notin \Omega \\ 2 & \text{otherwise} \end{cases}$$

which always makes the switched system asymptotically stable.

1.3.3 Well-definedness and Well-posedness

An important issue for the switched system is the existence and uniqueness of the solution. A switching path θ is said to be *well-defined on* $[t_1, t_2)$, if it is defined in $[t_1, t_2)$, and for all $t \in [t_1, t_2)$, both $\lim_{s \uparrow t} \theta(s)$ (for $t = t_1$, let

$\lim_{s\uparrow t_1} \theta(s) = \theta(t_1)$) and $\lim_{s\downarrow t} \theta(s)$ exist, and it has only finite jump instants in any finite time sub-interval of $[t_1, t_2)$. That is, the set of *jump times*

$$\{t \in [t_1, t_2): \lim_{s\uparrow t} \theta(s) \neq \lim_{s\downarrow t} \theta(s)\}$$

is finite for any finite time interval $[t_1, t_2)$.

Note that, for the continuous-time switching path, a jump time must be a discontinuous time, but the converse is not necessarily true. For example, the switching path

$$\theta(t) = \begin{cases} 1 & \text{if } t \in \mathbf{N} \\ 2 & \text{otherwise} \end{cases}$$

has infinite discontinuous times in $[t_0, \infty)$ for any $t_0 < \infty$. However, it has no jump time at all and hence is well-defined. On the other hand, the switching path

$$\theta(t) = \begin{cases} 1 & \text{if } t \in \{\frac{1}{k}\}_{k=1}^{\infty} \\ 2 & \text{otherwise} \end{cases}$$

is not well-defined on any interval containing zero because $\lim_{s\downarrow 0} \theta(s)$ does not exist.

In the book, we denote by $\mathcal{S}_{[t_0, t_1)}$(or $\mathcal{S}$ in short) the set of well-defined switching paths on $[t_0, t_1)$.

It is clear that the well-definedness excludes the possibility of the Zeno phenomenon (chattering) which is not desired in most situations. It is also clear that any discrete-time switching path is well-defined.

By means of the well-definedness of switching paths, we can further define the well-definedness of general switching signals. For this, fix a switched system and let σ be a switching signal of the system. The switching signal σ is said to be *well-defined at x_0 over $[t_0, t_1)$* (*w.r.t.* the switched system), if it is determinant at x_0 over $[t_0, t_1)$ and the switching path θ that uniquely generated by σ at x_0 over $[t_0, t_1)$ is well-defined on $[t_0, t_1)$. In addition, the switching signal is said to be *well-defined over $[t_0, t_1)$* (*w.r.t.* the switched system), if for any $x_0 \in \mathbf{R}^n$, it is well-defined at x_0 over $[t_0, t_1)$. The switching signal is said to be *(completely) well-defined*, if it is well-defined over any interval $[t_0, t_1)$ with $-\infty < t_0 < t_1 \leq \infty$.

Suppose that the subsystems' dynamics are given and switching signal σ is completely well-defined. Then, for any initial condition (t_0, x_0), the switching signal generates a unique switching path at x_0 over $[t_0, t_1)$ (for some $t_1 > t_0$), denoted by $\theta_{[t_0, t_1)}^{x_0}$. Therefore, the switching signal can be expressed by $\sigma(t) = \varphi(t; t_0, x_0)$, where

$$\varphi(t; t_0, x_0) = \theta_{[t_0, t_1)}^{x_0}(t) \quad \forall\, t \in [t_0, t_1).$$

In this book, whenever the subsystems' dynamics are given and clear from the context, we are quick to express a well-defined switching signal in either the standard form $\sigma(t+) = \psi(t, \sigma(t), x(t)/y(t), z(t))$ or in the form $\sigma(t) = \varphi(t; t_0, x_0)$.

On the other hand, a switched system is said to be *well-posed at x_0 over* $[t_0, t_1)$ *w.r.t.* switching signal σ, if for any given piecewise continuous and locally integrable input u, the switching signal σ is well-defined at x_0 over $[t_0, t_1)$ w.r.t. the switched system, and the switched system admits a unique solution via the switching signal at x_0 over $[t_0, t_1)$. Similarly, the switched system is said to be *well-posed over* $[t_0, t_1)$ *w.r.t.* the switching signal, if for any $x_0 \in \mathbf{R}^n$, the system is well-posed at x_0 over $[t_0, t_1)$ w.r.t. the switching signal; the switched system is said to be *(completely) well-posed w.r.t.* the switching signal, if for any time interval $[t_0, t_1)$ with $-\infty < t_0 < t_1 \le \infty$, the system is well-posed over $[t_0, t_1)$ w.r.t. the switching signal.

Given a switched system, if the switching signal is a switching path or a time-driven switching law, the well-posedness of the switched system and the well-definedness of the switching signal are decoupled and hence are independent of each other. However, for a switching signal in the event-driven form, its well-definedness implies that the state information is available on $[t_0, t_1)$, which means that the switched system admits at least one solution at x_0 over $[t_0, t_1)$. Note that in this case, the switched system is well-posed if, and only if, it admits only one solution. Accordingly, if the switched system satisfies the global Lipschitz condition, then, the well-posedness of the switched system is equivalent to the well-definedness of the switching signal. For the switched linear system, when the control input is globally integrable, then, the well-definedness of the switching signal implies the well-posedness of the switched system, hence they are equivalent to each other. In other words, for the switched linear system

$$\delta x(t) = A_\sigma x(t) + B_\sigma u(t)$$
$$y(t) = C_\sigma x(t) \tag{1.22}$$

where the control input is piecewise continuous and globally integrable, and the switching signal σ defined by

$$\sigma(t+) = \psi(t, \sigma(t), x(t)/y(t))$$

the following statements are equivalent:

(i) the system is well-posed at x_0 over $[t_0, t_1)$ under the switching signal;
(ii) the switching signal is well-defined at x_0 over $[t_0, t_1)$ w.r.t. the system; and
(iii) there is a unique pair (x, θ), where $x: [t_0, t_1) \mapsto \mathbf{R}^n$ is absolutely continuous and $\theta: [t_0, t_1) \mapsto M$ is well-defined, such that $x(t_0) = x_0$, and for almost any $t \in [t_0, t_1)$, we have

$$\delta x(t) = A_{\theta(t)} x(t) + B_{\theta(t)} u(t)$$
$$y(t) = C_{\theta(t)} x(t)$$
$$\theta(t+) = \psi(t, \theta(t), x(t)/y(t)).$$

In this book, we focus on designing switching signals such that they are well-defined and hence the switched linear systems are well-posed.

1.3.4 Switching Sequences

For a well-defined switching path θ defined on interval $[t_0, t_1)$, it has a finite number of jump instants in any finite length sub-interval of $[t_0, t_1)$. Any jump instant $t \in (t_0, t_1)$ is said to be a *switching time*. That is, a switching time t satisfies

$$\lim_{s \uparrow t} \theta(s) \neq \lim_{s \downarrow t} \theta(s).$$

As we have mentioned, for a continuous-time switching path, a switching time must be a discontinuous time, but the converse is not necessarily true. In fact, for any switched system, suppose that two switching paths are identical except on a set of isolated times, then, the two paths lead to the same system behavior. This implies that changing values of a switching path on isolated time instants does not affect the system dynamics. In the book, when we design a switching path, we usually make it to be either $\theta(t) = \theta(t+)$ in which the path is continuous from the right, or to be $\theta(t) = \theta(t-)$ in which the path is continuous from the left. Besides, we always assume that the switching path is continuous everywhere except at the switching times.

Note also that, for a well-defined path θ, the set of its switching times is a set of isolated instants. Let $s_1, s_2, \cdots, s_l$ be the ordered switching times in $[t_0, t_1)$ with

$$t_0 < s_1 < s_2 < \cdots < s_l < t_1.$$

Note that l is a nonnegative integer and is possibly infinite when $t_1 = \infty$. Let $s_0 = t_0$. The ordered sequence $\{s_0, s_1, \cdots, s_l\} = \{s_i\}_{i=0}^l$, is said to be the *switching time sequence of* θ on $[t_0, t_1)$. We denote by $TS_\theta^{[t_0,t_1)}$ this sequence and also TS_θ when the interval $[t_0, t_1)$ is clear from the context. Similarly, the index sequence $\{\theta(t_0+), \theta(s_1+), \cdots, \theta(s_l+)\} = \{\theta(s_i+)\}_{i=0}^l$ is said to be the *switching index sequence of* θ on $[t_0, t_1)$, and is denoted by $IS_\theta^{[t_0,t_1)}$ or IS_θ in short. The sequence of ordered pairs

$$\{(t_0, \theta(t_0+)), (s_1, \theta(s_1+)), \cdots, (s_l, \theta(s_l+))\} = \{(s_i, \theta(s_i+))\}_{i=0}^l$$

is said to be the *switching sequence of* θ over $[t_0, t_1)$, and will be denoted by $SS_\theta^{[t_0,t_1)}$ or SS_θ in short. Note that a switching sequence $\{(s_i, k_i)\}_{i=0}^l$ uniquely determines a switching path (up to possibly re-arranging the value at the switching times) by the relationship

$$\theta(t) = \begin{cases} k_0 & t \in [t_0, s_1) \\ k_1 & t \in [s_1, s_2) \\ \vdots \\ k_l & t \in [s_l, t_1). \end{cases}$$

Finally, let $h_i = s_{i+1} - s_i$ for $i = 0, \cdots, l-1$ and $h_l = t_1 - s_l$, then the sequence

$$\{(\theta(t_0+), h_0), \cdots, (\theta(s_l+), h_l)\}$$

is said to be the *switching duration sequence* of θ on $[t_0, t_1)$, and will be denoted by $DS_\theta^{[t_0, t_1)}$ or DS_θ in short. It is clear that the switching duration sequence is uniquely determined by the switching sequence, and vice versa.

The same sequences for a switching signal can be defined through the switching path generated by the switching signal. Suppose that switching signal σ is well-defined at x_0 over $[t_0, t_1)$, and $\theta_{[t_0, t_1)}$ is the switching path generated by the switching signal at x_0. The switching (time/index/duration) sequence at x_0 over $[t_0, t_1)$ is defined to be the switching (time/index/duration) sequence of the switching path $\theta_{[t_0, t_1)}$. To include the information of the initial condition, we add the initial state to the sequence. For example, a switching sequence is in the form

$$\{x_0, (s_0, k_0), (s_1, k_1), \cdots, (s_l, k_l)\}.$$

It should be noted that a well-defined switching law always generates a unique switching path at a given initial condition which is also well-defined. In general, different initial conditions may correspond to different switching paths and hence different switching sequences. This is a critical feature which makes the switched linear system essentially distinct from a linear time-varying system, where all initial states correspond to a single path. To further understand this point, let us examine a simple example.

Example 1.5. Consider a planar switched linear system given by

$$\dot{x}(t) = A_\sigma x(t) \quad \sigma \in \{1, 2\}$$

with

$$A_1 = \begin{bmatrix} 1 & 0 \\ 0 & -2 \end{bmatrix} \text{ and } A_2 = \begin{bmatrix} -2 & 0 \\ 0 & 1 \end{bmatrix}.$$

Let the switching signal be

$$\sigma(t) = \arg\min\{x^T(t) A_1 x(t), x^T(t) A_2 x(t)\}. \tag{1.23}$$

The switching signal is event-driven so it is initial-state-dependent. In general, the switching sequences may be finite or infinite for different initial conditions. It can be seen that, if the initial condition is $x(0) = x_0 = [a, 0]^T$ with $a \neq 0$, then the corresponding switching sequence is $\{x_0; (0, 2)\}$, which means that the second subsystem is always active hence there is no switch at all. Similarly, if the initial condition is $x(0) = x_0 = [0, b]^T$ with $b \neq 0$, then the corresponding switching sequence is $\{x_0; (0, 1)\}$, which means that the first subsystem is always active and there is no switch at all. In both cases, the switched system degenerates into linear time-invariant systems. For initial state $x(0) = x_0 = [a, b]^T$ with $ab \neq 0$, however, the switching sequence is always infinite.

This example clearly illustrates the fact that the switching sequence depends heavily on the initial state.

Finally, a determinant switching signal is said to be with *dwell time* τ, if $t_{i+1} - t_i \geq \tau$ for any two consecutive switching times t_i and t_{i+1}. Let $\mathcal{S}_\tau$ be the set of switching signals with dwell-time τ. It is clear that any switching signal with a positive dwell time is well-defined. The converse is not necessarily to be true, for instance, the switching path

$$\theta(t) = \begin{cases} 1 & \text{if } t \in [k - \frac{1}{k}, k + \frac{1}{k}) \text{ for some } k \in \mathbf{N}^+ \\ 2 & \text{otherwise} \end{cases}$$

is completely well-defined over $[0, \infty)$, but it does not permit a positive dwell time.

1.3.5 Solutions of Switched Linear Systems

For a nonlinear system, usually it is very hard (if not impossible) to explicitly express the solution in terms of the system parameters in an analytic way. On the other hand, for well-posed switched linear systems, this is always possible.

For clarity, let $\phi(t; t_0, x_0, u, \sigma)$ denote the state trajectory at time t of a continuous-time switched linear system

$$\dot{x}(t) = A_\sigma x(t) + B_\sigma u(t) \tag{1.24}$$

initialized at $x(t_0) = x_0$ with input u and switching signal σ.

Suppose that the switching signal is well-defined and its switching sequence is

$$\{x_0, (t_0, i_0), (t_1, i_1), \cdots, (t_l, i_l)\}.$$

As the i_0th subsystem is active during $[t_0, t_1)$, we have

$$\dot{x}(t) = A_{i_0} x(t) + B_{i_0} u(t) \quad x(t_0) = x_0 \quad t \in [t_0, t_1).$$

This is a linear differential equation with an initial condition, so its solution can be given explicitly by

$$\phi(t; t_0, x_0, u, \sigma) = e^{A_{i_0}(t-t_0)} x_0 + \int_{t_0}^t e^{A_{i_0}(t-\tau)} B_{i_0} u(\tau) d\tau \quad t \in [t_0, t_1)$$

and (by the continuity of the state trajectory)

$$x_1 = x(t_1) = \phi(t_1; t_0, x_0, u, \sigma) = e^{A_{i_0}(t_1-t_0)} x_0 + \int_{t_0}^{t_1} e^{A_{i_0}(t_1-\tau)} B_{i_0} u(\tau) d\tau.$$

During period $[t_1, t_2)$, the i_1th subsystem is active, thus, we have

$$\dot{x}(t) = A_{i_1} x(t) + B_{i_1} u(t) \quad x(t_1) = x_1 \quad t \in [t_1, t_2).$$

Again, the solution can be given explicitly by

$$\phi(t; t_0, x_0, u, \sigma) = e^{A_{i_1}(t-t_1)} x_1 + \int_{t_1}^{t} e^{A_{i_1}(t-\tau)} B_{i_1} u(\tau) d\tau$$

$$= e^{A_{i_1}(t-t_1)} e^{A_{i_0}(t_1-t_0)} x_0 + e^{A_{i_1}(t-t_1)} \int_{t_0}^{t_1} e^{A_{i_0}(t_1-\tau)} B_{i_0} u(\tau) d\tau$$

$$+ \int_{t_1}^{t} e^{A_{i_1}(t-\tau)} B_{i_1} u(\tau) d\tau \quad t \in [t_1, t_2)$$

and

$$x_2 = x(t_2) = \phi(t_2; t_0, x_0, u, \sigma) = e^{A_{i_1}(t_2-t_1)} x_1 + \int_{t_1}^{t_2} e^{A_{i_1}(t-\tau)} B_{i_1} u(\tau) d\tau.$$

Continuing with the above procedure, the solution for the switched system can be computed to be

$$\phi(t; t_0, x_0, u, \sigma) = e^{A_{i_k}(t-t_k)} e^{A_{i_{k-1}}(t_k-t_{k-1})} \cdots e^{A_{i_1}(t_2-t_1)} e^{A_{i_0}(t_1-t_0)} x_0$$

$$+ e^{A_{i_k}(t-t_k)} \cdots e^{A_{i_1}(t_2-t_1)} \int_{t_0}^{t_1} e^{A_{i_0}(t_1-\tau)} B_{i_0} u(\tau) d\tau$$

$$+ \cdots + e^{A_{i_k}(t-t_k)} \int_{t_{k-1}}^{t_k} e^{A_{i_{k-1}}(t_k-\tau)} B_{i_{k-1}} u(\tau) d\tau$$

$$+ \int_{t_k}^{t} e^{A_{i_k}(t-\tau)} B_{i_k} u(\tau) d\tau \quad t \in [t_k, t_{k+1}). \tag{1.25}$$

Let

$$\Psi(t, \sigma, x_0) = e^{A_{i_k}(t-t_k)} e^{A_{i_{k-1}}(t_k-t_{k-1})} \cdots e^{A_{i_0}(t_1-t_0)} \quad t \in [t_k, t_{k+1}].$$

It is clear that the state transition matrix is given by

$$\Phi(t_1, t_2, \sigma, x_0) = \Psi(t_1, \sigma, x_0) \left(\Psi(t_2, \sigma, x_0)\right)^{-1}.$$

The solution of the system can be re-written, in terms of the transition matrix, as

$$\phi(t; t_0, x_0, u, \sigma) = \Phi(t, t_0, \sigma, x_0) x_0 + \int_{t_0}^{t} \Phi(t, \tau, \sigma, x_0) u(\tau) d\tau.$$

From this expression, we can draw a few useful conclusions as follows:

(i) For a switched linear system, if the switching signal is well-defined and the input is globally integrable, then the system always permits a unique solution for the forward time space.

(ii) The solution is usually not continuously differentiable at the switching instants, even if the input is smooth.

(iii) The state transition matrix is a multiple multiplication of matrix function of the form e^{At}. Accordingly, properties of functions in this form play an important role in the analysis of switched linear systems.

Similarly, for a discrete-time switched linear system

$$x_{k+1} = A_\sigma x_k + B_\sigma u_k \tag{1.26}$$

the solution is

$$x_k = A_{\sigma(k-1)} \cdots A_{\sigma(1)} A_{\sigma(0)} x_0 + A_{\sigma(k-1)} \cdots A_{\sigma(1)} B_{\sigma(0)} u_0 + \cdots$$
$$+ A_{\sigma(k-1)} B_{\sigma(k-2)} u_{k-2} + B_{\sigma(k-1)} u_{k-1}. \tag{1.27}$$

The state transition matrix is

$$\Phi(k_1, k_2, \sigma) = A_{\sigma(k_1-1)} \cdots A_{\sigma(k_2)} \quad k_1 > k_2.$$

In terms of the transition matrix, the solution can be rewritten as

$$x_k = \Phi(k, 0, \sigma) x_0 + \sum_{j=0}^{k-1} \Phi(k, j, \sigma) u_j.$$

From the solution, we have:

(i) For a switched linear system, the system permits a unique solution for the forward time space. Hence, any discrete-time switched system is well-posed.

(ii) The state transition matrix is a multiple multiplication of matrices. Accordingly, properties of matrix multiplication play an important role in analyzing the switched system.

Finally, for both continuous-time and discrete-time switched linear systems, the state trajectory possesses several nice properties under mild conditions. As the properties are widely used in the following chapters, we state them separately as propositions for easy reference.

Proposition 1.6. *For the switched linear system, suppose that the switching signal is time-invariant. Then, for any given initial condition (t_0, x_0), input u, and time τ, we have*

$$\phi(t; t_0, x_0, u, \sigma) = \phi(t + \tau; t_0 + \tau, x_0, u', \sigma') \quad \forall\, t \geq t_0$$

where $u'(t) = u(t - \tau)$ for $t \geq t_0 + \tau$, and $\sigma' = \sigma^{\to h}$ is the time-transition of σ by h.

The proposition asserts that the state trajectory is time-transition invariant when the switching signal is time-invariant, hence we term it as the *time-transition invariance property*. This means that, if $x(\cdot)$ is a trajectory of the switched system with some initial condition (t_0, x_0), then, for any $\tau \in \mathbf{R}$, its time transition $y(\cdot) = x(\cdot + \tau)$ is also a trajectory of the switched system with the same initial state. In other words, the set of state trajectories is invariant under any time transition. Accordingly, the switched linear system behaves in a 'time-invariant' manner.

Proposition 1.7. *For the switched linear system, suppose that the switching signal is radial-invariant. Then, for any given initial condition (t_0, x_0), input u, and $\lambda \in \mathbf{R}$, we have*

$$\phi(t; t_0, \lambda x_0, u, \sigma) = \lambda \phi(t; t_0, x_0, u^\lambda, \sigma) \ \forall\, t \ge t_0 \ \lambda \in \mathbf{R}$$

where $(u^\lambda)(t) = \lambda u(t)$ for $t \ge t_0$.

The proposition asserts that the state trajectory is linear *w.r.t.* the initial state when the switching signal is radial-invariant, hence we term it as the *radial-transition invariance property*. This means that, if $x(\cdot)$ is a trajectory of the switched system with initial condition $x(t_0) = x_0$, then its radial transition $y(\cdot) = \lambda x(\cdot)$ is also a trajectory of the switched system with the initial state $y(t_0) = \lambda x_0$. In other words, the switched linear system behaves in a 'radially linear' manner.

For a completely well-posed switched linear system, let $\mathcal{U}_{[t_0,\infty)}$ denote the set of inputs which are piecewise continuous over $[t_0, \infty)$. For a well-defined switching path $\theta_{[t_0,\infty)} \in \mathcal{S}_{[t_0,\infty)}$, let

$$\mathcal{T}^\theta_{[t_0,\infty)}(x_0) = \{\phi(t; t_0, x_0, u, \theta) \colon \forall\, t \in [t_0, \infty), \ u \in \mathcal{U}_{[t_0,\infty)}\}.$$

This set includes the states attainable from $x(t_0) = x_0$ via the switching path θ over $[t_0, \infty)$. As a switching path is both time-invariant and radial-invariant, Proposition 1.6 implies that set $\mathcal{T}^\theta_{[t_0,\infty)}(x_0)$ is t_0-independent, *i.e.*,

$$\mathcal{T}^{\theta \to (s_2 - s_1)}_{[s_1,\infty)}(x_0) = \mathcal{T}^\theta_{[s_2,\infty)}(x_0) \ \ \forall\, s_1, s_2 \in \mathbf{R} \ \ x_0 \in \mathbf{R}^n.$$

Similarly, Proposition 1.7 implies that the set is radially linear *w.r.t.* the initial state, *i.e.*,

$$\mathcal{T}^\theta_{[t_0,\infty)}(\lambda x_0) = \lambda \mathcal{T}^\theta_{[t_0,\infty)}(x_0) \ \ \forall\, t_0, \lambda \in \mathbf{R} \ \ x_0 \in \mathbf{R}^n$$

where $\lambda \Omega = \{\lambda x \colon x \in \Omega\}$ for a set Ω. Furthermore, let

$$\mathcal{T}_{[t_0,\infty)}(x_0) = \cup_{\theta \in \mathcal{S}_{[t_0,\infty)}} \mathcal{T}^\theta_{[t_0,\infty)}(x_0)$$

and, for a set $\Omega \subseteq \mathbf{R}^n$

$$\mathcal{T}_{[t_0,\infty)}(\Omega) = \cup_{x \in \Omega} \mathcal{T}_{[t_0,\infty)}(x).$$

It can be seen that the sets are independent of t_0 and are radially linear. As a result, if Ω is a neighborhood of the origin (*i.e.*, contains the origin as an interior point), then

$$\mathcal{T}_{[t_0,\infty)}(\Omega) = \mathcal{T}_{[t_1,\infty)}(\mathbf{R}^n) \quad \forall \ t_0, t_1 \in \mathbf{R}.$$

On the other hand, as any well-defined switching signal generates a unique switching path at any $x_0 \in \mathbf{R}^n$ which is in $\mathcal{S}_{[t_0,\infty)}$, the set $\mathcal{T}_{[t_0,\infty)}(\Omega)$ in fact contains all the states attainable from Ω via any well-defined switching signal over $[t_0, \infty)$, including those that are not time-invariant and/or radial-invariant.

The above discussions reveal several fundamental features of switched linear systems. As an implication, for a property of the state that does not explicitly rely on the time, we have the following observations:

(i) if the property holds at some t_0, then it also holds at any other time; and
(ii) if such a property can be achievable via a well-defined switching signal, then it can also be achievable via a well-defined switching signal that is transition-invariant.

To make the above observations clearer, we discuss here a typical type of property that does not explicitly rely on the time. Let Ω_1 and Ω_2 be two subsets of $\mathbf{R}^n$, any property of the form

(the closure of) $\mathcal{T}_{[t_0,\infty)}(\Omega_1)$ contains/spans Ω_2

does not explicitly rely on the time t_0. This form includes a few special but important cases. To see this, let

$$\mathbf{S}_r = \{x \in \mathbf{R}^n \colon \|x\| = r\} \text{ and } \mathbf{B}_r = \{x \in \mathbf{R}^n \colon \|x\| \leq r\} \quad r \in \mathbf{R}_+.$$

Note that for an unforced switched linear system, the derivative of the state is bounded by a linear function of state. Note also that the origin is, or can be made to be, an equilibrium of the unforced/forced switched system. Taking into account the above observations, we can re-state the following well-known properties as:

(a) *(uniform) stability at t_0*: $(u \equiv 0)$ $\exists \ r < 1$ such that

$$x \in \mathbf{S}_r \implies \mathcal{T}_{[t_0,\infty)}(x) \in \mathbf{B}_1$$

(b) *asymptotic stability at t_0*: $(u \equiv 0)$ stability plus $0 \in \mathrm{cl}\mathcal{T}_{[t_0,\infty)}(x)$ for any $x \in \mathbf{S}_1$, where 'cl' denotes the closure of a set;
(c) *controllability at t_0*: $x \in \mathbf{S}_1 \implies 0 \in \mathcal{T}_{[t_0,\infty)}(x)$; and
(d) *reachability at t_0*: $\mathcal{T}_{[t_0,\infty)}(0) \supset \mathbf{S}_1$.

As a result, these properties do not rely on the initial time t_0. For instance, if the switched linear system is controllable at t_0, then it is also controllable at any other time instant. Other related properties, such as the (asymptotic) stabilizability and observability, also possess this initial-time-independent feature. In this book, when we formulate and discuss the notions, we simply assume that $t_0 = 0$ without loss of generality. Similarly, when we seek switching signals that make the switched system stable or controllable, we always seek from the set of well-defined switching paths or the set of well-defined and transition-invariant switching signals.

Finally, note that in Items (b)-(d), the unit sphere $\mathbf{S}_1$ can be replaced by any neighborhood of the origin, or any sphere that encircles the origin. This means that the local properties (asymptotic stability, *etc.*) are in fact equivalent to the global ones. Accordingly, we do not distinguish these local properties from their global counterparts.

1.4 Notes and References

The switched system is a special form of the hybrid dynamic system which contains both continuous dynamics and discrete elements. Hybrid systems have attracted considerable attention from both the control and computer communities. However, as general hybrid models are quite complex, the study of hybrid systems is still in the elementary stage [18]. The reader is referred to the monographs [156, 117] for an introduction. There has also been an annual conference on hybrid systems since 1993 [70, 71].

The study of switched systems has a long history and could be traced back to the pioneering work on the Lyapunov approach for the absolute stability of Lur'e systems. It is well known that a Lur'e system is absolutely stable if there is a common Lyapunov function for all the extreme systems. This corresponds to the stability of a switched linear system with subsystems coinciding with the extreme systems. This idea had been extended later to address the stability of time-varying systems and the robustness of uncertain linear systems.

Since the 1990s, switched systems have attracted increasing attention and recent years have witnessed an enormous growth of interest in switched systems. The literature, especially that on the stability issues, grew at an exponential rate. Various mathematical tools, such as linear and multilinear algebra, nonsmooth analysis, and game theory, have been exploited in the study of switched systems. The reader is referred to the survey papers [31, 92, 99, 173, 102, 141] and the monograph [90] for recent development. For switched systems where the switching signals are governed by random processes (*e.g.*, jump linear systems), the reader is referred to [84, 149, 97] for surveys and [42, 29] for recent development. In addition, the advantages of multiple-controller switching in complex systems were exploited in [45, 109, 110, 98, 88, 87], and more tutorial and introductory material can be found in the special issues [3, 106, 2, 34].

2

Mathematical Preliminaries

2.1 Introduction

This chapter presents a number of useful mathematical concepts and tools, many of which implicitly or explicitly underlie the material to be covered in the main portion of the book. While much of the material is standard and can be found in classical textbooks, we also include a number of useful items that are not commonly found elsewhere. Thus, this chapter serves as a short review and as a convenient reference when necessary. In addition, this chapter forms the basis for the subsequent development.

2.2 Linear Spaces

A *linear (vector) space* consists of an additive group, of elements called *vectors*, and an underlying field of scalars. We consider only spaces over the field of real numbers, **R**. Linear spaces are denoted by script capitals $\mathcal{X}, \mathcal{Y}, \cdots$; their elements (vectors) by lower case Roman letters x, y, $\cdots$ and field elements (real numbers) by lower case Roman or Greek letters.

Suppose that $\mathcal{X}$ is a linear space and $x_1, \cdots, x_k$ its elements. The *span*, denoted by

$$\text{span}\{x_1, \cdots, x_k\}$$

is the set of all linear combinations of the x_i, with coefficients in **R**. Space $\mathcal{X}$ is said to be *finite-dimensional* if there exist a finite k and vectors $x_1, \cdots, x_k$ such that

$$\mathcal{X} = \text{span}\{x_1, \cdots, x_k\}.$$

Suppose that $\mathcal{X} \neq 0$, the *dimension of $\mathcal{X}$*, denoted by $\dim \mathcal{X}$, is the least k which this happens. Let $\dim 0 = 0$.

A set of vectors $\{x_i, i = 1, \cdots, k\}$ is said to be *(linearly) independent*, if the relation

$$\sum_{i=1}^{k} c_i x_i = 0 \quad c_i \in \mathbf{R}$$

implies $c_i = 0$ for all $i \in \bar{k} \stackrel{def}{=} \{1, \cdots, k\}$. If $\{x_i, i \in \bar{k}\}$ is independent and $x \in \text{span}\{x_i, i \in \bar{k}\}$, then the representation

$$x = c_1 x_1 + \cdots + c_m x_m$$

is unique for each $x \in \mathcal{X}$. For an n-dimensional space $\mathcal{X}$, there exists a set of independent vectors $\{x_i, i \in \bar{n}\}$, called a basis for $\mathcal{X}$, which spans the space

$$\mathcal{X} = \text{span}\{x_i, i \in \bar{n}\}.$$

Unless otherwise stated, all linear spaces in this book are finite dimensional.

A *(linear) subspace* $\mathcal{Y}$ of linear space $\mathcal{X}$ is a nonempty subset of $\mathcal{X}$ which is a linear space under the operations of vector addition and scalar multiplication inherited from $\mathcal{X}$. That is, for any vectors $y_1, y_2 \in \mathcal{Y}$ and any real scalars c_1 and c_2, we have $c_1 y_1 + c_2 y_2 \in \mathcal{Y}$. The notation $\mathcal{Y} \subseteq \mathcal{X}$ will henceforth mean that $\mathcal{Y}$ is a subspace of $\mathcal{X}$. Geometrically, a subspace is a hyperplane passing through the origin of $\mathcal{X}$. Thus, the zero vector belongs to any subspace.

Suppose that $\mathcal{Y}$ and $\mathcal{Z}$ are subspaces of $\mathcal{X}$, we define their *summation* and *intersection* by

$$\mathcal{Y} + \mathcal{Z} = \{y + z : y \in \mathcal{Y}, z \in \mathcal{Z}\}$$
$$\mathcal{Y} \cap \mathcal{Z} = \{x : x \in \mathcal{Y}, x \in \mathcal{Z}\}.$$

It is clear that the summation and intersection of subspaces are also subspaces. In contrast, unions of subspaces are generally not subspaces. Another important distinction between summation and union of subspaces lies in that, it is always possible for a linear space to be expressed as a summation of subspaces of strictly lesser dimensions, while a finite dimensional linear space can in no way be expressed as a finite (even countable) union of subspaces of strictly lesser dimensions. The latter property, known as the *Baire's Category Theorem* [169], will play a role in proving Theorems 4.30 and 4.31.

Given a linear space, the family of all its subspaces is partially ordered by subspace inclusion ($\subseteq$), and under the operation $+$ and $\cap$, is to form a lattice, namely, $\mathcal{Y} + \mathcal{Z}$ is the minimum subspace that contains both $\mathcal{Y}$ and $\mathcal{Z}$, while $\mathcal{Y} \cap \mathcal{Z}$ is the largest subspace contained in both $\mathcal{Y}$ and $\mathcal{Z}$.

A family of k subspaces $\mathcal{X}_1, \cdots, \mathcal{X}_k$ of $\mathcal{X}$ is said to be *independent*, if

$$\mathcal{X}_i \cap \left(\sum_{j \neq i} \mathcal{X}_j\right) = 0 \quad \forall\, i \in \bar{k}.$$

If the family $\{\mathcal{X}_i, i \in \bar{k}\}$ is independent, then every vector in $\sum_{i=1}^{k} \mathcal{X}_i$ has a unique representation

$$x = x_1 + \cdots + x_k \quad x_i \in \mathcal{X}_i \quad i \in \bar{k}.$$

The sum $\mathcal{X}_0 = \sum_{i=1}^{k} \mathcal{X}_i$ is called an *internal direct sum*, and is written as

$$\mathcal{X}_0 = \mathcal{X}_1 \oplus \cdots \oplus \mathcal{X}_k = \oplus_{i=1}^{k} \mathcal{X}_i.$$

If $\mathcal{Y}$ and $\mathcal{Z}$ are subspaces of $\mathcal{X}$, then there exist $\mathcal{Y}_1 \subseteq \mathcal{Y}$ and $\mathcal{Z}_1 \subseteq \mathcal{Z}$ such that

$$\mathcal{Y} + \mathcal{Z} = \mathcal{Y}_1 \oplus (\mathcal{Y} \cap \mathcal{Z}) \oplus \mathcal{Z}_1.$$

Let $\mathcal{X}_1$ and $\mathcal{X}_2$ be arbitrary linear spaces. The *external direct sum* of $\mathcal{X}_1$ and $\mathcal{X}_2$, denoted by $\mathcal{X}_1 \tilde{\oplus} \mathcal{X}_2$, is the linear space of all ordered pairs $\{(x_1, x_2) \colon x_1 \in \mathcal{X}_1, x_2 \in \mathcal{X}_2\}$, under componentwise addition and scalar multiplication. Under isomorphism $\approx$, we have

$$\tilde{\oplus}_{i=1}^{k} \mathcal{X}_i \approx \oplus_{i=1}^{k} \mathcal{X}_i.$$

We shall not distinguish between the two types of direct sum, and will denote both by $\oplus$.

Finally, let $\mathcal{Y} \subseteq \mathcal{X}$. Vectors $x, y \in \mathcal{X}$ are said to be *equivalent mod* $\mathcal{Y}$ if $x - y \in \mathcal{Y}$. We define the *quotient space* $\frac{\mathcal{X}}{\mathcal{Y}}$ as the set of equivalence classes

$$\bar{x} = \{y \in \mathcal{X} \colon y - x \in \mathcal{Y}\}, x \in \mathcal{X}.$$

The function $x \mapsto \bar{x}$ is a map $P \colon \mathcal{X} \mapsto \frac{\mathcal{X}}{\mathcal{Y}}$ called the *canonical projection* of $\mathcal{X}$ on $\frac{\mathcal{X}}{\mathcal{Y}}$.

2.3 Maps and Matrices

Let $\mathcal{X}$ and $\mathcal{Y}$ be linear spaces. A function $\varphi \colon \mathcal{X} \mapsto \mathcal{Y}$ is a linear transformation (map) if

$$\varphi(c_1 x_1 + c_2 x_2) = c_1 \varphi(x_1) + c_2 \varphi(x_2) \quad \forall x_1, x_2 \in \mathcal{X} \quad c_1, c_2 \in \mathbf{R}.$$

$\mathcal{X}$ is the *domain* of C and $\mathcal{Y}$ is the *codomain*. If $\mathcal{Y} = \mathcal{X}$, then the map is an *endomorphism* of $\mathcal{X}$.

Let $\{x_i, i \in \bar{n}\}$ be a basis for $\mathcal{X}$ and $\{y_j, j \in \bar{m}\}$ a basis for $\mathcal{Y}$. For any linear map $C \colon \mathcal{X} \mapsto \mathcal{Y}$, we have

$$C x_i = c_{1i} y_1 + \cdots + c_{mi} y_m \quad i \in \bar{n}$$

for uniquely determined elements $c_{ji} \in \mathbf{R}$. The array

$$\mathrm{Mat}\, C = \begin{bmatrix} c_{11} & \cdots & c_{1n} \\ & \ddots & \\ c_{m1} & \cdots & c_{mn} \end{bmatrix}$$

is the matrix of C w.r.t. the given basis pair. Usually, we do not distinguish between a linear map and its matrix representation, and write $C = \mathrm{Mat}\, C$. We denote by $\mathbf{R}^{m \times n}$ the class of all m by n matrices with real entries.

Let $C \colon \mathcal{X} \mapsto \mathcal{Y}$ be a map, $\mathcal{V} \subseteq \mathcal{X}$, and $\mathcal{W} \subseteq \mathcal{Y}$. The *image* of C on $\mathcal{V}$ is the set

$$C\mathcal{V} = \{Cx \colon x \in \mathcal{V}\} \subseteq \mathcal{Y}$$

while the *kernel* of C on $\mathcal{W}$ is the set

$$C^{-1}\mathcal{W} = \{x \in \mathcal{X} \colon Cx \in \mathcal{W}\} \subseteq \mathcal{X}.$$

Specifically, we denote $C\mathcal{X}$ by $\mathrm{Im}\, C$, and $C^{-1}0$ by $\mathrm{Ker}\, C$. Both $\mathrm{Im}\, C$ and $\mathrm{Ker}\, C$ are subspaces.

Let $\mathcal{X} = \mathcal{R}_1 \oplus \mathcal{R}_2$. Since the representation $x = r + s$, $r \in \mathcal{R}_1$ and $s \in \mathcal{R}_2$, is unique for each $x \in \mathcal{X}$, there is a function $x \mapsto r$, called the *projection* on $\mathcal{R}_1$ along $\mathcal{R}_2$. The projection is in fact an endomorphism $Q \colon \mathcal{X} \mapsto \mathcal{X}$ such that $\mathrm{Im}\, Q = \mathcal{R}_1$ and $\mathrm{Ker}\, Q = \mathcal{R}_2$. Similarly, there is a projection on $\mathcal{R}_2$ along $\mathcal{R}_1$, which corresponds to an endomorphism P. It is clear that

$$I_d = Q + P$$

where I_d stands for the identity map mapping any vector into itself.

We assume that the reader is familiar with the rules of matrix algebra. For a column (or row) vector a, $a(j)$ denotes its jth entry; for a matrix A, $A(i,j)$ denotes its (i,j)th entry. Given a real matrix A, denote its *rank* by $\mathrm{rank}\, A$, its *determinant* by $\det A$ (when A is square). The nth-order identity matrix is denoted by I_n. The *characteristic polynomial* of an nth-order square matrix A is the nth degree monic polynomial

$$\pi(A) = \det(sI_n - A).$$

The *spectrum* of A, denoted by $\lambda(A)$, is the set of n complex zeros of its characteristic polynomial, listed according to multiplicity. The elements of the spectrum are the *eigenvalues* of A. The *spectral radius* of A, written as $\mathrm{sr}\, A$, is the radius of the smallest disc centered at the origin in the complex plane that includes all the eigenvalues of A.

A matrix P is said to be *positive (negative) definite*, denoted by $P > 0$ $(P < 0)$, if it is symmetric and all its eigenvalues are positive (negative). Similarly, a semi-positive (semi-negative) definite matrix P is written as $P \geq 0$ $(P \leq 0)$. A matrix A is said to be *stable*, or *Hurwitz*, if its spectrum locates in the open left half of the complex plane. Similarly, matrix A is said to be *discrete-time stable* or *Schur*, if its spectrum locates in the open unit ball of the complex plane.

The inverse of a nonsingular matrix T is denoted by T^{-1}. The set of nonsingular matrices in $\mathbf{R}^{n \times n}$, denoted by $Gl(n)$, forms a group under matrix multiplication. Two matrices A and B are said to be *similar*, if $\exists\, T \in Gl(n)$ such that $A = T^{-1}BT$. The relation of similarity is an equivalence relation and two similar matrices represent the same linear map under different bases. Indeed, let $C \colon \mathbf{R}^n \mapsto \mathbf{R}^n$ be an endomorphism. Suppose that $\mathrm{Mat}\, C$ is a matrix representation of C under a basis of $\mathbf{R}^n$. Then, the matrix set

$$\{T(\mathrm{Mat}\, C)T^{-1} \colon T \in Gl(n)\}$$

defines all possible matrix representations under different bases (coordinates). The simplest representations, usually judged by the number of zero entries and the possession of triangular structures, are said to be the *canonical forms*. One canonical form for any linear map A, known as the *rational canonical form*, is the block diagonal form

$$\mathrm{Mat}\, A = \mathrm{diag}(A_1, \cdots, A_k) = \begin{bmatrix} A_1 & \cdots & 0 \\ & \ddots & \\ 0 & \cdots & A_k \end{bmatrix}$$

where $A_i, i \in \bar{k}$ is in the companion form

$$A_i = \begin{bmatrix} 0 & 1 & 0 & \cdots & 0 \\ 0 & 0 & 1 & \cdots & 0 \\ & & \ddots & & \\ 0 & 0 & 0 & \cdots & 1 \\ a_{i1} & a_{i2} & a_{i3} & \cdots & a_{ij_i} \end{bmatrix}.$$

Let $\| \cdot \|$ denote a norm of a vector or induced norm of a matrix. Well-known norms including the l_2 (Euclidean) norm and l_∞ (max) norm. For any matrix norm $\| \cdot \|$, we have

$$\|A\| \geq \mathrm{sr}(A) = \lim_{k \to \infty} \|A^k\|^{\frac{1}{k}} \quad \forall\, A \in \mathbf{R}^{n \times n}.$$

For a matrix exponential function e^{At}, we have

$$\|e^{At}\| \leq p(t)e^{\gamma t} \quad \forall\, t \in \mathbf{R}_+ \tag{2.1}$$

where $p(t)$ is a polynomial with a degree less than n, and $\gamma = \max\{\Re s \colon s \in \gamma(A)\}$ is the maximum real part of the spectrum of A. In general, γ is said to be the *convergence rate* of A as there is no smaller γ for Inequality (2.1) to hold. By (2.1), for any $\alpha > \gamma$, there is a real constant $\beta = \beta(\alpha)$ such that

$$\|e^{At}\| \leq \beta e^{\alpha t} \quad \forall\, t \in \mathbf{R}_+.$$

Another useful estimation is

$$\|e^{At}\| \le \sum_{k=0}^{\infty} \frac{(\|A\|t)^k}{k!} = e^{\|A\|t} \quad \forall \, t \in \mathbf{R}_+$$

where $k!$ is the factorial of k with $0! = 1$.

2.4 Invariant Subspaces and Controllable Subspaces

Let $A\colon \mathcal{X} \mapsto \mathcal{X}$ be an endomorphism and $\mathcal{Y}$ be a subspace of $\mathcal{X}$. $\mathcal{Y}$ is said to be A-*invariant* if $A\mathcal{Y} \subseteq \mathcal{Y}$. Denote $\mathcal{V} = \frac{\mathcal{X}}{\mathcal{Y}}$, and let $P\colon \mathcal{X} \mapsto \mathcal{V}$ be the canonical projection. Then, there exists a unique map $\bar{A}\colon \mathcal{V} \mapsto \mathcal{V}$ such that $\bar{A}P = PA$.

Suppose that $\mathcal{Y}$ is A-invariant and $\mathcal{Z}$ is any subspace such that $\mathcal{Y} \oplus \mathcal{Z} = \mathcal{X}$. Let $\{y_i, i \in \bar{l}\}$ and $\{z_j, j \in \bar{k}\}$ be bases of $\mathcal{Y}$ and $\mathcal{Z}$, respectively. It is clear that the vectors

$$\{y_1, \cdots, y_l, z_1, \cdots, z_k\}$$

form a basis for $\mathcal{X}$, and under this basis, we have

$$\text{Mat } A = \begin{bmatrix} A_1 & A_3 \\ 0 & A_2 \end{bmatrix} \quad A_1 \colon l \times l \quad A_2 \colon k \times k.$$

Let $A\colon \mathcal{X} \mapsto \mathcal{X}$ and $\mathcal{Y}$ be a subspace of $\mathcal{X}$. We denote the smallest A-invariant subspace containing $\mathcal{Y}$ by $\Gamma_A \mathcal{Y}$. We also term this subspace as the *(single) controllable subspace* of pair $(A, \mathcal{Y})$. This subspace can be described in terms of A and $\mathcal{Y}$ by

$$\Gamma_A \mathcal{Y} = \mathcal{Y} + A\mathcal{Y} + \cdots + A^{n-1}\mathcal{Y}. \tag{2.2}$$

The operation can be defined recursively as

$$\Gamma_{A_1} \Gamma_{A_2} \mathcal{Y} = \Gamma_{A_1}(\Gamma_{A_2}\mathcal{Y}).$$

Similarly, let $\mathbf{A} = \{A_1, \cdots, A_m\}$ be a set of maps. Subspace $\mathcal{Y}$ is said to be *(multiple)* $\mathbf{A}$-*invariant*, if it is A_i-invariant for all $i \in \bar{m}$. Suppose that $\mathcal{Y}$ is $\mathbf{A}$-invariant and $\mathcal{Z}$ is any subspace such that $\mathcal{Y} \oplus \mathcal{Z} = \mathcal{X}$. Let $\{y_i, i \in \bar{l}\}$ and $\{z_j, j \in \bar{k}\}$ be bases of $\mathcal{Y}$ and $\mathcal{Z}$, respectively. In the basis $\{y_1, \cdots, y_l, z_1, \cdots, z_k\}$ of $\mathcal{X}$, we have

$$\text{Mat } A_i = \begin{bmatrix} A_{i1} & A_{i3} \\ 0 & A_{i2} \end{bmatrix} \quad A_{i1} \colon l \times l \quad A_{i2} \colon k \times k \quad i \in \bar{m}.$$

That is, matrices A_i, $i \in \bar{m}$ are simultaneously block triangularizable with consistent block dimensions.

For map set $\mathbf{A} = \{A_1, \cdots, A_m\}$ and subspace $\mathcal{Y}$, let $\Gamma_{\mathbf{A}}\mathcal{Y}$ denote the smallest $\mathbf{A}$-invariant subspace containing $\mathcal{Y}$. This subspace is termed the *(multiple) controllable subspace* of $(\mathbf{A}, \mathcal{Y})$.

Let us define the nested subspaces

$$\mathcal{Y}_0 = \mathcal{Y}$$
$$\mathcal{Y}_{j+1} = \Gamma_{A_1}\mathcal{Y}_j + \cdots + \Gamma_{A_m}\mathcal{Y}_j \quad j = 0, 1, \cdots.$$

Then, we have

$$\Gamma_{\mathbf{A}}\mathcal{Y} = \sum_{k=0}^{\infty} \mathcal{Y}_k.$$

Note that if $\dim\mathcal{Y}_j = \dim\mathcal{Y}_{j+1}$, then $\mathcal{Y}_l = \mathcal{Y}_j$ for any $l > j$. This implies that $\mathcal{Y} = \mathcal{Y}_n$, where $n = \dim \mathcal{X}$. Directly in terms of $\mathbf{A}$ and $\mathcal{Y}$, we have

$$\Gamma_{\mathbf{A}}\mathcal{Y} = \sum_{\substack{j_1, \cdots, j_{n-1} \in \underline{\mathbf{n}} \\ i_0, \cdots, i_{n-1} \in \bar{m}}} A_{i_{n-1}}^{j_{n-1}} \cdots A_{i_1}^{j_1} \mathcal{Y} \tag{2.3}$$

where $\underline{\mathbf{n}} \overset{def}{=} \{0, 1, \cdots, n-1\}$. In (2.3), $\Gamma_{\mathbf{A}}\mathcal{Y}$ is the summation of $(mn)^n$ items. It requires large computational effort to calculate this subspace if m and n are relatively large. To overcome this, here we provide a procedure to determine the subspace efficiently.

Denote the nested subspaces as

$$\mathcal{W}_0 = \mathcal{Y}$$
$$\mathcal{W}_j = \mathcal{W}_{j-1} + \sum_{k=1}^{m} A_k \mathcal{W}_{j-1} \quad j = 1, 2, \cdots.$$

Let $\mathcal{W} = \sum_{j=0}^{\infty} \mathcal{W}_j$. Then, we have

$$\mathcal{W}_0 \subset \mathcal{W}_1 \subset \mathcal{W}_2 \subset \cdots \subset \mathcal{W}$$

and

$$\Gamma_{\mathbf{A}}\mathcal{Y} = \mathcal{W}.$$

Note that if $\mathcal{W}_j = \mathcal{W}_{j+1}$ for some j, then $\mathcal{W}_k = \mathcal{W}_j$ for any $k > j$ and further $\mathcal{W}_j = \mathcal{W}$. This fact, together with $\dim\mathcal{W} \leq n$, implies that $\mathcal{W}_{n-n_0} = \mathcal{W} = \Gamma_{\mathbf{A}}\mathcal{Y}$, where $n_0 = \dim\mathcal{W}_0$.

Denote

$$\rho = \min\{k: \mathcal{W}_k = \Gamma_{\mathbf{A}}\mathcal{Y}\} \leq n - n_0$$

and

$$n_k = \dim \mathcal{W}_k \quad k = 1, \cdots, \rho.$$

A basis of $\Gamma_{\mathbf{A}} \mathcal{Y}$ can be constructed according to the following procedure. First, choose a basis $\gamma_1, \cdots, \gamma_{n_0}$ of $\mathcal{W}_0$.

Next, since

$$\mathcal{W}_1 = \mathcal{W}_0 + \mathrm{span}\{A_k \gamma_j, \ k = 1, \cdots, m, \ j = 1, \cdots, n_0\}$$
$$= \mathrm{span}\{\gamma_1, \cdots, \gamma_{n_0}, A_k \gamma_j, \ k = 1, \cdots, m, \ j = 1, \cdots, n_0\}$$

we can find a basis $\gamma_1, \cdots, \gamma_{n_1}$ of $\mathcal{W}_1$ by searching the set

$$\{\gamma_1, \cdots, \gamma_{n_0}, A_k \gamma_j, \ k = 1, \cdots, m, \ j = 1, \cdots, n_0\}$$

from left to right.

Then, continuing with the process, we can find a basis

$$\gamma_1, \cdots, \gamma_{n_0}, \cdots \gamma_{n_{l-1}+1}, \cdots, \gamma_{n_l}$$

for $\mathcal{W}_l$. Since

$$\mathcal{W}_{l+1} = \mathcal{W}_l + \mathrm{span}\{A_j \gamma_k, \ j = 1, \cdots, m, \ k = n_{l-1} + 1, \cdots, n_l\}$$
$$= \mathrm{span}\{\gamma_1, \cdots, \gamma_{n_k}, A_j \gamma_k, \ j = 1, \cdots, m, \ k = n_{l-1} + 1, \cdots, n_l\}$$

by searching the set

$$\{\gamma_1, \cdots, \gamma_{n_l}, A_j \gamma_k, \ j = 1, \cdots, m, \ k = n_{l-1} + 1, \cdots, n_l\}$$

from left to right for linearly independent column vectors, we can find a basis

$$\gamma_1, \cdots, \gamma_{n_0}, \cdots, \gamma_{n_{l-1}+1}, \cdots, \gamma_{n_l}, \gamma_{n_l+1}, \cdots, \gamma_{n_{l+1}}$$

for $\mathcal{W}_{l+1}$.

Finally, we have

$$\Gamma_{\mathbf{A}} \mathcal{Y} = \mathrm{span}\{\gamma_1, \cdots, \gamma_{n_0}, \cdots, \gamma_{n_{\rho-1}+1}, \cdots, \gamma_{n_\rho}\}.$$

It involves not more than $m^2 n$ column vectors in the procedure, which is only a small fraction of the original quantity, $(mn)^n$.

From the above analysis, if $\gamma_j = A_{i_k} \cdots A_{i_1} b_{i_0}$, then

$$A_{i_l} \cdots A_{i_1} b_{i_0} \in \{\gamma_1, \cdots, \gamma_{n_\rho}\} \quad \forall \, l \in \underline{\mathbf{k}}.$$

As $n_\rho \leq n$, there are at most n different (indexes of) subsystems whose parameters appear in $\{\gamma_1, \cdots, \gamma_{n_\rho}\}$. This implies the following proposition.

Proposition 2.1. *Suppose that* $\mathbf{A} = \{A_i, i \in \bar{m}\}$ *and* $m > n = \dim \mathcal{X}$. *Then, for any subspace* $\mathcal{Y}$, *there exists a subset* $\bar{\mathbf{A}}$ *of* $\mathbf{A}$ *with less than or equal to* n *elements, such that*

$$\Gamma_{\mathbf{A}} \mathcal{Y} = \Gamma_{\bar{\mathbf{A}}} \mathcal{Y}. \tag{2.4}$$

A pair $(\mathbf{A}, \mathcal{Y})$ is said to be *reducible*, if there is a strict subset $\bar{\mathbf{A}}$ of $\mathbf{A}$ such that (2.4) holds. Otherwise, the pair is *irreducible*. Proposition 2.1 asserts that any pair $(\{A_i, i \in \bar{m}\}, \mathcal{Y})$ with $m > n$ is reducible.

2.5 Reachability of Linear Systems

Consider the linear time-invariant system given by

$$\dot{x}(t) = Ax(t) + Bu(t) \tag{2.5}$$

where $x \in \mathbf{R}^n$ is the state, $u \in \mathbf{R}^p$ is the input, $A \in \mathbf{R}^{n \times n}$ and $B \in \mathbf{R}^{n \times p}$.

Given input u and initial state $x(0) = x_0$, the solution of the dynamical system can be expressed as

$$x(t) = \phi(t; 0, x_0, u) = e^{At}x_0 + \int_0^t e^{A(t-\tau)}Bu(\tau)d\tau.$$

Let T be any positive real number. The reachable set of the system at time T is the set

$$R_T = \{\phi(T; 0, 0, u) \colon u \in C([0, T]; \mathbf{R}^p)\}$$

where $C([0, T]; \mathbf{R}^p)$ is the space of continuous functions from $[0, T]$ to $\mathbf{R}^p$. As

$$R_T = \{\int_0^T e^{A(T-\tau)}Bu(\tau)d\tau \colon u \in C([0, T]; \mathbf{R}^p)\}$$

it is clear that the reachable set is a subspace of $\mathbf{R}^n$.

To set up the connection between subspace R_T and the curve of subspaces $t \mapsto \mathrm{Im}(e^{At}B)$, define W_T to be the smallest subspace of $\mathbf{R}^n$ that contains the image of $e^{At}B$ for all $t \in [0, T]$. That is, W_T is the subspace spanned by the set of vectors

$$L_T = \{e^{At}Bz \colon t \in [0, T],\ z \in \mathbf{R}^p\}.$$

Proposition 2.2. $R_T = W_T$.

Proof. By making the change of variable $s = T - \tau$, we have

$$\phi(T; 0, 0, u) = \int_0^T e^{As}Bu(T - s)ds. \tag{2.6}$$

The integrand in (2.6) is in W_T. As W_T is a closed subspace of $\mathbf{R}^n$, we have

$$\phi(T; 0, 0, u) \in W_T \quad \forall\, u \in C([0, T]; \mathbf{R}^p).$$

This clearly implies that $R_T \subseteq W_T$.

To establish the reverse inclusion, we introduce the controllability grammian. The *controllability grammian* at time T, is given by

$$C_T = \int_0^T e^{A\tau}BB^T e^{A^T\tau}d\tau.$$

It is clear that C_T is symmetric and semi-positive definite. For each $v \in \mathbf{R}^n$, define the control function

$$u_v(t) = B^T e^{A^T(T-t)} v.$$

It can be seen that

$$\phi(T; 0, 0, u_v) = \int_0^T e^{A\tau} B B^T e^{A^T \tau} v d\tau = C_T v.$$

As a result, we have

$$\{\phi(T; 0, 0, u_v) : v \in \mathbf{R}^n\} = \operatorname{Im} C_T$$

which implies that $\operatorname{Im} C_T \subseteq R_T$. On the other hand, since C_T is symmetric, we have

$$\operatorname{Im} C_T = (\operatorname{Ker} C_T)^\perp. \tag{2.7}$$

Suppose that $v \in \operatorname{Ker}(C_T)$. Then, we have

$$0 = <v, C_T v> = \int_0^T <v, e^{A\tau} B B^T e^{A^T} v> d\tau$$

$$= \int_0^T <B^T e^{A^T \tau} v, B^T e^{A^T \tau} v> d\tau = \int_0^T \|B^T e^{A^T \tau} v\|^2 d\tau$$

where $< \cdot, \cdot >$ denotes the inner product of vectors in $\mathbf{R}^n$. From this, we can conclude that

$$B^T e^{A^T \tau} v = 0 \quad \tau \in [0, T]. \tag{2.8}$$

As a result, we have

$$0 = <B^T e^{A^T \tau} v, z> = <v, e^{A\tau} B z> \quad \forall \, z \in \mathbf{R}^p.$$

It follows that v is orthogonal to L_T. As a result, v is orthogonal to subspace W_T which is spanned by L_T. This means that

$$\operatorname{Ker}(C_T) \subseteq (W_T)^\perp$$

which in turn means that

$$W_T \subseteq \operatorname{Im} C_T \subseteq R_T. \quad \square$$

From the proof of Proposition 2.2, we in fact have

$$R_T = W_T = \operatorname{Im} C_T.$$

In addition, by repeatedly differentiating both sides of Equation (2.8) at $\tau = 0$, we have

$$B^T(A^T)^k v = 0 \quad \forall\, \tau \in [0, T] \quad k = 0, 1, \cdots.$$

This is equivalent to

$$v \in \cap_{k=0}^{\infty} \mathrm{Ker}\left(B^T(A^T)^k\right) = \cap_{k=0}^{n-1}\left(\mathrm{Im}(A^k B)\right)^{\perp}$$

$$= \left(\sum_{k=0}^{n-1} \mathrm{Im}\, A^k B\right)^{\perp} = (\Gamma_A \, \mathrm{Im}\, B)^{\perp}.$$

Together with the reasonings in the proof of Proposition 2.2, we have

$$\mathrm{Im}\, C_T = \Gamma_A \, \mathrm{Im}\, B.$$

The above analysis is summarized in the following lemma which will be frequently utilized in Chapter 4.

Lemma 2.3. *For any positive time T, the following subspaces always coincide with each other:*

(i)   the reachable set $R_T = \{\int_0^T e^{A(T-\tau)} Bu(\tau)d\tau \colon u \in C([0,T]; \mathbf{R}^p)\}$;
(ii) the image space of the controllability grammian

$$C_T = \int_0^T e^{A\tau} BB^T e^{A^T \tau} d\tau;$$

(iii) subspace $W_T = \mathrm{span}\{e^{At} Bz \colon t \in [0,T], z \in \mathbf{R}^p\}$; and
(iv) the smallest A-invariant subspace that contains $\mathrm{Im}\, B$, $\Gamma_A \, \mathrm{Im}\, B$.

2.6 Variety and Genericity

Let $A, B, \cdots$ be real matrices with known entries and $\mathcal{P}$ be a parameter space. Suppose that $\Pi(\mathcal{P}, A, B, \cdots)$ is some property which may be asserted about them. For some values in $\mathcal{P}$, the property may be true, while for other values, it may not. In some cases, it turns out that once a property is true for one parameter value, it is true for almost all parameter values. To make this idea precise, we borrow the terminology from algebraic geometry. Let

$$p = (p_1, \cdots, p_N) \in \mathbf{R}^N$$

and consider a polynomial $\varphi(p_1, \cdots, p_N)$ with coefficients in $\mathbf{R}$. An *algebraic variety* $V \in \mathbf{R}^N$ is defined to be the locus of common zeros of a finite number of polynomials $\varphi_1, \cdots, \varphi_k$:

$$V = \{p \colon \varphi_i(p_1, \cdots, p_N) = 0 \quad i \in \bar{k}\}.$$

V is *proper* if $V \neq \mathbf{R}^N$ and *nontrivial* if $V \neq \emptyset$, the empty set. A *property* Π is merely a function $\Pi \colon \mathbf{R}^N \mapsto \{0, 1\}$, where $\Pi(p) = 1$ (or 0) means Π

holds (or fails) at p. Let V be a proper variety. The property Π is said to be *(algebraically) generic w.r.t. V* if $\Pi(p) = 0$ only for $p \in V$. Property Π is *(algebraically) generic* if such a V exists.

For slightly broader notions, we define the analytic variety and genericity as follows. Suppose that Ω^N is the set of real analytic functions on $\mathbf{R}^N$. An *analytic variety* is defined to be the locus of the set of common zeros of a finite number of analytic functions

$$V = \{p \colon \phi_i(p_1, \cdots, p_N) = 0, \phi_i \in \Omega^N \quad i \in \bar{k}\}.$$

Let V be a proper analytic variety. The property Π is said to be *(analytically) generic w.r.t. V* provided that $\Pi(p) = 0$ only for $p \in V$. Property Π is *(analytically) generic* if such a V exists.

Any algebraic variety is also an analytic variety, and accordingly, any algebraically generic property is also analytically generic. As an illustrative example, a proper algebraic variety in $\mathbf{R}$ contains only a finite number of points, while a proper analytic variety may consist of an (countable) infinite number of (isolated) points.

In our approach, we do not distinguish between the algebraic variety and genericity from the analytic ones and we shall refer to them as variety and genericity.

2.7 Stability and Lyapunov Theorems

Consider the vector differential equation described by

$$\dot{x}(t) = f(t, x) \quad x(0) = x_0 \tag{2.9}$$

where $x(t) \in \mathbf{R}^n$, and $f \colon \mathbf{R}_+ \times \mathbf{R}^n \mapsto \mathbf{R}^n$ is piecewise smooth and globally Lipschitz. We further assume that the origin is an equilibrium, that is $f(t, 0) = 0$, for $t \geq 0$.

Let $\phi(t; t_0, x_0)$ denote the solution of (2.9) corresponding to the initial condition $x(t_0) = x_0$, evaluated at time t.

Stability concerns the behavior of the solution when $x_0 \neq 0$ but is close to it.

Definition 2.4. *The equilibrium 0 is said to be:*

- stable *if, for each $\epsilon > 0$ and each $t_0 \geq 0$, there exists a $\delta = \delta(\epsilon, t_0)$ such that*

$$\|x_0\| < \delta(\epsilon, t_0) \Longrightarrow \|\phi(t; t_0, x_0)\| < \epsilon \quad \forall\, t \geq t_0$$

- uniformly stable *if, for each $\epsilon > 0$, there exists a $\delta = \delta(\epsilon)$ such that*

$$\|x_0\| < \delta(\epsilon) \quad t_0 \geq 0 \Longrightarrow \|\phi(t; t_0, x_0)\| < \epsilon \quad \forall\, t \geq t_0$$

- attractive *if, for each $t_0 \geq 0$, there exists a $\delta = \delta(t_0) > 0$ such that*

$$\|x_0\| < \delta(t_0) \implies \|\phi(t_0 + t; t_0, x_0)\| \to 0 \text{ as } t \to \infty$$

- uniformly attractive *if there exists a $\delta > 0$ such that*

$$\|x_0\| < \delta, t_0 \geq 0 \implies \|\phi(t_0 + t; t_0, x_0)\| \to 0 \text{ as } t \to \infty \text{ uniformly in } x_0, t_0$$

- asymptotically stable *if it is both stable and attractive;*
- uniformly asymptotically stable *if it is uniformly stable and uniformly attractive; and*
- exponentially stable *if there exist real constants $r, \alpha, \beta > 0$ such that*

$$\|\phi(t_0 + t; t_0, x_0)\| \leq \beta e^{-\alpha t} \|x_0\| \quad \forall \, t, t_0 \geq 0 \quad \|x_0\| < r.$$

These stability concepts are local in nature, *i.e.*, they are concerned only with a small neighborhood of the equilibrium. Global versions can be defined by allowing the attractive properties to hold in a global sense. If the solution is radially linear, that is

$$\phi(t; t_0, \lambda x_0) = \lambda \phi(t; t_0, x_0) \quad \forall \, \lambda \in \mathbf{R} \quad t \geq t_0 \quad x_0 \in \mathbf{R}^n$$

then a local stability concept is equivalent to the corresponding global one. This is precisely the case for switched linear systems (*c.f.* Section 1.3.5), hence, we do not need to distinguish between the local and global stability concepts.

We say that the dynamical system is stable (attractive, *etc.*) if the origin is a stable (attractive, *etc.*) equilibrium point of the system.

The Lyapunov approach provides a rigorous method for addressing stability. Here, we review several concepts that are used in Lyapunov stability theory.

A function $\alpha \colon \mathbf{R}_+ \mapsto \mathbf{R}_+$ is of class $\mathcal{K}$ if it is continuous, strictly increasing, and $\alpha(0) = 0$.

A continuous function $V(x, t) \colon \mathbf{R}^n \times \mathbf{R}_+ \mapsto \mathbf{R}$ with $V(0, t) \equiv 0$ is:

- *(locally) positive definite* $(V(x, t) \succ 0)$ if there exist a constant $r > 0$, and a class $\mathcal{K}$ function $\alpha(\cdot)$, such that

$$V(x, t) \geq \alpha(\|x\|) \quad \forall \, t \geq 0 \quad x \in \mathbf{B}_r$$

- *(locally) semi-positive definite* $(V(x, t) \succeq 0)$ if there exists a constant $r > 0$, such that

$$V(x, t) \geq 0 \quad \forall \, t \geq 0 \quad x \in \mathbf{B}_r$$

and

- *decrescent* if there exist a constant $r > 0$, and a class $\mathcal{K}$ function $\beta(\cdot)$, such that

$$V(x, t) \leq \beta(\|x\|) \quad \forall \, t \geq 0 \quad x \in \mathbf{B}_r.$$

Given a continuously differentiable (C^1) function $V \colon \mathbf{R}^n \times \mathbf{R}_+ \mapsto \mathbf{R}$, the (Lie) derivative of V along system (2.9) is defined as

$$\dot{V}(x,t) = \frac{dV}{dt}(x,t) = \frac{\partial V}{\partial t}(x,t) + \left[\frac{\partial V}{\partial x}(x,t)\right]^T f(t,x).$$

Theorem 2.5. *(Lyapunov Theorem) The equilibrium 0 of system (2.9) is:*

(i) stable if there exists a C^1 $V(x,t) \succ 0$ such that $-\dot{V}(x,t) \succeq 0$;

(ii) uniformly stable if there exists a C^1 decrescent function $V(x,t) \succ 0$ such that $-\dot{V}(x,t) \succeq 0$;

(iii) asymptotically stable if there exists a C^1 $V(x,t) \succ 0$ such that $-\dot{V}(x,t) \succ 0$;

(iv) uniformly asymptotically stable if there exists a C^1 decrescent function $V(x,t) \succ 0$ such that $-\dot{V}(x,t) \succ 0$;

(v) exponentially stable if there exist a C^1 $V(x,t) \succ 0$, and positive real constants α, β, γ, and $p \geq 1$, such that, for all x and t, we have

$$\alpha\|x\|^p \leq V(x,t) \leq \beta\|x\|^p \ and \ \dot{V}(x,t) \leq -\gamma\|x\|^p;$$

and

(vi) unstable if there exist a C^1 decrescent function $V \colon \mathbf{R}^n \times \mathbf{R}_+ \mapsto \mathbf{R}$, and a time $t_0 \geq 0$, such that $\dot{V} \succ 0$, and for any sufficiently small positive real number r, there exists a non-origin point $x \in \mathbf{B}_r$ such that $V(x,t_0) \geq 0$.

Function $V(x,t)$ in Theorem 2.5 is called the *Lyapunov function*. The theorem provides sufficient conditions for the origin to be stable, *etc.*

As a special case of (2.9), we consider the linear time-varying system given by

$$\dot{x}(t) = A(t)x(t) \quad t \geq 0 \tag{2.10}$$

where entries of $A(t)$ are piecewise smooth functions of time.

It is clear that the origin is always an equilibrium of system (2.10). The solution of (2.10) is given by

$$\phi(t; t_0, x_0) = \Phi(t, t_0)x_0$$

where $\Phi(\cdot, \cdot)$ is the state transition matrix associated with $A(\cdot)$ and is the unique solution of equation

$$\frac{d}{dt}\Phi(t, t_0) = A(t)\Phi(t, t_0) \quad \Phi(t_0, t_0) = I_n \quad \forall \, t_0 \geq 0 \quad t \geq t_0.$$

With the aid of this explicit characterization of the solution, it is possible to derive some useful conditions for stability, as stated in the following result.

Theorem 2.6. *Linear time-varying system (2.10) is uniformly asymptotically stable if and only if it is exponentially stable.*

Next, consider the linear time-invariant system given by

$$\dot{x}(t) = Ax(t) \quad t \geq 0. \tag{2.11}$$

In this special case, Lyapunov theory is very complete, and we have the following theorem.

Theorem 2.7. *For linear time-invariant system (2.11), the following statements are equivalent:*

(i) the system is asymptotically stable;
(ii) the system is exponential stable;
(iii) matrix A is Hurwitz;
(iv) the Lyapunov equation

$$A^T P + PA = -Q \tag{2.12}$$

has a unique solution $P > 0$ for any $Q > 0$; and
(v) Equation (2.12) has a unique solution $P > 0$ for some $Q > 0$.

The last statement asserts the existence of a quadratic Lyapunov function

$$V(x) = x^T P x$$

where P is symmetric and positive definite.

For the discrete-time linear time-invariant system

$$x_{k+1} = Ax_k \quad k \in \mathbf{N}_+ \tag{2.13}$$

a similar result can be stated as follows.

Theorem 2.8. *For discrete-time linear time-invariant system (2.13), the following statements are equivalent:*

(i) the system is asymptotically stable;
(ii) the system is exponential stable;
(iii) matrix A is Schur;
(iv) the Lyapunov difference equation

$$P - A^T P A = Q \tag{2.14}$$

has a unique solution $P > 0$ for any $Q > 0$; and
(v) Equation (2.14) has a unique solution $P > 0$ for some $Q > 0$.

Finally, for the switched linear system

$$\delta x(t) = A_\sigma x(t) \tag{2.15}$$

(guaranteed) stability means that the system is stable with arbitrary switching. That is, for any switching path $\theta_{[0,\infty)} \in \mathcal{S}$, the time-varying system

$$\delta x(t) = A_{\theta(t)} x(t) \quad t \geq 0$$

is stable.

Suppose that there exists a common quadratic Lyapunov function $V(x) = x^T P x$ for the subsystems:

$$A_i^T P + P A_i \leq 0 \quad i \in M. \tag{2.16}$$

It can be seen that the Lyapunov function decreases along any state trajectory of the switched system. Accordingly, the switched system is (guaranteed) stable. In fact, if a switched linear system is (guaranteed) stable, then, there exists such a Lyapunov function, though not necessarily quadratic, as stated in the following result.

Theorem 2.9. *A necessary and sufficient condition for guaranteed stability of the switched linear system is the existence of a smooth function of state which is a common Lyapunov function for all the subsystems.*

Another stability of switched systems is the switched stability. Suppose that the switching signal is given and known, for instance,

$$\sigma(t) = \varphi(t, t_0, x_0) \tag{2.17}$$

where φ is a known function. As we mentioned before, we can always assume that $t_0 = 0$ without loss of generality. For a fixed x_0, define

$$f(t, x) = A_{\varphi(t,0,x_0)} x.$$

Then, the switched system becomes

$$\delta x(t) = f(t, x(t)) \quad x(0) = x_0. \tag{2.18}$$

Switched system (2.15) with the specified switching signal (2.17) is said to be switched (asymptotically, exponentially, *etc.*) stable if system (2.18) is (asymptotically, exponentially, *etc.*) stable as defined in Definition 2.4.

Note that the switched stability is defined only when the switching signal is specified. In the book, when we talk about stability of a switched system with a specified switching signal, we always refer to switched stability.

2.8 Campbell-Baker-Hausdorff Formula and Average Systems

In this section, we briefly review the average method based on a formula from Lie algebra known as the Campbell-Baker-Hausdorff (CBH) formula.

Given two real square matrices, A and B, matrix $e^A e^B$ is always nonsingular. Hence, we can find a complex matrix C such that

$$e^A e^B = e^C. \tag{2.19}$$

Generally speaking, there is no guarantee that we can find such a matrix which is real. However, if $\|A\| + \|B\| \leq \ln 2$, then a real C satisfying (2.19) exists, and is given by the convergent infinite expression (the CBH formula)

$$C = A + B + \frac{1}{2}[A, B] + \frac{1}{12}[[A, B], B] + \frac{1}{12}[[B, A], A] + \cdots$$

where $[A, B] = AB - BA$ is the *commutator product* of A and B.

Based on the CBH formula, we can prove the following lemma.

Lemma 2.10. *Let $A_1, \cdots, A_m$ be matrices in $\mathbf{R}^{n \times n}$. Then, there is a positive real number η such that*

$$\exp(A_m t) \exp(A_{m-1} t) \cdots \exp(A_1 t) = \exp\left(\left(\sum_{i=1}^m A_i\right) t + \Upsilon_t t^2\right)$$

for any $t \leq \eta$, where entries of matrix Υ_t are analytic and bounded. Moreover, an upper (norm) bound of Υ_t can be explicitly estimated.

This lemma provides a basis for the average method described below.

Suppose that we have a set of linear time-invariant systems

$$\dot{x}(t) = A_i x(t) \quad i \in \bar{m}.$$

We then integrate them in a multi-rate manner as follows. Let $T > 0$ be a base period and $\alpha_1, \cdots, \alpha_m$ be positive real numbers with $\sum_{i=1}^m \alpha_i = 1$. Let $A(\cdot)$ be the periodic matrix function of time, with period T: $A(t+T) = A(t)$, $\forall\, t \geq 0$, and

$$A(t) = \begin{cases} A_1 & t \in [0, \alpha_1 T) \\ A_2 & t \in [\alpha_1 T, (\alpha_1 + \alpha_2) T) \\ \vdots \\ A_m & t \in [(\sum_{i=1}^{m-1} \alpha_i) T, T). \end{cases}$$

For the linear periodic system

$$\dot{x}(t) = A(t) x(t) \quad t \geq 0 \tag{2.20}$$

let the time-invariant system

$$\dot{x}(t) = A_0 x(t) \quad A_0 = \sum_{i=1}^m \alpha_i A_i \tag{2.21}$$

be its *average system*. Note that the *average matrix* A_0 is a linear convex combination of A_i, $i \in \bar{m}$.

The state transition matrix Φ of system (2.20) satisfies

$$\Phi(T, 0) = \exp(\alpha_m A_m T) \exp(\alpha_{m-1} A_{m-1} T) \cdots \exp(\alpha_1 A_1 T)$$

while the state transition matrix Ψ of system (2.21) satisfies

$$\Psi(T, 0) = \exp(A_0 T).$$

According to Lemma 2.10, we have

$$\lim_{T \to 0+} \exp(\alpha_m A_m T) \exp(\alpha_{m-1} A_{m-1} T) \cdots \exp(\alpha_1 A_1 T) = \exp(A_0 T).$$

This leads to the following observation.

Lemma 2.11. *Suppose that the average system satisfies*

$$\|\Psi(t, t_0)\| \le \beta e^{\delta(t - t_0)} \quad \forall \, t \ge t_0$$

for some β and δ. Then, for any $\epsilon > 0$, there exist positive real numbers κ and ρ such that

$$\|\Phi(t, t_0)\| \le \kappa e^{(\delta + \epsilon)(t - t_0)} \quad \forall \, t \ge t_0$$

for time-varying system (2.20) with $T \le \rho$.

The lemma asserts that the convergence rate of the time-varying system (2.20) can arbitrarily approach that of the average system (2.21) by means of high frequency switching. In particular, if the average system (2.21) is exponentially stable, then the time-varying system (2.20) is also exponentially stable if period T is sufficiently small.

Finally, it should be noticed that the average approach applies only to continuous-time systems.

2.9 Differential Inclusions

Let $\mathcal{X}$ and $\mathcal{Y}$ be two normed spaces. A *set-valued map* F from $\mathcal{X}$ to $\mathcal{Y}$ is a map that associates a set $F(x) \subseteq \mathcal{Y}$ with any $x \in \mathcal{X}$. The set $\{x \in \mathcal{X} : F(x) \ne \emptyset\}$ is called the *domain* of F, and the set $\cup_{x \in \mathcal{X}} F(x)$ is the *image* of F. The *graph* of F, denoted by $\operatorname{gr} F$, is defined as

$$\operatorname{gr} F = \{(x, y) \in \mathcal{X} \times \mathcal{Y} : y \in F(x)\}.$$

A differential inclusion is described by

$$\dot{x}(t) \in F(x(t)) \tag{2.22}$$

where $F\colon \mathbf{R}^n \mapsto \mathbf{R}^n$ is a set-valued map. A function $x\colon [t_0, t_1] \mapsto \mathbf{R}^n$ is said to be a trajectory of the differential inclusion, if it is absolutely continuous and satisfies (2.22) almost everywhere in $[t_0, t_1]$.

Note that the differential inclusions model a wide class of dynamical systems, including conventional control systems and switched systems. Indeed, the control system

$$\dot{x}(t) = f(t, x(t), u(t)) \quad u(t) \in U(t)$$

can be associated with the differential inclusion

$$\dot{x}(t) \in \cup_{u \in U(t)} f(t, x(t), u).$$

Similarly, a switched system

$$\dot{x}(t) = f_\sigma(t, x(t), u(t)) \quad u(t) \in U(t) \quad \sigma \in M$$

can be associated with

$$\dot{x}(t) \in \cup_{i \in M} \cup_{u \in U(t)} f_i(t, x(t), u).$$

In general, we are interested in the class of differential inclusions (2.22) where $F(x)$ is closed and convex for all $x \in \mathbf{R}^n$, as these systems possess good properties and are relatively easy to analyze. For a differential inclusion which does not belong to this class, we can generalize the system to

$$\dot{x}(t) \in \operatorname{clco} F(x(t)) \tag{2.23}$$

where $\operatorname{clco} F$ stands for the closed convex hull of set F. System (2.23) is said to be a *relaxed system* of system (2.22). It is clear that any trajectory of a differential inclusion is also a trajectory of its relaxed system. In general, the reverse is not true. However, under mild assumptions, each trajectory of the relaxed system can be approximated by a trajectory of the original system. This is guaranteed by the well-known Filippov-Wazewski Theorem. The following result states a special case of the theorem.

Lemma 2.12. *Fix a finite time T, a vector $\xi \in \mathbf{R}^n$, and let $z\colon [0, T] \mapsto \mathbf{R}^n$ be a solution of the differential inclusion*

$$\dot{z}(t) \in \operatorname{clco}\{A_1 z(t), \cdots, A_m z(t)\} \quad z(0) = \xi. \tag{2.24}$$

Let $r\colon [0, T] \mapsto \mathbf{R}$ be a continuous function satisfying $r(t) > 0$ for all $t \in [0, T]$. Then, there exists a solution $x\colon [0, T] \mapsto \mathbf{R}^n$ of

$$\dot{x}(t) \in \{A_1 x(t), \cdots, A_m x(t)\} \quad x(0) = \xi$$

such that

$$\|z(t) - x(t)\| \le r(t) \quad \forall \, t \in [0, T].$$

By this lemma, in a finite time interval, any trajectory of the relaxed system can be approximated by a trajectory of the original switched linear system with the same initial condition. For an infinite time interval, this property does not hold in general. However, if we allow some relaxation in the initial condition, the property does hold for the infinite-time interval, as stated in the following result.

Lemma 2.13. *Fix $\xi \in \mathbf{R}^n$ and let $z\colon [0, \infty) \mapsto \mathbf{R}^n$ be a solution of*

$$\dot{z}(t) \in \mathrm{clco}\{A_1 z(t), \cdots, A_m z(t)\} \quad z(0) = \xi.$$

Let $r\colon [0, \infty) \mapsto \mathbf{R}$ be a continuous function satisfying $r(t) > 0$ for all $t \geq 0$. Then, there exist an η with $\|\eta - \xi\| \leq r(0)$, and a solution $x\colon [0, \infty) \mapsto \mathbf{R}^n$ of

$$\dot{x}(t) \in \{A_1 x(t), \cdots, A_m x(t)\} \quad x(0) = \eta$$

such that

$$\|z(t) - x(t)\| \leq r(t) \quad \forall\, t \in [0, \infty).$$

This lemma sets up a connection between stability of a switched linear system and stability of its relaxed system. Indeed, suppose that the switched linear system is stable (asymptotically stable, exponentially stable, resp.), then by the lemma, the differential inclusion (2.24) is also stable (asymptotically stable, exponentially stable, resp.). As the reverse holds trivially, stability is in fact equivalent pairwise.

2.10 Lie Product and Chow's Theorem

Suppose that Ω is an open set in $\mathbf{R}^n$. Let f and g be mappings (vector fields) from Ω to $\mathbf{R}^n$. Hence, the mappings can be represented as

$$f(x) = \begin{pmatrix} f_1(x_1, \cdots, x_n) \\ f_2(x_1, \cdots, x_n) \\ \vdots \\ f_n(x_1, \cdots, x_n) \end{pmatrix} \text{ and } g(x) = \begin{pmatrix} g_1(x_1, \cdots, x_n) \\ g_2(x_1, \cdots, x_n) \\ \vdots \\ g_n(x_1, \cdots, x_n) \end{pmatrix}.$$

The vector field f is said to be *analytic*, if each f_i is a real analytic function. For an analytic vector field f on Ω and $x_0 \in \Omega$, there is an integral curve ξ defined on an open interval (t_1, t_2) of $\mathbf{R}$ with $0 \in (t_1, t_2)$ and $\xi(0) = x_0$. That is, ξ is a solution of the differential equation

$$\dot{\xi}(t) = f(\xi(t)) \quad \xi(0) = x_0 \quad t \in (t_1, t_2).$$

For notational convenience, denote $\xi(t)$ by $\Phi_t^f(x_0)$ for $t \in (t_1, t_2)$. When f is a linear vector field $f(x) = Ax$, it is clear that $\Phi_t^f(x_0) = e^{At}x_0$.

The *Lie product (or bracket) of f and g*, denoted by $[f, g]$, is the vector field defined by

$$[f, g](x) = \frac{\partial f}{\partial x}g(x) - \frac{\partial g}{\partial x}f(x) \quad x \in \Omega$$

where $\frac{\partial f}{\partial x}$ is the Jacobian matrix of mapping f:

$$\frac{\partial f}{\partial x} = \begin{pmatrix} \frac{\partial f_1}{\partial x_1} & \cdots & \frac{\partial f_1}{\partial x_n} \\ & \ddots & \\ \frac{\partial f_n}{\partial x_1} & \cdots & \frac{\partial f_n}{\partial x_n} \end{pmatrix}.$$

When f and g are linear vector fields, i.e., $f(x) = Ax$, and $g(x) = Bx$, we have

$$[f, g](x) = [Ax, Bx] = (AB - BA)x.$$

It is clear that

$$[Ax, Bx] = [A, B]x$$

where $[A, B]$ is the commutator product of A and B.

The following simple proposition summarizes several basic properties of the Lie product.

Proposition 2.14. *The Lie product of vector fields possesses the following properties:*

*(i) it is bilinear over **R**, i.e., if f_1, f_2, g_1, g_2 are vector fields and r_1, r_2 are real numbers, then, we have*

$$[r_1 f_1 + r_2 f_2, g_1] = r_1[f_1, g_1] + r_2[f_2, g_1]$$
$$[f_1, r_1 g_1 + r_2 g_2] = r_1[f_1, g_1] + r_2[f_1, g_2];$$

(ii) it is skew commutative, i.e., $[f, g] = -[g, f]$; and
(iii) it satisfies the Jacobi identity, i.e., for any vector fields f_1, f_2 and f_3, we have

$$[f_1, [f_2, f_3]] + [f_2, [f_3, f_1]] + [f_3, [f_1, f_2]] \equiv 0.$$

These properties are basic requirements for a linear space to be considered a *Lie algebra*, which is defined below.

Definition 2.15. *A linear space V over **R** is a Lie algebra if, in addition to its linear space structure, it is possible to define a binary operation $V \times V \mapsto V$, denoted by $[\cdot, \cdot]$, which has the following properties:*

(i) the operation is skew commutative, i.e.,

$$[v, w] = -[w, v] \quad \forall \, v, w \in V$$

(ii) the operation is bilinear over **R**, *i.e.,*

$$[\alpha v_1 + \beta v_2, v_3] = \alpha[v_1, v_3] + \beta[v_2, v_3] \quad \forall \, v_1, v_2, v_3 \in V \quad \alpha, \beta \in \mathbf{R}$$

(iii) the operation satisfies the Jacobi identity, i.e.,

$$[v_1, [v_2, v_3]] + [v_2, [v_3, v_1]] + [v_3, [v_1, v_2]] = 0 \quad \forall \, v_1, v_2, v_3 \in V.$$

By the definition, it can be seen that the set of analytic vector fields is a Lie algebra under the Lie product.

In the nonlinear setting, as we distinguish between points and tangent vectors, the concept of linear subspace generalizes to two different ones. First, a k-dimensional *(regularly imbedded) submanifold Γ of Ω* is a subset of Ω such that around each point of Γ, there exists a coordinate neighborhood U such that $\Gamma \cap U$ is given by $\{x_j = c_j : j = 1, \cdots, n - k\}$, where $c_1, \cdots, c_{n-k}$ are constants. On $\Gamma \cap U$, we have local coordinates given by $(x_{n-k+1}, \cdots, x_n)$. Second, a *distribution Δ* on Ω is a mapping which assigns to each $x \in \Omega$ a subspace $\Delta(x)$ of $T_x \Omega$ in an analytic fashion, where $T_x \Omega$ is the tangent space to Ω at x. If each of these subspaces is of dimension k, then Δ is said to be of rank k. The connection between the two concepts is as follows. Λ is an *integral submanifold of Δ* if for every $x \in \Lambda$, $\Delta(x) = T_x \Omega$. In other words, $\Delta(x)$ is the tangent space to Ω at x.

Given a set of vector fields $f_1, \cdots, f_k$, let $\{f_1, \cdots, f_k\}_{LA}$ denote the Lie algebra generated by $f_1, \cdots, f_k$. That is, $\{f_1, \cdots, f_k\}_{LA}$ is the smallest Lie algebra that contains $f_1, \cdots, f_k$. This Lie algebra contains the linear combination of all the possible Lie product of the form

$$[g_1, [g_2, \cdots, [g_{s-1}, g_s]]] \quad g_i \in \{f_1, \cdots, f_k\} \quad i = 1, \cdots, s \quad 0 \le s < \infty.$$

It is clear that $\{f_1, \cdots, f_k\}_{LA}$ is a distribution.

With these preparatory concepts, we are in the position to state the main theorem of this section. The theorem is an extension of the well-known Chow's Theorem and hence is termed the Generalized Chow's Theorem.

Theorem 2.16. *Suppose that Ω is a pathwise connected open set in $\mathbf{R}^n$, and $L = \{f_1, \cdots, f_k\}$ is a set of analytic vector fields defined on Ω. Let $\mathcal{L} = \{f_1, \cdots, f_k\}_{LA}$ be the Lie algebra generated by L. For any $x \in \Omega$, let Λ_x be the largest integral submanifold of $\mathcal{L}$ passing through x. Then, for any $y \in \Lambda_x$, there exist a natural number s, a real number sequence $t_1, \cdots, t_s$, and an index sequence $i_1, \cdots, i_s$, such that*

$$y = \Phi_{t_s}^{f_{i_s}} \circ \cdots \circ \Phi_{t_1}^{f_{i_1}} \circ \Phi_{t_1}^{f_{i_1}}(x)$$

where '$\circ$' denotes the composition of functions.

2.11 Language and Directed Graph

We start with a finite but nonempty set of symbols, Ξ, called the *alphabet*. From the individual symbols we construct *strings*, which are finite sequences of the symbols from the alphabet. The *concatenation* of two strings w and v, denoted by wv, is the string obtained by appending the symbols of v to the right end of w, that is, if

$$w = a_1 a_2 \cdots a_i \text{ and } v = b_1 b_2 \cdots b_j$$

then, we have

$$wv = a_1 a_1 \cdots a_i b_1 b_2 \cdots b_j.$$

The *length* of string w, denoted by $|w|$, is the number of symbols in the string. A string with length zero is said to be an *empty string*. For a nonempty string w, any string of consecutive characters in w is said to be a *substring* of w.

If w is a string, then w^n stands for the string obtained by repeating w n times. As a special case, let w^0 be the empty string.

Consider a partially ordered set $\mathcal{Q}$ with a unique maximum element $\mathbf{1}$ and a unique minimum element $\mathbf{0}$. The set is said to be N-*Nether* if it does not contain any strictly decreasing sequence of length N. For a given finite set of maps

$$G = \{f_i \colon \mathcal{Q} \mapsto \mathcal{Q}\}_{i=1}^k$$

we define a language in alphabet $\{1, \cdots, k\}$ as

$$L_G = \{\omega = \omega_1 \cdots \omega_l \colon f_\omega(\mathbf{0}) < \mathbf{1}\}$$

where $f_\omega \overset{def}{=} f_{\omega_1} \circ f_{\omega_2} \circ \cdots \circ f_{\omega_l}$.

The pair $(\mathcal{Q}, G)$ is said to be a *monotone automaton* if all maps in G are monotone w.r.t. the order in $\mathcal{Q}$ and $f_i(\mathbf{1}) = \mathbf{1}$ for all $f_i \in G$.

The following lemma follows easily from the well-known Pumping Lemma.

Lemma 2.17. *Suppose that $(\mathcal{Q}, G)$ is a monotone automaton, and $\mathcal{Q}$ is N-Nether. Then, there exists a nature number $F(N, k)$, such that, for any $w \in L_G$ with $|w| > F(N, k)$, we can find strings x, y and z satisfying the following properties:*

(i) $w = xyz$;
(ii) $|xy| \leq F(N, k)$;
(iii) $|y| \geq 1$; and
(iv) for all $i \in N_+$, we have $xy^i z \in L_G$.

Finally, we briefly introduce the concept of directed graphs. A *directed graph*, or *digraph*, consists of a finite set V of points and a collection of ordered pairs of distinct points. Any such pair (u, v) is called an *arc* or *directed line* and usually denoted by uv.

A *(directed) walk* in a digraph is an alternating sequence of points and arcs

$$v_0, x_1, v_1, \cdots, x_k, v_k$$

in which each arc x_i is $v_{i-1}v_i$. The *length* of such a walk is k, the number of occurrences of arcs in it. A *closed walk* has the same first and last points, and a *spanning walk* contains all the points. A *circle* is a non-trivial closed walk with all points distinct (expect for the first and last).

2.12 Notes and References

In this chapter, we briefly reviewed some mathematical concepts which will be useful in the development of the following chapters. It is assumed that the reader already has some working knowledge in these areas.

The geometrical concepts of invariant subspace are standard and the material was taken mainly from Wonham's famous monograph [160]. The concept, properties and algorithm of multiple controllable subspaces are natural extensions from those of single controllable subspaces. These were adopted from the recent works of the authors [52, 142]. Proposition 2.2 is based on the results in [36].

The matrix notation adopted here is standard and can be found in most textbooks, *e.g.*, [50] and [66].

Genericity is a fundamental concept in system science. The notion of (algebraic) genericity here was taken from [160]; see also the recent survey paper [35] for related issues. The analytic variety and genericity were defined by the authors for notational convenience. Nevertheless, the notions have a sound mathematical base which can be found, for example, in the classical work [78].

The stability theory here is standard and the reader is referred to [57] and [157] for further references. Theorem 2.9 was proven in [93, 30].

The Campbell-Baker-Hausdorff formula, also known as Baker-Campbell-Hausdorff formula, has several variations in different mathematical fields. Here, we adopted the version from [41] and [152]. The key point is to merge a multiplication of matrix exponentials into one matrix exponential. As the state transition matrix of a switched linear system is a multiplication of matrix exponentials, the formula is useful in addressing the stability issues for switched linear systems.

The differential inclusion material introduced in Section 2.9 was taken from [125]. Lemma 2.12 was reported in [47], and Lemma 2.13 was adopted from [72].

The Lie product introduced in Section 2.10 is very elementary and can be found in any standard Lie algebra textbook. For an advanced reference on the concepts and their extensive applications to nonlinear control theory, the reader is referred to the textbook [74]. The Generalized Chow's Theorem, Theorem 2.16, was taken from the recent work [22].

Finally, the section on automata and graphs, Section 2.11, is also standard and can be found in, *e.g.*, [95] and [58]. Lemma 2.17 was taken from [56].

3

Stabilizing Switching for Autonomous Systems

3.1 Introduction

In this chapter, we consider the switched linear autonomous system given by

$$\delta x(t) = A_\sigma x(t) \qquad (3.1)$$

where $x(t) \in \mathbf{R}^n$ is the state, $\sigma \in M \overset{def}{=} \{1, \cdots, m\}$ is the piecewise constant switching signal to be designed, $A_k \in \mathbf{R}^{n \times n}$, $k \in M$ are real constant matrices, and δ is the derivative operator in continuous time and the shift forward operator in discrete time.

Our primary aim is to design, if possible, switching signals to make the switched systems stable.

Definition 3.1. *System (3.1) is said to be* (asymptotically, exponentially) *stabilizable, if there is a switching signal σ such that the system is well-posed and uniformly (asymptotically, exponentially) stable. Such a switching signal σ is said to be a* stabilizing switching signal *for system (3.1).*

Suppose that σ is completely well-defined. Then, it can be written as $\sigma(t) = \varphi(t; t_0, x_0)$, where (t_0, x_0) is the initial condition and $\varphi(\cdot; t_0, x_0)$ is the switching path generated by the switching signal at x_0 over $[t_0, \infty)$. The switching signal is said to be *consistent (to the initial state)*, if it is independent of the initial state, *i.e.*,

$$\varphi(t; t_0, x_1) = \varphi(t; t_0, x_2) \quad \forall \, t \geq t_0 \quad x_1, x_2 \in \mathbf{R}^n.$$

As stated in Section 1.3.5, when we address the stabilizability of switched linear systems, without loss of generality, we can focus on the design of transition-invariant switching signals with $t_0 = 0$. As a consequence, a consistent switching signal is in fact a switching path

$$\varphi(t; t_0, x_0) = \varphi(t) \quad \forall \, t \geq 0.$$

If the switched system is stabilizable by means of a consistent switching signal, then it is consistently stabilizable as defined in the following.

Definition 3.2. *System (3.1) is said to be* consistently (asymptotically, exponentially) stabilizable, *if there is a consistent switching signal σ such that the system is well-posed and uniformly (asymptotically, exponentially) stable.*

For comparison, we sometimes refer to the stabilizability defined in Definition 3.1 as *pointwise stabilizability*. It is obvious that consistent stabilizability implies pointwise stabilizability, but the converse is not true in general. For example, the switched linear system $\Sigma(A_i)_{\bar{2}}$ with

$$A_1 = \begin{bmatrix} 1 & -2 \\ 2 & 1 \end{bmatrix} \text{ and } A_2 = \begin{bmatrix} 3 & -2 \\ 1 & -1 \end{bmatrix} \tag{3.2}$$

is pointwise stabilizable as shown in Example 1.1. However, by Theorem 3.4, which will be presented later, this switched system is not consistently stabilizable.

We also need the notion of quadratic stabilizability which is a special case of asymptotic stabilizability.

Definition 3.3. *System (3.1) is said to be* quadratically stabilizable, *if there exist a switching signal σ, and a positive definite quadratic function $V(x) = x^T P x$, such that the system is well-posed, and $-V(x) \succ 0$.*

It is clear that the quadratic function $V(x)$ is a Lyapunov function of the switched system, and we have

$$\phi^T(t; 0, x_0, \sigma)(A_{\sigma(t)}^T P + P A_{\sigma(t)})\phi(t; 0, x_0, \sigma) < 0 \quad \forall\, t \geq t_0 \quad x_0 \neq 0$$

where $\phi(t; 0, x_0, \sigma)$ denotes the solution of (3.1) at t with initial condition $x(0) = x_0$ via σ.

Besides stability, there are other additional performance specifications which should be met.

An important issue for switching design is to reduce the switching frequency to an acceptable level. Take digital networks for example. The digital data must be transferred in real-time. This sets a data rate limit, which in turn limits the allowable switching frequency. Bearing this in mind, the switching signal should be designed to prevent the actuator from fast switching, or chattering, which can not only increase the necessary data rate, but also damage the system. However, the design of low frequency switching is very challenging in general, even for simple systems such as linear time-invariant systems [73, 107].

Another critical issue for switching design is to enhance the robustness against system uncertainties and perturbations. As disturbance exists almost everywhere, a switching signal cannot work properly if it is not robust. Moreover, for a state-feedback switching signal, the resultant switching paths may differ from each other for the nominal system and the perturbed system with the same initial condition. This poses an additional challenge because a well-defined switching signal for the nominal system may result in the ill-posed chattering phenomenon for the perturbed system.

In short, a 'good' switching signal should satisfy the following basic criteria:

- it makes the switched system stable;
- it can avoid fast switching, preferably with a guaranteed positive dwell time;
- it is robust to (time-varying and nonlinear) system perturbations; and
- it uses measurable system information only.

This chapter aims to provide a design methodology for a good stabilizing switching signal. We first present some general results that provide basic observations on the ability and limitations of switching design. Then, we analyze and design the periodic switchings, the state/output-feedback switchings, and the combined switchings for stability and robustness. We focus on continuous-time systems, except in Section 3.7 where discrete-time systems are addressed.

3.2 General Results

3.2.1 Algebraic Criteria

For a linear time-invariant system, it is well known that the system is stable when its poles are located in the open left half of the complex plane. For stabilizability of switched linear systems, we have a similar criteria as follows.

Theorem 3.4. *Suppose that the switched linear system $\Sigma(A_i)_M$ is consistently stabilizable. Then, there is a $k \in M$ such that*

$$\sum_{i=1}^{n} \lambda_i(A_k) \leq 0$$

where $\lambda_i(A)$, $1 \leq i \leq n$ are the eigenvalues of matrix A. Furthermore, if the system is consistently asymptotically stabilizable, then the inequality is strict.
Proof. Let σ be a consistent switching signal that stabilizes the switched system. Suppose that the switching duration sequence of σ is

$$DS_\sigma = \{(i_0, h_0), (i_1, h_1), \cdots\}.$$

If the sequence is finite, *i.e.*, there involve only finite switches in σ, then, it can be seen that the last active subsystem must be stable and the theorem follows immediately. If the sequence is infinite, it follows from the well-posedness of σ that there involve only finite switches in any finite time. As a consequence, $\sum_{i=1}^{l} h_i \to \infty$ as $l \to \infty$. According to Definition 3.2, by setting $\varepsilon = 1$, there exists a $\delta > 0$ such that

$$\|x_0\| \leq \delta \Longrightarrow \|\phi(t; 0, x_0, \sigma)\| \leq 1 \quad \forall\, t \geq t_0.$$

In particular,

$$\|e^{A_{i_s}h_s}\cdots e^{A_{i_1}h_1}e^{A_{i_0}h_0}x_0\| \leq 1 \quad \forall \, x_0 \in \mathbf{B}_\delta \quad s = 0,1,\cdots.$$

As a consequence, all entries of the matrices

$$e^{A_{i_0}h_0}, e^{A_{i_1}h_1}e^{A_{i_0}h_0}, \cdots, e^{A_{i_s}h_s}\cdots e^{A_{i_1}h_1}e^{A_{i_0}h_0}, \cdots \tag{3.3}$$

must be bounded by $\dfrac{1}{\delta}$. Suppose that

$$\varrho = \min_{k\in M}\left\{\sum_{i=1}^{n}\lambda_i(A_k)\right\} > 0.$$

Then, we have

$$\det e^{A_k h} = \exp\left(h\sum_{i=1}^{n}\lambda_i(A_k)\right) \geq e^{\varrho h} \quad k \in M \quad h > 0.$$

As a result,

$$\det e^{A_{i_s}h_s}\cdots e^{A_{i_1}h_1}e^{A_{i_0}h_0} \geq e^{\varrho \sum_{j=0}^{s} h_j} \to \infty \quad \text{as} \quad s \to \infty.$$

This contradicts the boundedness of entries of the matrices. This establishes the former part of the theorem. The latter part can be proven in a similar manner. $\square$

Theorem 3.5. *Suppose that the switched linear system $\Sigma(A_i)_M$ is pointwise stabilizable. Then, there is a $k \in M$ such that*

$$\min\left\{\lambda_1(A_k + A_k^T), \cdots, \lambda_n(A_k + A_k^T)\right\} \leq 0. \tag{3.4}$$

Furthermore, if the system is pointwise asymptotically stabilizable, then the inequality is strict.

Proof. We proceed by contradiction.

Suppose that all the eigenvalues of $A_k + A_k^T$ are positive. This implies that

$$A_k + A_k^T > 0 \quad k \in M.$$

As the index M is finite, there is a positive real number ϵ such that

$$A_k + A_k^T \geq \epsilon I_n \, . \quad k \in M.$$

Let $V(x) = x^T x$. It is easily seen that

$$\dot{V}\big|_{A_k x} \geq \epsilon V(x) \quad k \in M.$$

According to Theorem 2.5, each non-trivial trajectory diverges to infinity via any switching signal, hence the system is unstable.

The latter part can be proven in the same way. $\square$

Remark 3.6. The above two necessary conditions are by no means sufficient except for the simplest case of one dimensional systems. Nevertheless, an advantage of the conditions is that they are easily verifiable.

As an application, it can be easily verified that system (3.2) does not satisfy Theorem 3.4, hence, it is not consistently stabilizable.

Corollary 3.7. *For any second-order switched linear system consisting of one subsystem* $A_1 = \begin{bmatrix} 0 & 1 \\ -1 & 0 \end{bmatrix}$, *the system is pointwise asymptotically stabilizable if and only if*

$$\min_{k \in M} \min_{j \in \bar{n}} \{\lambda_j (A_k + A_k^T)\} < 0. \tag{3.5}$$

Proof. The *if* part follows immediately from Theorem 3.5.

According to (3.5), there exists another subsystem, say, A_2, such that $A_2 + A_2^T$ is not semi-positive definite. As a result, there is a sector

$$\Lambda = \{x \in \mathbf{R}^2 : k_2 x_1^2 \le x_1 x_2 \le k_1 x_1^2\}$$

with $k_2 < k_1$, such that

$$x^T (A_2 + A_2^T) x < 0 \quad \forall \, x \in \Lambda \quad x \ne 0.$$

On the other hand, the first subsystem can rotate a state to any direction without changing its 2-norm. Based on these observations, we assign a switching law σ which is the concatenation of $\sigma_1 \equiv 1$ with $\sigma_2 \equiv 2$ via Λ. That is, if the state is in Λ, let the second subsystem be active, otherwise, let the first subsystem be active. It can be seen that this switching strategy makes the switched system asymptotically stable. $\square$

This corollary establishes the simple geometric fact: If one subsystem is purely rotative, then the switched system is stabilizable when another subsystem is contractible along certain direction.

3.2.2 Equivalence of the Stabilization Notions

For a linear time-varying system, a well-known property is that uniformly asymptotic stabilizability implies (hence is equivalent to) exponential stabilizability (*c.f.* Theorem 2.6). A problem naturally arises: Does this equivalence still hold for switched linear systems? The following result provides an affirmative answer to the problem. To establish the equivalence, we need the concept of switched convergence.

Definition 3.8. *System (3.1) is said to be* switched convergent, *if for each state* $x_0 \in \mathbf{R}^n$, *there is a switching signal* σ_{x_0}, *such that the state trajectory initialized at* $x(0) = x_0$ *converges to the origin, that is*

$$\lim_{t \to \infty} \phi(t; 0, x_0, \sigma_{x_0}) = 0.$$

Theorem 3.9. *The following statements are equivalent*:

(i) the switched system is asymptotically stabilizable;
(ii) the switched system is exponentially stabilizable; and
(iii) the switched system is switched convergent.

Proof. It is obvious that $(ii) \Longrightarrow (i) \Longrightarrow (iii)$. Thus, we only need to prove $(iii) \Longrightarrow (ii)$.

First, by switched convergence, we know that, for each state x on the unit sphere $\mathbf{S}_1$, there exist a time t_x, and a switching path $\sigma_x \colon [0, t_x] \mapsto M$, such that

$$\phi(t_x; 0, x, \sigma_x) \in \mathbf{B}_{\frac{1}{4}}. \tag{3.6}$$

Suppose that the switching time sequence of σ_x is $t_1, \cdots, t_k$ with

$$t_0 = 0 < t_1 < \cdots < t_k < t_{k+1} \overset{def}{=} t_x.$$

Then, the corresponding state transition matrix is

$$\Phi(t, 0, \sigma_x) = e^{i_j(t-t_j)} e^{i_{j-1}(t_j - t_{j-1})} \cdots e^{i_0(t_1 - t_0)}$$
$$t \in [t_j, t_{j+1}] \quad j = 0, 1, \cdots, k$$

where $i_0, \cdots, i_k$ are the corresponding switching indices. In terms of the transition matrix, we can re-write Equation (3.6) as

$$\Phi(t_x, 0, \sigma_x)x \in \mathbf{B}_{\frac{1}{4}}.$$

As a result, there is a neighborhood N_x of x such that

$$\Phi(t_x, 0, \sigma_x)y \in \mathbf{B}_{\frac{1}{2}} \quad \forall\, y \in N_x.$$

Next, let x vary along the unit sphere, it is obvious that

$$\cup_{x \in \mathbf{S}_1} N_x \supseteq \mathbf{S}_1.$$

As the unit sphere is a compact set in $\mathbf{R}^n$, by the Finite Covering Theorem, there exist a finite number l, and a set of states $x_1, \cdots, x_l$ on the unit sphere, such that

$$\cup_{i=1}^l N_{x_i} \supseteq \mathbf{S}_1.$$

Accordingly, we can partition the unit sphere into l regions $R_1, \cdots, R_l$, such that

(a) $\cup_{i=1}^l R_i = \mathbf{S}_1$, and $R_i \cap R_j = \emptyset$ for $i \neq j$; and
(b) for each i, $1 \leq i \leq l$, we have

$$\Phi(t_{x_i}, 0, \sigma_{x_i})y \in \mathbf{B}_{\frac{1}{2}} \quad \forall\, y \in R_i.$$

In view of Item (b), for each $i = 1, \cdots, l$ and $x \in R_i$, we re-define t_x and σ_x by

$$t_x = t_{x_i} \quad \text{and} \quad \sigma_x = \sigma_{x_i}.$$

Let $T = \max_{i=1}^l t_{x_i}$, and $\eta = \max_{i \in M} \|A_i\|$. It is clear that

$$\|\Phi(t, 0, \sigma_x)\| \le e^{\eta T} \quad \forall x \in \mathbf{S}_1 \quad t \le t_x.$$

Then, for any $x_0 \ne 0$, construct a switching path $\theta_{x_0} : [0, \infty) \mapsto M$ as follows. Define recursively a sequence of states

$$z_0 = x_0$$
$$z_{k+1} = \phi(t_{\frac{z_k}{\|z_k\|}}; 0, z_k, \sigma_{\frac{z_k}{\|z_k\|}}) \quad k = 0, 1, \cdots.$$

As $z_k / \|z_k\| \in \mathbf{S}_1$ for all $k = 1, 2, \cdots$, from the previous derivations, it follows that the signal $\sigma_{\frac{z_k}{\|z_k\|}}(\cdot)$ is known over $[0, t_{\frac{z_k}{\|z_k\|}}]$. Let

$$\theta_{x_0}(t) = \begin{cases} \sigma_{\frac{z_0}{\|z_0\|}}(t) & t \in [0, t_{\frac{z_0}{\|z_0\|}}) \\ \sigma_{\frac{z_1}{\|z_1\|}}(t - t_{\frac{z_0}{\|z_0\|}}) & t \in [t_{\frac{z_0}{\|z_0\|}}, t_{\frac{z_0}{\|z_0\|}} + t_{\frac{z_1}{\|z_1\|}}) \\ \vdots & \\ \sigma_{\frac{z_k}{\|z_k\|}}(t - \sum_{i=0}^{k-1} t_{\frac{z_i}{\|z_i\|}}) & t \in [\sum_{i=0}^{k-1} t_{\frac{z_i}{\|z_i\|}}, \sum_{i=0}^{k} t_{\frac{z_i}{\|z_i\|}}) \\ \vdots & \end{cases}$$

which is defined over $[0, \infty)$. In this way, each non-origin state x_0 is assigned a switching path $\theta_{x_0} : [0, \infty) \mapsto M$. For $x_0 = 0$, assign θ_{x_0} to be any switching path.

Finally, we prove that each state trajectory under the above switching path is exponentially convergent. To see this, let

$$\alpha = \ln 2 / T \quad \text{and} \quad \beta = 2e^{\eta T}.$$

It follows from Proposition 1.7 that

$$\phi(t; 0, \lambda x_0, \theta_{x_0}) = \lambda \phi(t; 0, x_0, \theta_{x_0}) \quad \forall \, t, x_0 \quad \lambda \in \mathbf{R}.$$

Accordingly, we have

$$\|z_{k+1}\| \le \frac{\|z_k\|}{2} \quad k = 0, 1, \cdots.$$

On the other hand, as $t_x \le T$ for all $x \in \mathbf{S}_1$, we have

$$\|\phi(t; 0, x_0, \theta_{x_0})\| \le e^{\eta T} \|\phi(\sum_{i=0}^{k-1} t_{\frac{z_i}{\|z_i\|}}; 0, x_0, \theta_{x_0})\|$$
$$\forall \, t \in [\sum_{i=0}^{k-1} t_{\frac{z_i}{\|z_i\|}}, \sum_{i=0}^{k} t_{\frac{z_i}{\|z_i\|}}) \quad k = 0, 1, \cdots.$$

Combining the above reasonings gives

$$\|\phi(t; 0, x_0, \theta_{x_0})\| \le \beta \exp(-\alpha t) \|x_0\| \quad \forall\, x_0 \in \mathbf{R}^n \quad t \ge 0. \tag{3.7}$$

As the constants α and β are independent of x_0 and θ_{x_0}, Inequality (3.7) shows that the switched system is exponentially stabilizable. $\square$

Remark 3.10. The relationship between switched convergence and exponential stability is an important issue for dynamic systems. The equivalence between them has been established for linear differential inclusions [101]. As a direct consequence, for a switched linear system with arbitrary switching signals, convergence implies exponential stability. Theorem 3.9 ensures that, if a switched system is switched convergent, then the system is also exponentially stabilizable. In Section 6.3, this theorem will be utilized to address the infinite-time horizon optimal switching problem.

3.2.3 Periodic and Synchronous Switchings

Switching path $\theta_{[0,\infty)}$ is said to be *periodic*, if there exists a positive time T such that

$$\theta(t + T) = \theta(t) \quad \forall\, t \ge 0.$$

Switching path σ is said to be *synchronous*, if there exist a base rate ω, and a sequence of natural numbers $\{\mu_1, \mu_2, \cdots\}$, such that the switching time sequence is

$$\{0, \mu_1\omega, \mu_2\omega, \cdots\}.$$

Periodic and synchronous switching signals are interesting from the viewpoint of implementation.

Theorem 3.11. *If a switched system is consistently asymptotically stabilizable, then, there is a periodic and synchronous switching path which asymptotically stabilizes the switched system.*

Proof. If there is a subsystem, say, A_k, that is asymptotically stable, then the constant switching signal $\sigma \equiv k$ works. Otherwise, suppose that a switching signal σ with duration sequence

$$DS_\sigma = \{(i_0, h_0), (i_1, h_1), \cdots\}$$

asymptotically stabilizes the switched system. It is obvious that this switching signal must involve infinite switches. From the proof of Theorem 3.4, matrix sequence (3.3) converges to the zero matrix. Consequently, there is a finite number N such that

$$\|e^{A_{i_N} h_N} \cdots e^{A_{i_1} h_1} e^{A_{i_0} h_0}\| < 1. \tag{3.8}$$

Let us define a function $g \colon \mathbf{R}^{N+1} \mapsto \mathbf{R}_+$ by

$$g(s_0, s_1, \cdots, s_N) = \| e^{A_{i_N} s_N} \cdots e^{A_{i_1} s_1} e^{A_{i_0} s_0} \|.$$

It can be seen that g is a continuous function of its arguments. Since

$$g(h_0, h_1, \cdots, h_N) < 1$$

there is a neighborhood Λ of $(h_0, h_1, \cdots, h_N)^T$ in $\mathbf{R}^{N+1}$ such that

$$g(z) < 1 \quad \forall \, z \in \Lambda.$$

Choose a $z_0 = (r_0, r_1, \cdots, r_N)^T$ from Λ, where r_j is a rational number for any $j = 0, 1, \cdots, N$. It can be verified that the periodic and synchronous switching path θ with duration sequence

$$DS_\theta = \{(i_0, r_0), \cdots, (i_N, r_N), (i_0, r_0), \cdots, (i_N, r_N), \cdots\} \tag{3.9}$$

asymptotically stabilizes the switched system. $\square$

Estimation (3.8) is very important in analyzing the convergence of the systems. It establishes the contractibility uniformly for all initial states.

Corollary 3.12. *For a switched linear system, the following statements are equivalent*:

(i) the system is consistently asymptotically stabilizable;
(ii) the system is consistently exponentially stabilizable;
(iii) the system is periodically and synchronously asymptotically stabilizable;
(iv) there exist a natural number l, an index sequence $i_1, \cdots, i_l$, and a positive real number sequence $h_1, \cdots, h_l$, such that matrix $e^{A_{i_l} h_l} \cdots e^{A_{i_1} h_1}$ is Schur; and
(v) for any real number $s \in (0, 1)$, there exist a natural number $l = l(s)$, an index sequence $i_1, \cdots, i_l$, and a positive real number sequence $h_1, \cdots, h_l$, such that

$$\| e^{A_{i_l} h_l} \cdots e^{A_{i_1} h_1} \| \le s. \tag{3.10}$$

Proof. From the proof of Theorem 3.11, (i) implies that, there is a finite number N, such that

$$\| e^{A_{i_N} h_N} \cdots e^{A_{i_1} h_1} e^{A_{i_1} h_1} \| = \gamma < 1$$

for some sequences $i_1, \cdots, i_N$ and $h_1, \cdots, h_N$. Let $l = kN$, where k is to be determined later. Define

$$i_{j+\mu N} = i_j \text{ and } h_{j+\mu N} = h_j \quad j = 1, \cdots, N \quad \mu = 1, \cdots, k-1.$$

It can be seen that

$$\| e^{A_{i_l} h_l} \cdots e^{A_{i_1} h_1} e^{A_{i_1} h_1} \| = (\| e^{A_{i_N} h_N} \cdots e^{A_{i_1} h_1} e^{A_{i_1} h_1} \|)^k = \gamma^k.$$

Accordingly, for any $s \in (0, 1)$, by letting $k \ge \frac{\ln s}{\ln \gamma}$, Inequality (3.10) holds. This means that $(i) \implies (v)$. In the same manner, we can prove that $(iv) \implies (v)$. On the other hand, from the proof of Theorem 3.9, we have $(iv) \implies (iii)$. Other implications are trivial and hence the theorem follows. $\square$

3.2.4 Special Systems

A switched linear system $\Sigma(A_i)_M$ is said to be *triangular* if each subsystem is in the triangular form:

$$A_k = \begin{bmatrix} a_{1,1}^k & \cdots & a_{1,n}^k \\ & \ddots & \\ 0 & \cdots & a_{n,n}^k \end{bmatrix} \quad k \in M.$$

Triangular systems are interesting because they have a simple structure, and many non-triangular systems can be made to be triangular by means of equivalent transformations (simultaneous triangularization) [113]. Here, we present a stabilizability criterion for triangular systems.

Suppose that $A = (a_{i,j})_{n \times n}$. Let $A^d = \text{diag}(a_{1,1}, \cdots, a_{n,n})$ denote the matrix obtained from A by replacing all the off-diagonal entries with zeros, and $l(A)$ the row vector $(-a_{1,1}, \cdots, -a_{n,n})$. Matrix $L \in \mathbf{R}^{m \times n}$ is said to be *semi-positive* provided that there is an $x \in \mathbf{R}^n$ with $x \geq 0$, such that $Lx > 0$, the inequalities denoting entrywise inequality.

Theorem 3.13. *The following statements are equivalent*:

(i) triangular system $\Sigma(A_i)_M$ is consistently asymptotically stabilizable;
(ii) diagonal system $\Sigma(A_i^d)_M$ is consistently asymptotically stabilizable;
(iii) diagonal system $\Sigma(A_i^d)_M$ is pointwise asymptotically stabilizable; and
(iv) matrix $L = \left[l(A_1)^T, \cdots, l(A_m)^T \right]^T$ is semi-positive.

Proof. By Corollary 3.12, the system is consistently asymptotically stabilizable if and only if there exist a natural number k, and sequences $i_1, \cdots, i_k$ and $h_1, \cdots, h_k$, such that matrix $e^{A_{i_k} h_k} \cdots e^{A_{i_1} h_1}$ is Schur. The equivalence between (i) and (ii) comes from the fact that the eigenvalues of $e^{A_{i_k} h_k} \cdots e^{A_{i_1} h_1}$ are the same as those of $e^{A_{i_k}^d h_k} \cdots e^{A_{i_1}^d h_1}$.

Next, we show that $(ii) \iff (iv)$. It is clear that (ii) is equivalent to the existence of sequences $i_1, \cdots, i_k$ and $h_1, \cdots, h_k$ such that matrix $e^{A_{i_k}^d h_k} \cdots e^{A_{i_1}^d h_1}$ is Schur. Since diagonal matrices are commutative, this is equivalent to the existence of nonnegative real numbers $t_1, \cdots, t_m$ such that matrix

$$e^{A_1^d t_1} \cdots e^{A_m^d t_m} = \exp(A_1^d t_1 + \cdots + A_m^d t_m)$$

is Schur. Accordingly, we have

$$A_1(j,j)t_1 + \cdots + A_m(j,j)t_m > 0 \quad \forall\, j = 1, \cdots, n$$

which implies that matrix L is semi-positive.

Finally, we establish by contradiction that $(iii) \implies (iv)$. Suppose that (iv) does not hold. This means that any matrix in the form $\exp(A_1^d t_1 + \cdots + A_m^d t_m)$ is not convergent (to the zero matrix). As a result, any matrix of the form

$e^{A^d_{i_k} h_k} \cdots e^{A^d_{i_1} h_1}$ is not convergent. As such a matrix is diagonal, by Theorem 3.5, the diagonal system $\Sigma(A^d_i)_M$ is not pointwise asymptotically stabilizable. □

This theorem converts the stabilizability of triangular systems into the semi-positiveness of a matrix, which can be verified effectively [75].

For the switched linear system with a stable convex combination, the following theorem shows that these systems are consistently stabilizable.

Theorem 3.14. *Suppose that there is a stable convex combination of A_k, $k \in M$. Then, the switched system is consistently asymptotically stabilizable.*
Proof. Suppose that $\sum_{k \in M} w_k A_k$ is Hurwitz for $w_1, \cdots, w_m$ with

$$w_k \geq 0 \quad k \in M \quad \text{and} \quad \sum_{k \in M} w_k = 1.$$

By Lemma 2.10, there is a positive real number ρ such that

$$\mathrm{sr}\left(\exp(w_m h A_m) \cdots \exp(w_1 h A_1)\right) < 1 \quad \forall \, h \leq \rho \tag{3.11}$$

where $\mathrm{sr}(A)$ denotes the spectral radius of matrix A. As a result, matrix $\exp(w_m \rho A_m) \cdots \exp(w_1 \rho A_1)$ is Schur. Define the periodic switching path θ as

$$DS_\theta = \{(1, w_1\rho), \cdots, (m, w_m\rho), (1, w_1\rho), \cdots, (m, w_m\rho), \cdots\}. \tag{3.12}$$

It can be seen that θ asymptotically stabilizes the switched system. □

3.2.5 Robustness Issues

In this subsection, we briefly discuss the robustness of switched linear systems. Suppose that system (3.1) undergoes small perturbations:

$$\dot{x}(t) = (A_\sigma + \varepsilon_\sigma B_\sigma)x(t) \tag{3.13}$$

where $B_k \in \mathbf{R}^{n \times n}$ for $k \in M$ are given and fixed, and ε_k for $k \in M$ are real numbers.

Theorem 3.15. *Suppose that nominal system (3.1) is asymptotically stabilizable. Then, there are positive numbers $\kappa_1, \cdots, \kappa_m$, such that the perturbed system (3.13) is also asymptotically stabilizable if*

$$|\varepsilon_k| \leq \kappa_k \quad k \in M.$$

Proof. By the proof of Theorem 3.9, we can partition the unit sphere into a finite set of regions $R_1, \cdots, R_l$, such that

(a) $\cup_{i=1}^l R_i = \mathbf{S}_1$, and $R_i \cap R_j = \emptyset$ for $i \neq j$;

(b) for each i, $1 \leq i \leq l$, we have a time t_{x_i} and a switching path σ_{x_i}, such that

$$\Phi(t_{x_i}, 0, \sigma_{x_i})y \in \mathbf{B}_{\frac{1}{2}} \quad \forall \, y \in R_i$$

and

(c) there is a time T, such that $t_{x_i} \leq T$ for all $i = 1, \cdots, l$.

Now fix an $i \in \bar{l}$. Suppose that the switching duration sequence for σ_{x_i} in $[0, t_{x_i})$ is

$$\{(j_{i1}, h_{i1}), \cdots, (j_{ik_i}, h_{ik_i})\}.$$

Then, we have

$$\left\| e^{A_{j_{ik_i}} h_{ik_i}} \cdots e^{A_{j_{i1}} h_{i1}} y \right\| \leq \frac{1}{2} \quad \forall \, y \in R_i.$$

Define a function

$$g_i(\varepsilon_1, \cdots, \varepsilon_m) = \sup_{y \in R_i} \left\| e^{(A_{j_{ik_i}} + \varepsilon_{j_{ik_i}} B_{j_{ik_i}}) h_{ik_i}} \cdots e^{(A_{j_{i1}} + \varepsilon_{j_{i1}} B_{j_{i1}}) h_{i1}} y \right\|.$$

It is clear that g_i is a continuous function of its arguments. As

$$g_i(0, \cdots, 0) \leq \frac{1}{2}$$

there are positive numbers $\kappa_{i1}, \cdots, \kappa_{im}$, such that

$$g_i(\varepsilon_1, \cdots, \varepsilon_m) \leq \frac{2}{3} \quad \forall \, |\varepsilon_i| < \kappa_{ij} \quad j \in M.$$

Let i vary and denote

$$\kappa_k = \min\{\kappa_{1k}, \cdots, \kappa_{lk}\} \quad k \in M.$$

Suppose that the perturbed system (3.13) is with

$$|\varepsilon_k| \leq \kappa_k \quad k \in M.$$

Let Φ' denote the transition matrix of system (3.13). It is clear that

$$\Phi'(t_{x_i}, 0, \sigma_{x_i})y \in \mathbf{B}_{\frac{2}{3}} \quad \forall \, y \in R_i \quad i = 1, \cdots, l.$$

This, together with the proof of Theorem 3.9, implies that the perturbed system is asymptotically stabilizable. $\square$

Remark 3.16. Theorem 3.15 establishes an important fact on robustness of switched linear systems. Indeed, it asserts that sufficiently small perturbations on the system matrices $A_1, \cdots, A_m$ will not turn an asymptotically stabilizable switched system into an unstable system. This is the generalization of its time-invariant counterpart, namely, each stable linear time-invariant system has a positive stability margin.

Remark 3.17. If we take a $n \times n$ real matrix as a point in the n^2-dimensional Euclidean space, switched system (3.1) is associated to a point in the mn^2-dimensional Euclidean space. Theorem 3.15 implies that, all asymptotically stabilizable n-th order switched systems with m subsystems form an open set in the mn^2-dimensional Euclidean space.

In the perturbed model (3.13), we fix B_k for $k \in M$ and take ε_k for $k \in M$ as the perturbed variables. An alternative way is to describe the perturbed system as

$$\dot{x}(t) = (A_\sigma + B_\sigma)x(t) \tag{3.14}$$

where $B_k \in \mathbf{R}^{n \times n}$ for $k \in M$ are the perturbations. In this situation, the robustness theorem can be stated as follows.

Theorem 3.18. *Suppose that nominal system (3.1) is asymptotically stabilizable. Then, there are positive real numbers $\kappa_1, \cdots, \kappa_m$, such that the perturbed system (3.14) is also asymptotically stabilizable if*

$$\|B_k\| \leq \kappa_k \quad k \in M.$$

For consistently asymptotically stabilizable switched systems, a similar robustness property holds as follows.

Theorem 3.19. *Suppose that nominal system (3.1) is consistently asymptotically stabilizable. Then, there are positive numbers $\kappa_1, \cdots, \kappa_m$, such that the perturbed system (3.13) is also consistently asymptotically stabilizable if*

$$|\varepsilon_k| \leq \kappa_k \quad k \in M.$$

Proof. If there is a subsystem, say, A_1, that is asymptotically stable, then, there is a positive number κ_1 such that $A_1 + \varepsilon_1 B_1$ is still asymptotically stable when $|\varepsilon_1| < \kappa_1$. In this case, the constant switching signal $\sigma \equiv 1$ asymptotically stabilizes the perturbed system if $|\varepsilon_1| < \kappa_1$ and $|\varepsilon_i| < \infty, i = 2, \cdots, m$.

Next, we assume that no subsystem is asymptotically stable. By the proof of Theorem 3.11, there exist a finite number N, a sequence of indices $i_1, \cdots, i_N$, and a sequence of time intervals $h_1, \cdots, h_N$, such that

$$\|e^{A_{i_N} h_N} \cdots e^{A_{i_1} h_1}\| = r < 1. \tag{3.15}$$

Let us define a function $\rho \colon \mathbf{R}^m \mapsto \mathbf{R}_+$ by

$$\rho(s_1, \cdots, s_m) = \|e^{(A_{i_N} + s_{i_N} B_{i_N}) h_N} \cdots e^{(A_{i_1} + s_{i_1} B_{i_1}) h_1}\|.$$

It can be seen that ρ is a continuous function of its arguments. Because $\rho(0, \cdots, 0) = r < 1$, there is a sequence of positive numbers $\kappa_1, \cdots, \kappa_m$ such that

$$\rho(s_1, \cdots, s_m) < 1 \quad \forall \, s_i \in [-\kappa_i, \kappa_i] \quad i \in M.$$

Accordingly, the periodic switching path θ with

$$DS_\theta = \{(i_1, h_1), \cdots, (i_N, h_N), (i_1, h_1), \cdots, (i_N, h_N), \cdots\}$$

asymptotically stabilizes the perturbed systems provided that $|\varepsilon_i| < \kappa_i$, $i \in M$. $\square$

Remark 3.20. A special case of interest is when $B_k = I_n$ for $k \in M$. Because I_n is commutative with any matrix, we can easily prove that the perturbed system (3.13) is consistently asymptotically stabilizable if

$$\varepsilon_{i_1} h_1 + \cdots + \varepsilon_{i_N} h_N < -\ln r. \tag{3.16}$$

Note that there is no lower bound for ε_i's.

Example 3.21. For the third-order switched linear system with two subsystems

$$A_1 = \begin{bmatrix} 1 & 0 & 0 \\ 0 & -1 & 0 \\ 0 & 0 & -2 \end{bmatrix} \text{ and } A_2 = \begin{bmatrix} -10 & -1 & 10 \\ 1 & 0 & 6 \\ -1 & -4 & 2 \end{bmatrix}$$

it can be verified that

$$\| \exp(0.4 A_2) \exp(A_1) \| \approx 0.7803 < 1.$$

Consequently, this switched system is consistently asymptotically stabilizable.

Let us consider the perturbed system (3.13) with $B_1 = B_2 = I_3$. From Remark 3.20, the perturbed system is consistently asymptotically stabilizable if

$$\varepsilon_1 + 0.4\varepsilon_2 < -\ln 0.7803 \approx 0.2481.$$

3.3 Periodic Switching

In this section, we investigate the stabilizing and robust design issues via periodic switching signals. For the switched linear autonomous system (3.1), we make the following assumption.

Assumption 3.1. *There is a convex combination of A_k, $k \in M$ which is Hurwitz.*

In view of this assumption, let $A_0 \overset{def}{=} \sum_{k \in M} w_k A_k$ be Hurwitz with $w_k \geq 0$ and $\sum_{k \in M} w_k = 1$.

Under this assumption, we are able to apply the average technique to approximate the switched system with a linear time-invariant system. According to Lemma 2.10, for a sufficiently small ρ, we have

$$\exp(w_m A_m \rho) \exp(w_{m-1} A_{m-1} \rho) \cdots \exp(w_1 A_1 \rho) = \exp\left(\rho(A_0 + \rho \Upsilon_\rho)\right).$$

Moreover, for a sufficiently small ρ, matrix $\bar{A} \stackrel{def}{=} A_0 + \rho \Upsilon_\rho$ is Hurwitz. Let us fix such a ρ. Define a periodic switching path

$$\sigma(t) = \begin{cases} 1 & \mod (t, \rho) \in [0, w_1 \rho) \\ 2 & \mod (t, \rho) \in [w_1 \rho, (w_1 + w_2)\rho) \\ \vdots & \\ m & \mod (t, \rho) \in [(\sum_{i=1}^{m-1} w_i)\rho, \rho) \end{cases} \quad \forall\, t \geq t_0 \qquad (3.17)$$

where $\mod (a, b)$ denotes the remainder of a divided by b.

Let $\{(t_0, i_0), (t_1, i_1), \cdots\}$ be the switching sequence of the switching path. Define the matrix function

$$\Psi(t, \sigma) = e^{A_{i_k}(t - t_k)} e^{A_{i_{k-1}}(t_k - t_{k-1})} \cdots e^{A_{i_0}(t_1 - t_0)} \quad t \in [t_k, t_{k+1}].$$

The state transition matrix can be expressed as

$$\Phi(s_1, s_2, \sigma) = \Psi(s_1, \sigma)\Psi(s_2, \sigma)^{-1} \quad \forall\, s_1, s_2 \geq t_0. \qquad (3.18)$$

The solution of system (3.1) is given by

$$x(t) = \Phi(t, t_0, \sigma)x_0 \quad \forall\, t \geq t_0.$$

We have the following estimation for the state transition matrix.

Lemma 3.22. *Transition matrix (3.18) is exponentially convergent, that is, there exist two positive numbers α and β such that*

$$\|\Phi(s_1, s_2, \sigma)\| \leq \beta \exp\left(-\alpha(s_1 - s_2)\right) \quad \forall\, s_1 \geq s_2.$$

Proof. Since

$$\Phi(\rho, 0, \sigma) = \exp(w_m A_m \rho) \cdots \exp(w_1 A_1 \rho) = \exp(\bar{A}\rho)$$

for any nonnegative integers $l_1 \leq l_2$, we have

$$\Phi(l_2 \rho, l_1 \rho, \sigma) = \exp(\bar{A}(l_2 - l_1)\rho).$$

As $\bar{A}$ is Hurwitz, there are positive numbers α and κ such that

$$\|\Phi(l_2 \rho, l_1 \rho, \sigma)\| \leq \kappa \exp\left(-\alpha(l_2 \rho - l_1 \rho)\right).$$

For any $s_1 \leq s_2$, let l_1, l_2 satisfy

$$l_1 \rho \leq s_1 < (l_1 + 1)\rho \quad (l_2 - 1)\rho < s_2 \leq l_2 \rho.$$

Simple calculation gives

$$\|\Phi(s_2, s_1, \sigma)\| \leq \|\Phi(l_1\rho, s_1, \sigma)\| \|\Phi(l_2\rho, l_1\rho, \sigma)\| \|\Phi(s_2, l_2\rho, \sigma)\|$$
$$\leq \kappa \exp\left(-\alpha(l_2\rho - l_1\rho)\right) \|\Phi(0, s_1 - l_1\rho, \sigma)\| \|\Phi(0, l_2\rho - s_2, \sigma)\|.$$

Denoting

$$\kappa_1 = \max_{0 \leq t \leq \rho} \|\Phi(0, t, \sigma)\|$$

the maximum is attainable because $\Phi(0, t, \sigma)$ is continuous from t. Combining the above inequalities gives

$$\|\Phi(s_1, s_2, \sigma)\| \leq \kappa_1^2 \kappa \exp\left(-\alpha(l_2\rho - l_1\rho)\right) \leq \kappa_1^2 \kappa \exp\left(-\alpha(s_2 - s_1)\right).$$

Let $\beta = \kappa_1^2 \kappa$ and the lemma follows. $\square$

Next, we consider the perturbed switched system given by

$$\dot{x}(t) = A_\sigma x(t) + f_\sigma(t) \quad x(t_0) = x_0 \tag{3.19}$$

where $f_k \colon \mathbf{R}_+ \mapsto \mathbf{R}^n$, $k \in M$ are piecewise continuous vector functions representing system perturbations or uncertainties.

Note that the above model allows different perturbations for different subsystems, thus reflecting many practical situations. For example, in the framework of multi-controller switching, different controllers may induce different types of noises. If the noises are controller-independent, we simply have $f_1 = f_2 = \cdots = f_m$.

Let

$$N = \sup_{t \geq t_0, k \in M} \{\|f_k(t)\|\}.$$

The perturbations $f_k(\cdot)$, $k \in M$ are said to be

i) *bounded*, if $N < \infty$;
ii) *convergent (to the origin)*, if $\|f_k(t)\| \to 0$ as $t \to \infty$ for all $k \in M$; and
iii) *exponentially convergent (to the origin)*, if $\|f_k(t)\| \leq \delta \exp(-\gamma t)$, $\forall\, t,\, k \in M$ for some positive real numbers δ and γ.

Theorem 3.23. *Suppose that Assumption 3.1 holds. Then, for system (3.19) with periodic switching signal (3.17), we have the following:*

(i) the system state is bounded if the perturbation is bounded;
(ii) the system state is bounded and convergent if the perturbation is bounded and convergent; and
(iii) the system state is exponentially convergent if the perturbation is exponentially convergent.

Proof. According to Lemma 3.22, there exist two positive numbers α and β such that

$$\|\Phi(s_1, s_2, \sigma)\| \leq \beta \exp\left(-\alpha(s_1 - s_2)\right) \quad \forall\, s_1 \geq s_2.$$

Denote $f(t) = f_{\sigma(t)}(t)$, $t \geq t_0$. First, suppose that $f(\cdot)$ is bounded. Then, we have

$$\|x(t)\| = \|\Phi(t, t_0, \sigma)x_0 + \int_{t_0}^{t} \Phi(t, \tau, \sigma)f(\tau)d\tau\|$$

$$\leq \|\Phi(t, t_0, \sigma)\|\|x_0\| + N \int_{t_0}^{t} \|\Phi(t, \tau, \sigma)\|d\tau$$

$$\leq \beta\|x_0\| + \frac{\beta}{\alpha}N.$$

Hence, the state is bounded.

Second, suppose that $f(\cdot)$ is bounded and convergent. Then, for any given positive number ϵ, we can find a time $T \geq t_0$ such that

$$\|x(t)\| \leq \epsilon \quad \forall\, t \geq T. \tag{3.20}$$

Indeed, for any given positive number ϱ, there exists a $T_1 \geq t_0$ such that

$$\|f(t)\| \leq \varrho \quad \forall\, t \geq T_1.$$

Therefore, we have

$$\|x(t)\| = \|\Phi(t, t_0, \sigma)x_0 + \int_{t_0}^{t} \Phi(t, \tau, \sigma)f(\tau)d\tau\|$$

$$\leq \beta \exp\left(-\alpha(t - t_0)\right)\|x_0\| + N \int_{t_0}^{T_1} \beta \exp\left(-\alpha(t - \tau)\right)d\tau$$

$$+\varrho \int_{T_1}^{t} \beta \exp\left(-\alpha(t - \tau)\right)d\tau$$

$$\leq \beta \exp\left(-\alpha(t - t_0)\right)\|x_0\| + \frac{\beta}{\alpha}N \exp\left(-\alpha(t - T_1)\right) + \frac{\beta}{\alpha}\varrho.$$

Choose T and ϱ to satisfy

$$\exp(-\alpha T)\|x_0\| \leq \frac{\epsilon}{3} \quad \frac{\beta}{\alpha}N \exp\left(-\alpha(T - T_1)\right) \leq \frac{\epsilon}{3} \text{ and } \varrho \leq \frac{\epsilon\alpha}{3\beta}.$$

With these, inequality (3.20) follows. From the arbitrariness of ϵ, the convergence of the state follows.

Third, suppose that $\|f(t)\| \leq \delta \exp(-\gamma t)$, $\forall\, t$. Then, we have

$$\|x(t)\| \leq \beta \exp\left(-\alpha(t - t_0)\right)\|x_0\| + \int_{t_0}^{t} \beta \exp\left(-\alpha(t - \tau)\right)\delta \exp(-\gamma\tau)d\tau.$$

Simple derivation gives

$$\|x(t)\| \leq \begin{cases} (\beta\|x_0\| + \frac{\beta\delta}{\gamma-\alpha})\exp\left(-\alpha(t - t_0)\right) & \text{if } \gamma > \alpha \\ (\beta\|x_0\| + \frac{\beta\delta}{\alpha-\gamma})\exp\left(-\gamma(t - t_0)\right) & \text{if } \gamma < \alpha \\ (\beta\|x_0\| + \frac{\beta\delta}{\varepsilon e})\exp\left(-(\alpha - \varepsilon)(t - t_0)\right) & \text{if } \gamma = \alpha \end{cases}$$

where $e = \exp(1)$ and ε is any sufficiently small positive number. As a result, the state is exponentially convergent. This completes the proof of the theorem. $\square$

Remark 3.24. This theorem establishes several nice robustness properties for a class of switched linear systems. In particular, the bounded perturbation (implying) bounded state property is desirable in many practical situations. Moreover, it can be seen from the proof that the bound of the state can be explicitly estimated. Therefore, given any allowable state bound, we can estimate a bound of perturbations which makes the system state bounded within the allowed level.

Remark 3.25. It can be seen from the proof that the decay rate of the state relies heavily on the decay rate of the transition matrix, which in turn relies on the choice of the switching signal. A question naturally arises: How can we find a periodic switching signal with the largest possible decay rate for the transition matrix? Intuitively, we need to find a natural number l, an index sequence $i_0, i_1, \cdots, i_l$ and a duration sequence $h_0, h_1, \cdots, h_l$, such that

$$\frac{\ln \| \exp(A_{i_l} h_l) \cdots \exp(A_{i_1} h_1) \exp(A_{i_0} h_0) \|}{\sum_{j=0}^{l} h_j} \to \min.$$

Because this problem can be quite involved, we leave it open for further investigation.

3.4 State-feedback Switching

3.4.1 State-space-partition-based Switching

In this subsection, we formulate a state-feedback switching signal based on an appropriate partition of the state space.

First, suppose that Assumption 3.1 holds and the average matrix

$$A_0 = \sum_{k \in M} w_k A_k$$

is Hurwitz. Solving the Lyapunov equation

$$A_0^T P + P A_0 = -I_n$$

for symmetric matrix P, we obtain a positive definite solution P. Denote

$$Q_k = A_k^T P + P A_k \quad k \in M.$$

Then, fix a set of real numbers $r_i \in (0,1)$, $i \in M$. For any initial state $x(t_0) = x_0$, set

$$\sigma(t_0) = \arg\min\{x_0^T Q_1 x_0, \cdots, x_0^T Q_m x_0\}$$

where $\arg\min$ stands for the index which attains the minimum among M. If there are more than one such index, we simply choose the minimum index.

The first switching time instant is determined by

$$t_1 = \inf\left\{t > t_0 \colon\ x^T(t)Q_{\sigma(t_0)}x(t) > -r_{\sigma(t_0)}x^T(t)x(t)\right\}.$$

If the set is empty, then, let $t_1 = \infty$. Otherwise, define the switching index as

$$\sigma(t_1) = \arg\min\{x^T(t_1)Q_1 x(t_1), \cdots, x^T(t_1)Q_m x(t_1)\}.$$

Finally, we define the switching time/index sequences recursively by

$$t_{k+1} = \inf\left\{t > t_k \colon\ x^T(t)Q_{\sigma(t_k)}x(t) > -r_{\sigma(t_k)}x^T(t)x(t)\right\}$$
$$\sigma(t_{k+1}) = \arg\min\left\{x^T(t_{k+1})Q_1 x(t_{k+1}), \cdots, x^T(t_{k+1})Q_m x(t_{k+1})\right\}$$
$$k = 1, 2, \cdots. \tag{3.21}$$

Lemma 3.26. *Under the above switching law, system (3.1) is well-posed and quadratically stable.*

Proof. We first prove the well-posedness of the switching signal.

Suppose that t_k and t_{k+1} are two consecutive switching time instants. Let $i = \sigma(t_k+)$. It follows from the switching signal that

1) $x^T(t_k)Q_i x(t_k) = \min_{j\in M}\{x^T(t_k)Q_j x(t_k)\}$; and
2) $x^T(t_{k+1})Q_i x(t_{k+1}) \geq -r_i x^T(t_{k+1})x(t_{k+1})$.

As $\sum_{j\in M} w_j Q_j = -I_n$ and $\sum_{j\in M} w_j = 1$, it follows from Item 1) that

$$x^T(t_k)Q_i x(t_k) \leq -x^T(t_k)x(t_k). \tag{3.22}$$

For notational convenience, denote $x_k = x(t_k)$, $x_{k+1} = x(t_{k+1})$, and let ϑ be any real number greater than 1.

First, consider the case

$$\|x(t)\| \leq \vartheta\|x_{k+1}\| \quad \forall\, t \in [t_k, t_{k+1}]. \tag{3.23}$$

In this case, define a function

$$g(t) = x^T(t)(Q_i + I_n)x(t) \quad t \in [t_k, t_{k+1}].$$

It follows from (3.22) and Item 2) that

$$g(t_k) \leq 0 \quad g(t_{k+1}) \geq (1 - r_i)x_{k+1}^T x_{k+1}. \tag{3.24}$$

Simple computation gives

$$\frac{dg}{dt}(t) = x^T(t)\left(A_i^T(Q_i + I_n) + (Q_i + I_n)A_i\right)x(t).$$

Denote $\nu_i = \|A_i^T(Q_i + I_n) + (Q_i + I_n)A_i\|$. By inequality (3.23), we have

$$|\frac{dg}{dt}(t)| \leq \vartheta^2 \nu_i x_{k+1}^T x_{k+1} \quad \forall \, t \in [t_k, t_{k+1}].$$

Combining this with (3.24) yields

$$\vartheta^2 \nu_i (t_{k+1} - t_k) \geq (1 - r_i)$$

which implies that

$$t_{k+1} - t_k \geq (1 - r_i)/(\vartheta^2 \nu_i).$$

Second, suppose that (3.23) does not hold. This means that there is a $t^* \in [t_k, t_{k+1})$ satisfying

$$\|x(t^*)\| > \vartheta \|x_{k+1}\|. \tag{3.25}$$

From system equation (3.1), we have

$$x(t^*) = \exp\left(A_i(t^* - t_{k+1})\right) x_{k+1}.$$

From (3.25) and the fact that

$$\|\exp\left(A_i(t^* - t_{k+1})\right)\| \leq \exp\left(\|A_i\|(t_{k+1} - t^*)\right)$$

it follows that

$$t_{k+1} - t_k \geq t_{k+1} - t^* > \ln\vartheta/\|A_i\|.$$

From the above reasonings, we have

$$t_{k+1} - t_k \geq \eta \overset{def}{=} \sup_{\vartheta > 1} \min_{i \in M} \left((1 - r_i)/(\vartheta^2 \nu_i), \ln\vartheta/\|A_i\|\right).$$

Hence, the switching signal has a positive dwell time η, and hence it is well-defined.

To prove the quadratic stability of the switched system, let us consider the Lyapunov function candidate $V(x) = x^T P x$. Its derivative along the system trajectory is

$$\frac{dV}{dt}(x(t)) = x^T(t)Q_{\sigma(t)}x(t) \leq -r_{\sigma(t)}x^T(t)x(t) \leq -rx^T(t)x(t)$$

where $r = \min\{r_1, \cdots, r_m\}$. As a result, the quadratic function $V(x)$ is a Lyapunov function of the system and the theorem follows from the Lyapunov Theorem, Theorem 2.5. $\square$

Although the lemma guarantees the well-definedness of the switching signal, it cannot prevent fast switching and chattering in the event that the system undergoes perturbations. This is justified by the following example.

Example 3.27. Consider the perturbed system given by

$$\dot{x}(t) = A_\sigma x(t) + f_\sigma(t) \quad \sigma \in \{1, 2\}$$

$$A_1 = \begin{bmatrix} -2 & 0 \\ 0 & 1 \end{bmatrix} \text{ and } A_2 = \begin{bmatrix} 1 & 0 \\ 0 & -2 \end{bmatrix}$$

$$f_1(t) = \begin{bmatrix} -1 \\ 1 \end{bmatrix} \exp(-0.1t) \text{ and } f_2(t) = \begin{bmatrix} 1 \\ -1 \end{bmatrix} \exp(-0.1t) \quad (3.26)$$

where f_1 and f_2 are perturbations associated to the first and second subsystems, respectively.

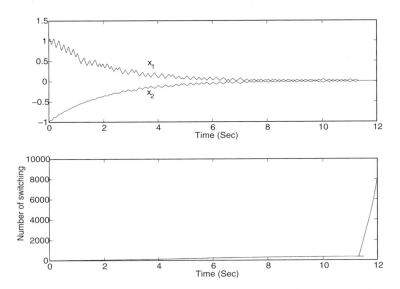

Fig. 3.1. State trajectory and switching number of system (3.26)

As $A_1 + A_2 = -I_2$, we can choose $w_1 = w_2 = \frac{1}{2}$. Simple computation gives $P = \frac{1}{2}I_2$. Suppose that we use the state-feedback switching signal (3.21) with $r_1 = r_2 = 0.4$. Figure 3.1 depicts the state trajectories and the number of switches, respectively, when the system initializes at $x(0) = [1, -1]^T$. It can be seen that the chattering phenomenon occurs when $t > 11.45$. In fact, as the state trajectory converges to the origin, the information of the state is 'merged' by the perturbations, that is, $\frac{\|f_\sigma(t)\|}{\|x(t)\|} \to \infty$. Because the state direction and the perturbation direction are always opposite, chattering occurs.

3.4.2 A Modified Switching Law

In this subsection, we consider the perturbed switched linear system (3.19) where the switching signal is chosen to be in state-feedback form.

For a switched system with nonlinear perturbations, the design of a state-feedback stabilizing switching law is quite sophisticated. Indeed, a state-feedback switching law usually generates different switching paths with respect to different system perturbations. In other words, the perturbations make the switching paths 'drift' away. Because of this, the perturbation analysis for time-varying systems is not applicable.

To ensure well-posedness of the perturbed switched systems, we modify the switching law given in the previous subsection as follows.

Fix a positive real number v. For perturbed switched system (3.19), define the following state-feedback switching signal.

For any initial state $x(t_0) = x_0$, set

$$\sigma(t_0) = \arg\min\{x_0^T Q_1 x_0, \cdots, x_0^T Q_m x_0\}.$$

If there are more than one such index, we select the minimum one.

The subsequent switching time/index sequences are defined recursively by

$$t_{k+1} = \inf\left\{t > t_k : \ x^T(t)Q_{\sigma(t_k)}x(t) > -r_{\sigma(t_k)}x^T(t)x(t), \ \|x(t)\| > v\right\}$$
$$\sigma(t_{k+1}) = \arg\min\{x^T(t_{k+1})Q_1 x(t_{k+1}), \cdots, x^T(t_{k+1})Q_m x(t_{k+1})\}$$
$$k = 0, 1, \cdots. \tag{3.27}$$

The switching signal is the same as that in (3.21), except that we fix a level set for switching, that is, no switch occurs within the v-neighborhood of the origin.

Under this switching law, the perturbed system possesses nice robustness properties as shown in the following theorem.

Theorem 3.28. *For system (3.19), suppose that Assumption 3.1 holds. If the perturbations are bounded, then the state-feedback switching signal (3.27) is well-defined and the system state is bounded.*

Proof. First, we prove that the switching law is well-defined by showing that the switching signal has a positive dwell time.

Suppose that t_k and t_{k+1} are two consecutive switching time instants. It follows from the switching signal that

1) $x^T(t_k)Q_{\sigma(t_k)}x(t_k) = \min_{i \in M}\{x^T(t_k)Q_i x(t_k)\}$;
2) $x^T(t_{k+1})Q_{\sigma(t_k)}x(t_{k+1}) \geq -r_{\sigma(t_k)}x^T(t_{k+1})x(t_{k+1})$; and
3) $\|x(t_{k+1})\| \geq v$.

As $\sum_{i \in M} w_i Q_i = -I_n$ and $\sum_{i \in M} w_i = 1$, it can be seen that Item 1) implies that

4) $x^T(t_k)Q_{\sigma(t_k)}x(t_k) \leq -x^T(t_k)x(t_k)$.

For notational convenience, denote

$$x_k = x(t_k) \quad x_{k+1} = x(t_{k+1}) \text{ and } j = \sigma(t_k).$$

Let us consider the case

$$\|x(t)\| \leq 2\|x_{k+1}\| \quad \forall \ t \in [t_k, t_{k+1}]. \tag{3.28}$$

In this case, define a function

$$g(t) = x^T(t)Q_j x(t) + x^T(t)x(t) \quad t \in [t_k, t_{k+1}].$$

It follows from Items 2) and 4) that

$$g(t_k) \leq 0 \text{ and } g(t_{k+1}) \geq (1 - r_j)x_{k+1}^T x_{k+1}. \tag{3.29}$$

Simple computation gives

$$\frac{dg}{dt}(t) = x^T(t)\left(A_j^T(Q_j + I_n) + (Q_j + I_n)A_j\right)x(t) + 2(f_j(t))^T(Q_j + I_n)x(t).$$

Let us denote

$$\nu_1 = \|A_j^T(Q_j + I_n) + (Q_j + I_n)A_j\|$$

and

$$\nu_2 = \|Q_j + I_n\| \text{ and } \nu_3 = \sup_{t \in [t_k, t_{k+1}]} \|f_j(t)\|.$$

In view of Item 3) and Inequality (3.28), we have

$$\|\frac{dg}{dt}(t)\| \leq 4\nu_1 x_{k+1}^T x_{k+1} + 4\nu_2\nu_3\|x_{k+1}\|$$

$$\leq (4\nu_1 + 4\nu_2\frac{\nu_3}{\upsilon})x_{k+1}^T x_{k+1} \quad \forall \ t \in [t_k, t_{k+1}].$$

Combining this with (3.29) yields

$$(4\nu_1 + 4\nu_2\frac{\nu_3}{\upsilon})(t_{k+1} - t_k) \geq (1 - r_j)$$

which implies that

$$t_{k+1} - t_k \geq (1 - r_j)/(4\nu_1 + 4\nu_2\frac{\nu_3}{\upsilon}).$$

Next, suppose that (3.28) does not hold. This means that there is a $t^* \in [t_k, t_{k+1})$ satisfying

$$\|x(t^*)\| > 2\|x_{k+1}\|. \tag{3.30}$$

From the system equation (3.19), we have

$$x_{k+1} = \exp\left(A_j(t_{k+1} - t^*)\right)x(t^*) + \int_{t^*}^{t_{k+1}} \exp\left(A_j(t_{k+1} - \tau)\right)f_j(\tau)d\tau$$

which is equivalent to

$$x(t^*) = \exp\left(A_j(t^* - t_{k+1})\right) x_{k+1} - \int_{t^*}^{t_{k+1}} \exp\left(A_j(t^* - \tau)\right) f_j(\tau) d\tau.$$

As $\exp(A_j t)$ is continuous and $\|\exp(A_j 0)\| = 1$, there is a positive number ν_4 such that

$$\|\exp(A_j t)\| \le \frac{3}{2} \quad \forall\, t \in [-\nu_4, 0].$$

Suppose that

$$t_{k+1} - t^* \le \min\left(\nu_4, \frac{\upsilon}{3N}\right)$$

where N is the upper bound of perturbations. Then, we have

$$\|x(t^*)\| \le \frac{3}{2}\|x_{k+1}\| + \frac{3}{2}(t_{k+1} - t^*)N \le 2\|x_{k+1}\|$$

which contradicts (3.30). As a consequence, we have

$$t_{k+1} - t_k \ge t_{k+1} - t^* > \min\left(\nu_4, \frac{\upsilon}{3N}\right).$$

The above reasonings show that for any consecutive switching time instants t_k and t_{k+1}, we have

$$t_{k+1} - t_k \ge \min_{j \in M}\left((1 - r_j)/(4\nu_1 + 4\nu_2 \frac{\nu_3}{\upsilon}), \nu_4, \frac{\upsilon}{3N}\right) \tag{3.31}$$

which sets a lower bound for the dwell time of the switching signal. This ensures that the switching signal is well-defined.

Finally, we show that the system state is bounded. In fact, we can prove by contradiction that

$$\|x(t)\| \le \left(\frac{\lambda_{max}}{\lambda_{min}}\right)^{\frac{1}{2}} \max\left(\|x_0\|, \frac{2\|P\|}{r}N, \upsilon\right) \stackrel{def}{=} \mu \quad \forall\, t \ge t_0 \tag{3.32}$$

where λ_{max} and λ_{min} are the maximum and minimum eigenvalues of matrix P, and $r = \min(r_1, \cdots, r_m)$. Indeed, suppose that $\|x(t_*)\| > \mu$ for some $t_* > t_0$. Then, let

$$\mu_1 = \max\left(\|x_0\|, \frac{2\|P\|}{r}N, \upsilon\right) = \left(\frac{\lambda_{min}}{\lambda_{max}}\right)^{\frac{1}{2}} \mu.$$

From $\|x_0\| \le \mu_1$ and the continuity of the state, there exists a time instant $s \in [t_0, t_*)$ satisfying

$$\|x(s)\| = \mu_1 \quad \text{and} \quad \|x(t)\| > \mu_1 \quad \forall\, t \in (s, t_*]. \tag{3.33}$$

Consider the Lyapunov function candidate given by

$$V(x) = x^T P x.$$

Its derivative along the system trajectory can be computed as

$$\frac{dV}{dt}(x(t)) = x^T(t)Q_{\sigma(t)}x(t) + 2(f_\sigma(t))^T P x(t).$$

As $\|x(t)\| \geq \mu_1 \geq v$, $\forall\, t \in [s, t_*]$, it follows from the switching signal that

$$x^T(t)Q_{\sigma(t)}x(t) \leq -r_{\sigma(t)}x^T(t)x(t) \quad \forall\, t \in [s, t_*].$$

This, together with (3.33), implies that

$$\frac{dV}{dt} \leq 0 \quad \forall\, t \in [s, t_*].$$

Consequently, we have

$$V(x(t)) \leq V(x(s)) \leq \lambda_{max}\|x(s)\|^2 = \lambda_{max}\mu_1^2 \quad \forall\, t \in [s, t_*]$$

which in turn implies that

$$x^T(t)x(t) \leq \frac{V(x(t))}{\lambda_{min}} = \frac{\lambda_{max}}{\lambda_{min}}\mu_1^2 = \mu^2 \quad \forall\, t \in [s, t_*].$$

Note that the last inequality contradicts the assumption that $\|x(t_*)\| > \mu$. As a result, Inequality (3.32) holds for all $t \geq t_0$. This completes the proof of the theorem. $\square$

Remark 3.29. In switching signal (3.27), the parameters v and r_k play twofold roles in determining the performance of the perturbed system. Generally speaking, the role of v is to prevent the switching signals from chattering. On one hand, a sufficiently small v may result in a sufficiently small dwell time and a sufficiently high switching frequency, which is usually undesirable in practice. On the other hand, a larger v may lead to a larger system state bound, which is also undesirable. Similarly, smaller r_k, $k \in M$ may lead to a larger system state bound, while larger r_k, $k \in M$ may result in a smaller dwell time. The choice of these parameters should be balanced according to the performance requirement.

If the perturbations are convergent, then the system state can be made to converge to any given small neighborhood of the origin. This can be seen from the following result.

Theorem 3.30. *For system (3.19), suppose that Assumption 3.1 holds. If the perturbations are bounded and convergent, then, there is a time instant $T \geq t_0$ such that*

$$\|x(t)\| \leq \left(\frac{\lambda_{max}}{\lambda_{min}}\right)^{\frac{1}{2}} \upsilon \quad \forall\, t \geq T. \tag{3.34}$$

Proof. As in the proof of Theorem 3.28, we consider the Lyapunov function candidate $V(x) = x^T P x$ and its derivative along the system trajectory

$$\frac{dV}{dt} = x^T(t) Q_{\sigma(t)} x(t) + 2(f_\sigma(t))^T P x(t).$$

Let $\eta = \|P\|$, $r = \min(r_1, \cdots, r_m)$, and $f(t) = f_\sigma(t)$. Whenever

$$\|x(t)\| \geq \max\left(\upsilon, \frac{4\eta}{r}\|f(t)\|\right) \overset{def}{=} \psi(t)$$

it follows from the switching law that

$$\frac{dV}{dt} \leq -r x^T(t) x(t) + 2 f^T(t) P x(t) \leq -\frac{r}{2} x^T(t) x(t) \leq -\frac{r}{2\lambda_{max}} V(x(t)).$$

Therefore, for any $t \geq t_0$, at least one of the following two inequalities holds

a) $\|x(t)\| \leq \psi(t)$; and

b) $\dfrac{d\ln(V(x(t)))}{dt} \leq -\dfrac{r}{2\lambda_{max}}.$

Fix a time instant $s \geq t_0$. Suppose that Item a) does not hold at $t = s$, then either

$$\|x(t)\| \geq \psi(t) \quad \forall\, t \in [t_0, s]$$

or there is an $s_* \in [t_0, s)$ such that

$$\|x(s_*)\| \leq \psi(s_*) \text{ and } \|x(t)\| > \psi(t) \quad \forall\, t \in (s_*, s].$$

In the former, integrating $\frac{d\ln(V(x(t)))}{dt}$ from t_0 to s gives

$$\ln(V(x(s))) - \ln(V(x(t_0))) \leq -\frac{r}{2\lambda_{max}}(s - t_0)$$

which implies that

$$\|x(s)\|^2 \leq \frac{\lambda_{max}}{\lambda_{min}} \exp\left(-\frac{r}{2\lambda_{max}}(s - t_0)\right) \|x_0\|^2.$$

Similarly, in the latter, we have

$$\|x(s)\|^2 \leq \frac{\lambda_{max}}{\lambda_{min}} \exp\left(-\frac{r}{2\lambda_{max}}(s - s_*)\right) \psi(s_*)^2.$$

To summarize, for any $s \geq t_0$, we have

$$\|x(s)\|^2 \leq \max \left(\psi(s)^2, \frac{\lambda_{max}}{\lambda_{min}} \exp\left(-\frac{r}{2\lambda_{max}}(s - t_0) \right) \|x_0\|^2 , \right.$$

$$\left. \sup_{s_* < s} \left\{ \frac{\lambda_{max}}{\lambda_{min}} \exp\left(-\frac{r}{2\lambda_{max}}(s - s_*) \right) \psi(s_*)^2 \right\} \right). \tag{3.35}$$

As $f(t)$ is convergent, there is a time $T_1 \geq t_0$ such that

$$\|f(t)\| \leq \frac{r}{4\eta} \upsilon \quad \forall \, t \geq T_1.$$

Recall that N is an upper bound of $\|f(\cdot)\|$. If $N \leq \frac{r\upsilon}{4\eta}$, then let $T_2 = 0$. Otherwise, let $T_2 = \frac{4\lambda_{max}}{r} \ln\left(\frac{4N\eta}{r\upsilon} \right)$. Finally, let $T_3 = \frac{4\lambda_{max}}{r} \ln\left(\frac{\|x_0\|}{\upsilon} \right)$. With these, it can be seen that, for any $t \geq \max(T_1 + T_2, T_3) \overset{def}{=} T$, we have

$$\psi(t) = \upsilon \quad \exp\left(-\frac{r}{2\lambda_{max}}(t - t_0) \right) \|x_0\|^2 \leq \upsilon^2$$

and

$$\sup_{s_* < t} \left\{ \exp\left(-\frac{r}{2\lambda_{max}}(t - s_*) \right) \psi(s_*)^2 \right\} \leq \upsilon^2.$$

It follows from (3.35) that

$$\|x(t)\|^2 \leq \frac{\lambda_{max}}{\lambda_{min}} \upsilon^2 \quad \forall \, t \geq T.$$

This completes the proof. $\square$

Utilizing this theorem, we can design a switching signal to make the perturbed system 'practically' convergent by choosing a suitable υ. Indeed, given a level of practical tolerance, ϵ, let $\upsilon = (\frac{\lambda_{min}}{\lambda_{max}})^{\frac{1}{2}} \epsilon$, then the switching signal will bring the system state into the ϵ-neighborhood of the origin in a finite time. Due to the arbitrariness of the ϵ, the switched system can be steered into any pre-assigned neighborhood of the origin with a well-defined switching law.

3.4.3 Observer-based Switching

In this subsection, we explore the possibility of designing a switching law based on measured output instead of the state information.

Consider the switched linear system described by

$$\begin{cases} \dot{x}(t) = A_\sigma x(t) + f_\sigma(t) \\ y(t) = C_\sigma x(t) \end{cases} \tag{3.36}$$

where $x(t), \sigma$ are the same as in (3.1), $y(t) \in \mathbf{R}^q$ is the measured output, $f_i(t)$, $i \in M$ are piecewise continuous vector functions denoting for system perturbations, and A_i, C_i, are matrices of compatible dimensions.

For system (3.36), we consider the state estimator (observer) given by

$$\dot{\hat{x}} = A_\sigma \hat{x} + L_\sigma [y - C_\sigma \hat{x}] \tag{3.37}$$

where y and σ are the output and switching signal of system (3.36), respectively, and gain matrices $L_1, \cdots, L_m \in \mathbf{R}^{n \times q}$ are to be determined later. Note that the observer itself is a switched linear system.

Assumption 3.2. *There exist a positive definite matrix P_1, and matrices Y_i, such that*

$$A_i^T P_1 + P_1 A_i - C_i^T Y_i^T - Y_i C_i < 0 \quad \forall \, i \in M.$$

Let $L_i = P_1^{-1} Y_i$. The above assumption asserts that there exists a common quadratic Lyapunov function for switched system

$$\dot{x} = (A_\sigma - L_\sigma C_\sigma) x. \tag{3.38}$$

Indeed, let $V(x) = x^T P_1 x$. Its derivative along system (3.38) is

$$\frac{dV}{dt}(x(t)) = x^T(t) \left(A_\sigma^T P_1 + P_1 A_\sigma - C_\sigma^T Y_\sigma^T - Y_\sigma C_\sigma \right) x(t)$$

which is a negative definite function of state. This implies that system (3.38) is stable under arbitrary switching (*c.f.* Theorem 2.9). Furthermore, let $\Phi_\sigma(t, t_0)$ be the state transition matrix of Equation (3.38). Then, it can be proven that

$$\|\Phi_\sigma(t, t_0)\| \le \left(\frac{\lambda_2}{\lambda_1} \right)^{\frac{1}{2}} e^{-\frac{\gamma}{2\lambda_1}(t - t_0)} \tag{3.39}$$

where λ_1 and λ_2 denote the minimum and maximum eigenvalues of matrix P_1, respectively, and γ denotes the minimum eigenvalue among $Q_i \overset{def}{=} -(A_i^T P_1 + P_1 A_i - C_i^T Y_i^T - Y_i C_i)$ for $i \in M$.

To stabilize system (3.36) via observer-based switching, we propose a modified switching strategy based on switching law (3.27).

Fix a positive real number v. Suppose that $\hat{x}$ is initialized at $\hat{x}(t_0) = \hat{x}_0$. Set

$$\sigma(t_0) = \arg\min\{\hat{x}_0^T Q_1 \hat{x}_0, \cdots, \hat{x}_0^T Q_m \hat{x}_0\}.$$

If there are more than one such index, just pick the minimum one.

The subsequent switching time/index sequences are defined recursively by

$$t_{k+1} = \inf \left\{ t > t_k : \ \hat{x}^T(t) Q_{\sigma(t_k)} \hat{x}(t) > -r_{\sigma(t_k)} \hat{x}^T(t) \hat{x}(t), \|\hat{x}(t)\| > v \right\}$$
$$\sigma(t_{k+1}) = \arg\min\{\hat{x}^T(t_{k+1}) Q_1 \hat{x}(t_{k+1}), \cdots, \hat{x}^T(t_{k+1}) Q_m \hat{x}(t_{k+1})\}$$
$$k = 0, 1, \cdots. \tag{3.40}$$

The switching signal is the same as that in (3.27), except that the observer $\hat{x}$ is used instead of the real state x.

Theorem 3.31. *For system (3.36), suppose that Assumptions 3.1 and 3.2 hold. Then, the dynamic-output-feedback switching law (3.40) is well-defined, and we have*

(i) the perturbed system is bounded if the perturbations are bounded; and
(ii) the perturbed system is practically convergent if the perturbations are convergent.

Proof. Define the difference between the real state and the estimated state

$$\tilde{x} = x - \hat{x}.$$

Subtracting (3.37) from (3.36), we obtain

$$\dot{\tilde{x}}(t) = (A_\sigma - L_\sigma C_\sigma)\tilde{x}(t) + f_\sigma(t). \tag{3.41}$$

Combining (3.37) with (3.41), we have

$$\dot{\tilde{x}} = (A_\sigma - L_\sigma C_\sigma)\tilde{x} + f_\sigma \tag{3.42}$$

$$\dot{\hat{x}} = A_\sigma \hat{x} + L_\sigma C_\sigma \tilde{x}. \tag{3.43}$$

Let $\Phi_\sigma(t, t_0)$ be the state transition matrix of the nominal system (3.38). The solution of Equation (3.42) can be given by

$$\tilde{x}(t) = \Phi_\sigma(t, t_0)\tilde{x}(t_0) + \int_{t_0}^{t} \Phi_\sigma(\tau, t_0)f_\sigma(\tau)d\tau.$$

It follows from Assumption 3.2 that (3.39) holds. From (3.39), it can be seen that $\tilde{x}$ is bounded if the perturbations f_i, $i \in M$ are bounded, and $\tilde{x}$ is convergent if the perturbations f_i, $i \in M$ are convergent.

For system (3.43), take $\dot{\hat{x}} = A_\sigma \hat{x}$ as the nominal system and $L_i C_i \tilde{x}$, $i \in M$ as the exoteric perturbations. The theorem follows directly from Theorems 3.28 and 3.30. □

Example 3.32. For comparison, we continue to address system (3.26) in Example 3.27.
Set $v = 0.2$, and simulate the state trajectory and switching signal under the modified switching law (3.27). This is shown in Figure 3.2. It is clear that no chattering occurs in this case, albeit at the expense of the state becoming practically convergent rather than exponentially convergent.

Next, suppose that the full state information is not available. Assume that the measured output is $y = x_2$ for the first subsystem and $y = x_1$ for the second subsystem. Let

$$L_1 = \begin{bmatrix} 0 \\ 2 \end{bmatrix} \text{ and } L_2 = \begin{bmatrix} 1 \\ 0 \end{bmatrix}.$$

It can be verified that Assumption 3.2 holds. Figure 3.3 shows the state and observer trajectories for the switched system under the observer-based switching law (3.40) with $v = 0.2$ and $\hat{x}(t_0) = [0, 0]^T$. The corresponding switching signal is depicted in Figure 3.4.

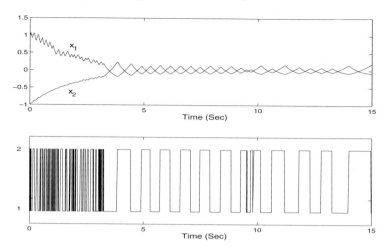

Fig. 3.2. State trajectory and switching signal under switching signal (3.27)

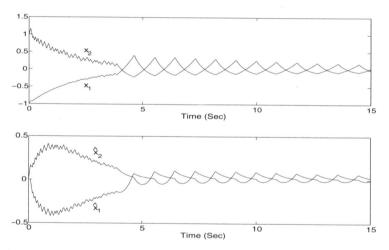

Fig. 3.3. State and estimator trajectories under switching signal (3.40)

3.5 Combined Switching

In this section, we first focus on the switched systems with two subsystems (*i.e.*, $m = 2$). Then, we extend it to the general case in the last subsection.

3.5.1 Switching Strategy Description

For a switched linear system with two subsystems, suppose that none of A_1 and A_2 is Hurwitz but Assumption 3.1 holds. Fix w_1, w_2, r_1, r_2 as described in Section 3.4.1.

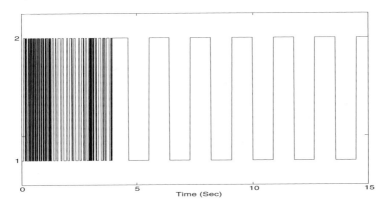

Fig. 3.4. Switching signal under the observer-based switching signal

Let $A_0 = w_1 A_1 + w_2 A_2$, and $P > 0$ be such that $A_0^T P + P A_0 = -I_n$. From Lemma 2.10, we can find a $\tau > 0$ such that

$$(A_0 + \Upsilon_t t)^T P + P(A_0 + \Upsilon_t t) < 0 \quad \forall \, t \leq \tau.$$

As a result, we have

$$(\exp(w_2 A_2 \tau) \exp(w_1 A_1 \tau))^T P (\exp(w_2 A_2 \tau) \exp(w_1 A_1 \tau))$$
$$\leq (1 - \delta) P < P \tag{3.44}$$

where $\delta \in (0, 1)$.

Let $\tau_1 = w_1 \tau$, $\tau_2 = w_2 \tau$, and $Q_i = A_i^T P + P A_i$, $i = 1, 2$.

The following switching strategy combines the ideas from the state-feedback switching signal (3.21) and from the average method introduced in Section 3.3.

Suppose that the system has the initial state $x(t_0) = x_0$. Set

$$\sigma(t_0) = \arg\min\{x_0^T Q_1 x_0, x_0^T Q_2 x_0\}.$$

The subsequent switching time/index sequences are defined recursively by

$$t_{k+1} = \begin{cases} \inf\left\{t > t_k: \ x^T(t) Q_1 x(t) > -r_1 x^T(t) x(t)\right\} + \tau_1 & \text{if } \sigma(t_k) = 1 \\ \inf\left\{t > t_k + \tau_2: \ x^T(t) Q_2 x(t) > -r_2 x^T(t) x(t)\right\} & \text{if } \sigma(t_k) = 2 \end{cases}$$

$$\sigma(t_{k+1}) = \begin{cases} 2 \text{ if } \sigma(t_k) = 1 \\ 1 \text{ if } \sigma(t_k) = 2 \end{cases} \quad k = 0, 1, \cdots . \tag{3.45}$$

According to this strategy, when the first subsystem is active, it should be kept active for the additional dwell-time τ_1 after the state-feedback switching time is due. On the other hand, if the second subsystem is active, it should be kept active for the dwell time τ_2, then the state-feedback switching mechanism decides the next switching time. Suppose that t_k is a switching time with

$\sigma(t_k) = 1$, it can be seen from the strategy that, the period $[t_k, t_{k+2})$ can be divided into two phases: in $[t_k, t_{k+1} - \tau_1) \cup [t_{k+1} + \tau_2, t_{k+2})$, the system is governed by the state-feedback switching mechanism, while in $[t_{k+1} - \tau_1, t_{k+1} + \tau_2)$, the system is governed by the time-driven mechanism. That is, the average period (*i.e.*, time duration for a cyclic switching) is the summation of the two phases. In this way, the resultant switching frequency is lower than that of a single mechanism.

In the above strategy, the state-feedback switching mechanism works in a nontrivial way as shown in the following theorem.

Theorem 3.33. *There is a positive real number η, such that the dwell time of the first subsystem is not less than $\bar{\tau}_1 \overset{def}{=} \eta + \tau_1$.*
Proof. Suppose that t_k and t_{k+1} are two consecutive switching time instants with $\sigma(t_k) = 1$. Let $\bar{t}_{k+1} = t_{k+1} - \tau_1$. It follows from the switching signal that

1) $x^T(t_k)Q_2x(t_k) \geq -r_2 x^T(t_k)x(t_k)$; and
2) $x^T(\bar{t}_{k+1})Q_1x(\bar{t}_{k+1}) \geq -r_1 x^T(\bar{t}_{k+1})x(\bar{t}_{k+1})$.

As $w_1 Q_1 + w_2 Q_2 = -I_n$ and $w_1 + w_2 = 1$, it follows from Item 1) that

$$(1 - w_2)x^T(t_k)Q_1x(t_k) = -x^T(t_k)x(t_k) - w_2 x^T(t_k)Q_2x(t_k)$$
$$\leq -(1 - w_2 r_2)x^T(t_k)x(t_k)$$

which further implies that

3) $x^T(t_k)Q_1x(t_k) \leq -x^T(t_k)x(t_k)$.

For notational convenience, let $x_k = x(t_k)$, $x_{k+1} = x(\bar{t}_{k+1})$, and ϑ be any real number greater than 1.

First, consider the case

$$\|x(t)\| \leq \vartheta\|x_{k+1}\| \quad \forall\, t \in [t_k, \bar{t}_{k+1}]. \tag{3.46}$$

In this case, define a function

$$g(t) = x^T(t)(Q_1 + I_n)x(t) \quad t \in [t_k, \bar{t}_{k+1}].$$

It follows from Items 2) and 3) that

$$g(t_k) \leq 0 \quad g(\bar{t}_{k+1}) \geq (1 - r_1)x_{k+1}^T x_{k+1}. \tag{3.47}$$

Simple computation gives

$$\frac{dg}{dt}(t) = x^T(t)\left(A_1^T(Q_1 + I_n) + (Q_1 + I_n)A_1\right)x(t).$$

Denote $\nu_1 = \|A_1^T(Q_1 + I_n) + (Q_1 + I_n)A_1\|$. By inequality (3.46), we have

$$|\frac{dg}{dt}(t)| \leq \vartheta^2 \nu_1 x_{k+1}^T x_{k+1} \quad \forall\, t \in [t_k, \bar{t}_{k+1}].$$

Combining this with (3.47) yields

$$\vartheta^2 \nu_1 (\bar{t}_{k+1} - t_k) \geq (1 - r_1)$$

which implies that

$$\bar{t}_{k+1} - t_k \geq (1 - r_1)/(\vartheta^2 \nu_1).$$

Second, suppose that (3.46) does not hold. This means that there is a $t^* \in [t_k, \bar{t}_{k+1})$ satisfying

$$\|x(t^*)\| > \vartheta \|x_{k+1}\|. \tag{3.48}$$

From the system equation, we have

$$x(t^*) = \exp\left(A_1(t^* - \bar{t}_{k+1})\right) x_{k+1}.$$

It follows from (3.48) and

$$\|\exp\left(A_1(t^* - \bar{t}_{k+1})\right)\| \leq \exp\left(\|A_1\|(\bar{t}_{k+1} - t^*)\right)$$

that

$$\bar{t}_{k+1} - t_k \geq \bar{t}_{k+1} - t^* > \ln \vartheta / \|A_1\|.$$

From the above reasonings, we have

$$\bar{t}_{k+1} - t_k \geq \eta \stackrel{def}{=} \sup_{\vartheta > 1} \min\left((1 - r_1)/(\vartheta^2 \nu_1), \ln \vartheta / \|A_1\|\right). \quad \square$$

Note that, function $(1 - r_1)/(\vartheta^2 \nu_1)$ monotonically decreases as ϑ increases, and $\ln \vartheta / \|A_1\|$ monotonically increases as ϑ increases. Hence, η is the intersection value.

Theorem 3.34. *The switched system is exponentially stable under the switching signal (3.45).*

Proof. Consider the Lyapunov function candidate given by

$$V(x) = x^T P x. \tag{3.49}$$

Its derivative along the system trajectory is

$$\frac{dV}{dt}(x(t)) = x^T(t) Q_{\sigma(t)} x(t).$$

First, let λ_1 and λ_n be the minimum and maximum eigenvalues of matrix P, respectively. Suppose that t_k is any switching time with $\sigma(t_k) = 1$. We examine the consecutive switching intervals $[t_k, t_{k+1})$ and $[t_{k+1}, t_{k+2})$. According to the switching signal, we have

$$x^T(t)Q_{\sigma(t)}x(t) \leq -r_{\sigma(t)}x^T(t)x(t) \leq -\frac{r_{\sigma(t)}}{\lambda_n}V(x(t))$$

$$t \in [t_k, t_{k+1} - \tau_1) \cup [t_{k+1} + \tau_2, t_{k+2}). \tag{3.50}$$

Hence, $V(x(t))$ decays exponentially in the above time intervals. On the other hand, from (3.44), it follows that

$$V(x(t_{k+1} + \tau_2)) < (1 - \delta)V(x(t_{k+1} - \tau_1)).$$

Within the interval $[t_{k+1} - \tau_1, t_{k+1} + \tau_2]$, we have

$$\|x(s_2)\| \leq e^{\max(\|A_1\|,\|A_2\|)(s_2-s_1)}\|x(s_1)\| \quad \forall \, s_1 \leq s_2$$

which implies that

$$V(x(s_2)) \leq \lambda_n x^T(s_2)x(s_2) \leq \frac{\lambda_n}{\lambda_1}e^{\max(\|A_1\|,\|A_2\|)(s_2-s_1)}V(x(s_1)) \quad \forall \, s_1 \leq s_2.$$

Let

$$\alpha = \min\left(\frac{r_1}{\lambda_n}, \frac{r_2}{\lambda_n}, \frac{-\ln(1-\delta)}{\tau}\right)$$

and

$$\beta = \frac{\lambda_n}{\lambda_1}\exp\left((\max(\|A_1\|,\|A_2\|) + \alpha)\tau\right).$$

The above analysis shows that

$$V(x(t_{k+2})) = V(x(t_k))e^{-\alpha(t_{k+2}-t_k)}$$

and

$$V(x(s_2)) \leq \beta V(x(s_1))e^{-\alpha(s_2-s_1)} \quad \forall \, t_k \leq s_1 \leq s_2 \leq t_{k+2}. \tag{3.51}$$

Note that if $\sigma(t_0) = 2$, then (3.51) holds in $[t_0, t_0 + \tau_2)$ and (3.50) holds in $[t_0 + \tau_2, t_1)$.

Next, for any two time instants s_1 and s_2 with $t_0 \leq s_1 < s_2$, suppose that

$$s_1 \in [t_{k_1}, t_{k_1+2}) \quad s_2 \in [t_{k_2}, t_{k_2+2}).$$

Then, we have

$$V(x(s_2)) \leq \beta e^{-\alpha(s_2-t_{k_2})}V(x(t_{k_2})) \leq \beta e^{-\alpha(s_2-t_{k_1+2})}V(x(t_{k_1+2}))$$
$$\leq \beta^2 e^{-\alpha(s_2-s_1)}V(x(s_1)).$$

As a result, the Lyapunov function decays exponentially, and hence the switched system is exponentially stable. $\square$

Remark 3.35. From the proof, it is clear that, when the state-feedback mechanism works, the Lyapunov function candidate strictly decreases along the state trajectory. While in the time-driven period, the Lyapunov function decreases at discrete time instants. In between these time instants, the Lyapunov function may strictly increase along the state trajectory. Therefore, the state trajectory does not necessarily admit a quadratic Lyapunov function. That is, there may not exist a quadratic Lyapunov function $W(x)$ such that

$$W(x(t_2)) \le W(x(t_1)) \quad \forall \ t_2 > t_1.$$

This is an essential feature of the switching strategy. Later, we will show that the strategy can be extended to cope with more general classes of switched systems, including systems which are not quadratically stabilizable.

3.5.2 Robustness Properties

Consider a perturbed switched linear system given by

$$\dot{x}(t) = A_\sigma x(t) + f_\sigma(t) \quad x(t_0) = x_0 \tag{3.52}$$

where $\sigma \in \{1,2\}$, and $f_i : [t_0, \infty) \mapsto \mathbf{R}^n$, $k = 1,2$ are piecewise continuous vector functions representing system perturbations or uncertainties.

Theorem 3.36. *For perturbed system (3.52), suppose that Assumption 3.1 holds. Then, under the switching law (3.45), we have*

(i) the system state is bounded if the perturbations are bounded;
(ii) the system state is bounded and convergent if the perturbations are bounded and convergent; and
(iii) the system state is exponentially convergent if the perturbations are exponentially convergent.

Proof. In terms of the notation defined in Section 3.5.1, we further define the function

$$f(t) = f_\sigma(t) \quad t \ge t_0$$

and denote

$$r = \min(r_1, r_2) \quad p = \|P\| \quad q = \max(\|Q_1\|, \|Q_2\|)$$

$$\theta = \min\left(\frac{r}{4\lambda_n}, -\frac{\ln(1 - \frac{\delta}{2})}{2\tau}\right) \quad \varrho = e^{\|A_1\|\tau_1} e^{\|A_2\|\tau_2}$$

and

$$\varpi = \left(\frac{\lambda_n}{\lambda_1}\right)^{\frac{1}{2}} \exp\left(\left(\frac{2q+r}{2\lambda_1} + 2\theta\right)\tau\right).$$

We estimate an upper bound for the state norm in terms of the initial state and the perturbations. Define a function

$$\mu_1(t) = \max\left(\|x_0\|e^{-\theta(t-t_0)}, \frac{4p}{r}\|f(t)\|, \frac{8p\varrho^2}{\delta\lambda_1}\int_t^{t+\tau}\|f(\varsigma)\|d\varsigma\right) \quad t \geq t_0.$$

It can be seen that $\mu_1(\cdot)$ is piecewise continuous. Re-define the function if necessary at the discontinuous points such that the function is continuous from the left. We are to prove by contradiction that

$$\|x(t)\| \leq \mu(t) \overset{def}{=} \varpi \sup_{\varsigma\in[t_0,t]} \mu_1(\varsigma)e^{-\theta(t-\varsigma)} \quad \forall\, t \geq t_0. \tag{3.53}$$

For this, suppose that

$$\|x(t_*)\| > \mu(t_*) \tag{3.54}$$

for some $t_* > t_0$. From $\|x_0\| \leq \mu_1(t_0)$ and the continuity of the state, there exists a time instant $s \in [t_0, t_*)$ such that

$$\|x(s)\| \leq \mu_1(s) \text{ and } \|x(t)\| \geq \mu_1(t) \quad \forall\, t \in (s, t_*]. \tag{3.55}$$

Consider the Lyapunov function candidate given by

$$V(x) = x^T P x.$$

Its derivative along the system trajectories can be computed to be

$$\frac{dV}{dt}(x(t)) = x^T(t)Q_{\sigma(t)}x(t) + 2f^T(t)Px(t).$$

Suppose that the switching time sequence is $\{t_0, t_1, t_2, \cdots\}$. Any time instant s falls into either of the two cases:

(a) there is a nonnegative integer k such that $\sigma(t_k) = 1$ and $s \in [t_k, t_{k+2})$; or
(b) $\sigma(t_0) = 2$ and $s \in [t_0, t_1)$.

For Case (a), it follows from (3.55) that

$$\frac{dV}{dt}(x(t)) \leq -r_\sigma x^T(t)x(t) + 2p\|f(t)\|\|x(t)\| \leq -\frac{r}{2}x^T(t)x(t) \leq -\frac{r}{2\lambda_n}V(x(t))$$
$$\forall\, t \in (s, t_*] \cap ([t_k, t_{k+1} - \tau_1) \cup [t_{k+1} + \tau_2, t_{k+2})) \tag{3.56}$$

and

$$\frac{dV}{dt}(x(t)) \leq qx^T(t)x(t) + 2p\|f(t)\|\|x(t)\| \leq (q + \frac{r}{2})x^T(t)x(t)$$
$$\leq \frac{2q+r}{2\lambda_1}V(x(t)) \quad \forall\, t \in (s, t_*] \cap [t_{k+1} - \tau_1, t_{k+1} + \tau_2]. \tag{3.57}$$

In addition, if

$$[t_{k+1} - \tau_1, t_{k+1} + \tau_2] \in [s, t_*]$$

then, we have

$$x(t_{k+1} + \tau_2) = e^{A_2 \tau_2} e^{A_1 \tau_1} x(t_{k+1} - \tau_1) + e^{A_2 \tau_2} \int_{t_{k+1}-\tau_1}^{t_{k+1}} e^{A_1(t_{k+1}-\zeta)} f(\zeta) d\zeta$$

$$+ \int_{t_{k+1}}^{t_{k+1}+\tau_2} e^{A_2(t_{k+1}+\tau_2-\zeta)} f(\zeta) d\zeta$$

which implies that

$$V(x(t_{k+1} + \tau_2)) \le x^T(t_{k+1} - \tau_1) e^{A_1^T \tau_1} e^{A_2^T \tau_2} P e^{A_2 \tau_2} e^{A_1 \tau_1} x(t_{k+1} - \tau_1)$$

$$+ 2p\varrho^2 \|x(t_{k+1} - \tau_1)\| \int_{t_{k+1}-\tau_1}^{t_{k+1}+\tau_2} \|f(\zeta)\| d\zeta + p\varrho^2 \left(\int_{t_{k+1}-\tau_1}^{t_{k+1}+\tau_2} \|f(\zeta)\| d\zeta \right)^2$$

$$\le (1 - \delta) V(x(t_{k+1} - \tau_1)) + \frac{\delta}{4} \lambda_1 \mu_1 (t_{k+1} - \tau_1) \|x(t_{k+1} - \tau_1)\|$$

$$+ \frac{\delta^2}{64p} \lambda_1^2 (\mu_1 (t_{k+1} - \tau_1))^2$$

$$\le (1 - \frac{\delta}{2}) V(x(t_{k+1} - \tau_1)) \le e^{-2\theta\tau} V(x(t_{k+1} + \tau_2))$$

where the relationships $\mu_1(t_{k+1} - \tau_1) \le \|x(t_{k+1} - \tau_1)\|$, $p \ge \lambda_1$, and $\delta \in (0, 1)$ are used. Combining this with (3.56), we have

$$V(x(t_{k+2})) \le e^{-2\theta(t_{k+2}-t_k)} V(x(t_k)).$$

On the other hand, it follows from (3.56) and (3.57) that, for all $s_1, s_2 \in (s, t_*] \cap [t_k, t_{k+2})$, we have

$$V(x(s_2)) \le \exp \left((\frac{2q + r}{2\lambda_1} + 2\theta) \tau \right) \exp(-2\theta(s_2 - s_1)) V(x(s_1)). \quad (3.58)$$

If $t_* > t_{k+2}$, let l be the largest natural number such that $t_* > t_{k+2l}$. A similar argument gives

$$V(x(t_{k+2l})) \le \exp(-2\theta(t_{k+2l} - t_{k+2})) V(x(t_{k+2})) \quad (3.59)$$

and

$$V(x(t_*)) \le \exp \left(\frac{2q + r}{2\lambda_1} + 2\theta) \tau \right) \exp(-2\theta(t_* - t_{k+2l})) V(x(t_{k+2l})). \quad (3.60)$$

Combining (3.58) and (3.59) with (3.60), we have

$$V(x(t_*)) \leq \exp\left(2(\frac{2q+r}{2\lambda_1} + 2\theta)\tau\right)\exp(-2\theta(t_* - s))V(x(s)). \quad (3.61)$$

For Case (b), we can proceed in the same way, using $[t_0, t_0 + \tau_2)$ instead of $[t_{k+1} - \tau_1, t_{k+1} + \tau_2)$ and $[t_0 + \tau_2, t_1)$ instead of $[t_k, t_{k+1} - \tau_1) \cup [t_{k+1} + \tau_2, t_{k+2})$. The same reasonings show that (3.61) also holds in this case.

To summarize, (3.61) holds for both cases. Therefore, we have

$$\|x(t_*)\|^2 \leq \frac{1}{\lambda_1}V(x(t_*))$$

$$\leq \frac{1}{\lambda_1}\exp\left(2(\frac{2q+r}{2\lambda_1} + 2\theta)\tau\right)\exp(-2\theta(t_* - s))V(x(s))$$

$$\leq \frac{\lambda_n}{\lambda_1}\exp\left(2(\frac{2q+r}{2\lambda_1} + 2\theta)\tau\right)\exp(-2\theta(t_* - s))\mu_1^2(s)$$

$$\leq \mu^2(t_*)$$

which contradicts the assumption (3.54). As a result, the estimation (3.53) always holds.

Now, we are ready to derive (i)-(iii) from (3.53).

(i) Suppose that $\|f(t)\| \leq N_f$ for all $t \geq t_0$. It is clear that

$$\mu_1(t) \leq \max\left(\|x_0\|, \frac{4p}{r}N_f, \frac{8p\varrho^2}{\delta\lambda_1}\tau N_f\right) \quad t \geq t_0.$$

Therefore, the state is bounded by

$$\|x(t)\| \leq \varpi \max\left(\|x_0\|, \frac{4p}{r}N_f, \frac{8p\varrho^2}{\delta\lambda_1}\tau N_f\right) \stackrel{def}{=} N_x.$$

(ii) Suppose that $f(t)$ is bounded and convergent. For any given positive real number $\epsilon < N_f$, there is a time T_1 such that

$$\|x_0\| \leq \frac{\epsilon}{\varpi}e^{\theta(t-t_0)} \quad \|f(t)\| \leq \min\left(\frac{r}{4p}, \frac{\delta\lambda_1}{8p\varrho^2\tau}\right)\frac{\epsilon}{\varpi} \quad \forall\, t \geq T_1.$$

Let $T_2 = \frac{1}{\theta}\ln\frac{N_x}{\epsilon}$ and $T = T_1 + T_2$. Simple analysis gives

$$\mu(t) \leq \varpi \max\left(\sup_{s\in[t_0,T_1]}\mu_1(s)e^{-\theta(t-T_1)}, \sup_{s\in[T_1,t]}\mu_1(t)\right) \leq \epsilon \quad \forall\, t \geq T.$$

From the arbitrariness of ϵ, the convergence of the state follows.

(iii) Suppose that

$$\|f(t)\| \leq \beta e^{-\alpha(t-t_0)} \quad t \geq t_0$$

for some positive real numbers α and β. Then, simple calculation yields

$$\mu(t) \leq \varpi \max\left(\|x_0\|, \frac{4p}{r}\beta, \frac{8p\varrho^2\beta}{\delta\lambda_1\alpha}\right)e^{-\min(\theta,\alpha)(t-t_0)} \quad t \geq t_0$$

which shows that the system state is exponentially convergent. $\square$

Remark 3.37. It can be seen from the proof that the bound of the state can be explicitly estimated. Moreover, we can also estimate the eventual bound of the state. Indeed, suppose that the eventual bound of the perturbation is N_{fe}, that is

$$\limsup_{t \to \infty} \|f(t)\| = N_{fe}.$$

Then, from (3.53), it is clear that the eventual bound of the state is given by

$$\limsup_{t \to \infty} \|x(t)\| \leq N_{xe} = \max\left(\frac{4p}{r} N_{fe}, \frac{8p\varrho^2}{\delta\lambda_1}\tau N_{fe}\right).$$

Consequently, given any allowed state (eventual) bound, we can estimate a (eventual) bound of perturbations that makes the system state (eventually) bounded within the allowed level.

3.5.3 Observer-based Switching

In this subsection, we explore the possibility of designing switching signals based on the measured output instead of the state information.

Consider the switched linear system described by

$$\begin{cases} \dot{x}(t) = A_\sigma x(t) + f_\sigma(t) \\ y(t) = C_\sigma x(t) + h_\sigma(t) \end{cases} \tag{3.62}$$

where $x(t), \sigma$ and $f_i(t)$ are the same as in (3.52), $y(t) \in \mathbf{R}^q$ is the measured output, $h_i(t)$, $i = 1, 2$ are piecewise continuous vector functions that stand for output perturbations, and A_i, C_i, $i = 1, 2$ are matrices of compatible dimensions. We assume that information of the state and perturbations is not available on-line.

System (3.62) represents a switched linear system with multiple sensor devices. The description includes multi-sensor scheduling as a special case [117].

Assumption 3.3. *Both the matrix pairs (C_1, A_1) and (C_2, A_2) are observable.*
Note that this assumption is very mild as the property of observability is generic for all matrix pairs [160].

For system (3.62), consider the state estimator given by

$$\dot{\hat{x}} = A_\sigma \hat{x} + L_\sigma [y(t) - C_\sigma \hat{x}] \tag{3.63}$$

where $y(t)$ and σ are the output and switching signal of system (3.62), respectively, and matrices $L_1, L_2 \in \mathbf{R}^{n \times q}$ are to be determined later.

Fix w_1, w_2, r_1, r_2 and τ as described in Section 3.5.1. Let $\tau_1 = w_1\tau$ and $\tau_2 = w_2\tau$. In addition, fix a real number $\varphi \in (0, 1)$.

Choose gain matrices $L_i, i = 1, 2$ such that

$$\|e^{(A_i - L_i C_i)t}\| < \varphi \quad \forall \, \tau_i \leq t < 2\tau_i \quad i = 1, 2. \tag{3.64}$$

Note that the above choice is always possible because of the observability assumption. Intuitively, for any given $t \geq 0$, the following inequality holds

$$\|e^{(A_i - L_i C_i)t}\| \leq p(\rho)e^{\rho_n t}$$

where $\rho = \{\rho_1 \leq \cdots \leq \rho_n\}$ denotes the real parts of the eigenvalues of matrix $(A_i - L_i C_i)$, and $p(\cdot)$ is a polynomial function. Note that ρ can be arbitrarily assigned by choosing appropriate gain matrices L_i. When ρ approaches $-\infty$ proportionately, $p(\rho)$ increases polynomially but $e^{\rho_n t}$ decreases exponentially, hence $\|e^{(A_i - L_i C_i)t}\|$ decreases and can be made arbitrarily small. See [176] and [89] for more details.

With these preparations, we are ready to formulate the observer-based switching law as follows.

Suppose that $\hat{x}$ initializes at $\hat{x}(t_0) = \hat{x}_0$. Set

$$\sigma(t_0) = \arg\min\{\hat{x}_0^T Q_1 \hat{x}_0, \hat{x}_0^T Q_2 \hat{x}_0\}.$$

The subsequent switching time/index sequences are defined recursively by

$$t_{k+1} = \begin{cases} \inf\left\{t > t_k: \ \hat{x}^T(t)Q_1\hat{x}(t) > -r_1\hat{x}^T(t)\hat{x}(t)\right\} + \tau_1 & \text{if } \sigma(t_k) = 1 \\ \inf\left\{t > t_k + \tau_2: \ \hat{x}^T(t)Q_2\hat{x}(t) > -r_2\hat{x}^T(t)\hat{x}(t)\right\} & \text{if } \sigma(t_k) = 2 \end{cases}$$

$$\sigma(t_{k+1}) = \begin{cases} 2 \text{ if } \sigma(t_k) = 1 \\ 1 \text{ if } \sigma(t_k) = 2 \end{cases} \quad k = 0, 1, \cdots. \tag{3.65}$$

This switching strategy is exactly the same as (3.45), except that the state $x(t)$ is substituted by the estimator $\hat{x}(t)$. The following result establishes the fact that it inherits the nice robustness properties.

Theorem 3.38. *For system (3.62), suppose that Assumptions 3.1 and 3.3 hold. Then, under the switching strategy (3.65), we have*

(i) the system state and the estimator are bounded if the perturbations are bounded;

(ii) the system state and the estimator are bounded and convergent if the perturbations are bounded and convergent; and

(iii) the system state and the estimator are exponentially convergent if the perturbations are exponentially convergent.

Proof. Define the difference between the real state and the estimated state by

$$\tilde{x} = x - \hat{x}.$$

Subtracting (3.63) from (3.62), we obtain

$$\dot{\tilde{x}} = (A_\sigma - L_\sigma C_\sigma)\tilde{x} + f_\sigma(t) - L_\sigma h_\sigma(t). \tag{3.66}$$

On the other hand, it follows from (3.63) that

$$\dot{\hat{x}} = A_\sigma \hat{x} + L_\sigma C_\sigma \tilde{x} + L_\sigma h_\sigma(t). \tag{3.67}$$

For this equation, take $\dot{\hat{x}} = A_\sigma \hat{x}$ as the nominal system and $L_\sigma C_\sigma \tilde{x}(t)$ as an exoteric perturbation, along with $L_\sigma h_\sigma(t)$. By Theorem 3.36, we only need to prove that $\tilde{x}(t)$ is bounded (convergent, exponentially convergent) if the perturbations f_σ and h_σ are bounded (convergent, exponentially convergent).

Suppose that the switching time sequence is $t_0 < t_1 < t_2 < \cdots$ and the switching index sequence is $j_0 = \sigma(t_0+), j_1 = \sigma(t_1+), j_2 = \sigma(t_2+), \cdots$. It follows from the switching strategy (3.65) that, for any consecutive switching times t_k and t_{k+1}, we have

$$t_{k+1} - t_k \geq \begin{cases} \tau_1 & \text{if } j_k = 1 \\ \tau_2 & \text{if } j_k = 2. \end{cases} \tag{3.68}$$

By (3.64), we can find a positive real number α such that

$$\|e^{(A_i - L_i C_i)t}\| \leq e^{-\alpha t} \quad \forall\, t \geq \tau_i. \tag{3.69}$$

Indeed, let

$$\alpha = \min\left(-\frac{\ln \varphi}{2\tau_1}, -\frac{\ln \varphi}{2\tau_2}\right).$$

For any $t \in [\tau_i, 2\tau_i)$, we have

$$\|e^{(A_i - L_i C_i)t}\| \leq \varphi \leq e^{-\alpha t}.$$

For any $t \geq 2\tau_i$, there exists a natural number l such that $t \in [(l+1)\tau_i, (l+2)\tau_i)$. As a result, we have

$$\|e^{(A_i - L_i C_i)t}\| \leq \|e^{(A_i - L_i C_i)(t - l\tau_i)}\|\|e^{(A_i - L_i C_i)l\tau_i}\| \leq e^{-\alpha(t - l\tau_i)}\varphi^l \leq e^{-\alpha t}.$$

Define matrix function

$$\Psi(t, \sigma) = e^{(A_{j_k} - L_{j_k} C_{j_k})(t - t_k)} \cdots e^{(A_{j_1} - L_{j_1} C_{j_1})(t_2 - t_1)} e^{(A_{j_0} - L_{j_0} C_{j_0})(t_1 - t_0)}$$
$$t_k < t \leq t_{k+1}.$$

The state transition matrix of the error system (3.66) can be expressed as

$$\Phi(s_1, s_2, \sigma) = \Psi(s_1, \sigma)\Psi(s_2, \sigma)^{-1} \quad \forall\, s_1, s_2 \geq t_0.$$

The solution of system (3.66) is given by

$$\tilde{x}(t) = \Phi(t, t_0, \sigma)\tilde{x}_0 + \int_{t_0}^{t} \Phi(t, \zeta, \sigma)(f_\sigma(\zeta) - L_\sigma h_\sigma(\zeta))d\zeta \quad \forall\, t \geq t_0.$$

From (3.68) and (3.69), it follows that

$$\|\Phi(s_1, s_2, \sigma)\| \leq \beta e^{-\alpha(s_1 - s_2)} \quad \forall\ s_1 \geq s_2 \geq t_0$$

where $\beta = \exp\left(\|(A_1 - L_1 C_1)\| \tau_1\right) \exp\left(\|(A_2 - L_2 C_2)\| \tau_2\right)$. Therefore,

$$\|\tilde{x}(t)\| \leq \beta e^{-\alpha(t - t_0)} \|\tilde{x}_0\| + \beta \int_{t_0}^{t} e^{-\alpha(t - \zeta)} \|(f_\sigma(\zeta) - L_\sigma h_\sigma(\zeta))\| d\zeta \quad \forall\ t \geq t_0.$$

From this we can prove that $\tilde{x}(t)$ is bounded/convergent/exponentially-convergent for bounded/convergent /exponentially-convergent perturbations. We omit the details for briefness. $\square$

According to Theorem 3.38, the observer-based switching strategy works well in the event that the system undergoes unmeasurable perturbations. Note that we can also allow perturbations for the estimator in the scheme. That is, instead of (3.63), the estimator is given by

$$\dot{\hat{x}} = A_\sigma \hat{x} + L_\sigma [y(t) - C_\sigma \hat{x}] + g_\sigma(t)$$

where $g_i(t)$, $i = 1, 2$ are piecewise continuous vector functions representing perturbations induced by the estimator. In this case, it is clear that Theorem 3.38 still holds.

Combining the above proof with that of Theorem 3.36, an explicit bound estimation can be obtained for the system state and estimator. Note that, due to the extra perturbations from the error system, the estimated state bound may be quite larger than that of the system under the state-feedback switching strategy.

3.5.4 Extensions

In this subsection, we discuss the possibility of generalizing the results to a broader class of switched linear systems. The systems are described by

$$\begin{cases} \dot{x}(t) = A_\sigma x(t) + f_\sigma(t) \\ y(t) = C_\sigma x(t) + h_\sigma(t) \end{cases} \tag{3.70}$$

where the system structure is the same as in (3.62), but with an arbitrarily large number of subsystems rather than only two.

In the scheme, a key role of Assumption 3.1 is to guarantee (3.44), that is, there exists a (quadratic) Lyapunov-like function which decreases at discrete times. Thus, it is natural to substitute Assumption 3.1 by (3.44). In fact, the systems satisfying (3.44) form exactly the class of consistently asymptotically stabilizable systems (c.f. Section 3.2).

In view of Corollary 3.12, for a consistently asymptotically stabilizable system $\Sigma(A_i)_M$, there exist a natural number ι, an index set $j_1, \cdots, j_\iota \in M$, and a positive real number set $\tau_1, \cdots, \tau_\iota$, such that matrix $e^{A_{j_\iota} \tau_\iota} \cdots e^{A_{j_1} \tau_1}$

is Schur stable. In addition, we can assume, without loss of generality, that $j_1 \neq j_\iota$. In fact, if $j_1 = j_\iota$, we can prove that $e^{A_{j_{\iota-1}} \tau_{\iota-1}} \cdots e^{A_{j_1}(\tau_1 + \tau_k)}$ is also asymptotically stable. In addition, it is clear that we can assume $j_i \neq j_{i+1}$ for $i = 1, \cdots, \iota - 1$.

In the following, we consider the class of systems which are consistently asymptotically stabilizable. We start from the equivalent assumption.

Assumption 3.4. *There exist a natural number ι, an index set $j_1, \cdots, j_\iota \in M$ with $j_i \neq j_{i+1}$ and $j_1 \neq j_\iota$, and a positive real number set $\tau_1, \cdots, \tau_\iota$, such that matrix $e^{A_{j_\iota} \tau_\iota} \cdots e^{A_{j_1} \tau_1}$ is Schur.*

Note that Assumption 3.1 implies Assumption 3.4, but the converse is generally not true. This means that the class of systems which satisfies Assumption 3.4 is broader than that satisfying Assumption 3.1. In particular, Assumption 3.1 implies quadratic stabilizability but Assumption 3.4 does not exclude non-quadratic stabilizability (an example is given in Example 3.41 in Section 3.6).

Given a matrix $Q > 0$, solve for the discrete-time Lyapunov equation

$$(e^{A_{j_\iota} \tau_\iota} \cdots e^{A_{j_1} \tau_1})^T P e^{A_{j_\iota} \tau_k} \cdots e^{A_{j_1} \tau_1} - P + Q = 0$$

we have a unique symmetric solution $P > 0$.

Concerning the system output, we make the following assumption.

Assumption 3.5. *Matrix pair (C_i, A_i) is observable for all $i \in M$.*

As mentioned in the previous subsection, fix a real number $\varphi \in (0, 1)$, we can find gain matrices $L_i, i \in M$ such that

$$\|e^{(A_i - L_i C_i)t}\| < \varphi \quad \forall \, \tau_i \leq t < 2\tau_i \quad i \in M. \tag{3.71}$$

We construct the following state estimator

$$\dot{\hat{x}} = A_\sigma \hat{x} + L_\sigma [y(t) - C_\sigma \hat{x}] + g_\sigma(t) \tag{3.72}$$

where $g_i(t)$, $i \in M$ are the possible perturbations induced by the state estimator.

Fix $r_i \in (0, 1)$ for $i \in M$ and let $\tau_0 = \tau_\iota$. The generalized switching strategy is formulated as follows.

1. **Initialization.** Set

$$\sigma(t_0) = \begin{cases} j_1 & \text{if } \hat{x}_0^T Q_{j_1} \hat{x}_0 \leq \hat{x}_0^T Q_{j_\iota} \hat{x}_0 \\ j_\iota & \text{otherwise} \end{cases}$$

and

$$\iota_0 = \begin{cases} 1 & \text{if } \sigma(t_0) = j_1 \\ 0 & \text{if } \sigma(t_0) = j_\iota. \end{cases}$$

2. **Recursion.** Define recursively the consecutive time and index as

$$
t_{k+1} = \begin{cases} \inf \left\{ t > t_k : \ \hat{x}^T(t) Q_{\sigma(t_k)} \hat{x}(t) > -r_{\sigma(t_k)} \hat{x}^T(t) \hat{x}(t) \right\} + \tau_1 \\ \qquad \text{if mod } (k, \iota) \in \{0, \iota - 1\} \text{ and } \sigma(t_k) = j_1 \\ \inf \left\{ t > t_k + \tau_\iota : \ \hat{x}^T(t) Q_{\sigma(t_k)} \hat{x}(t) > -r_{\sigma(t_k)} \hat{x}^T(t) \hat{x}(t) \right\} \\ \qquad \text{if mod } (k, \iota) \in \{0, \iota - 1\} \text{ and } \sigma(t_k) = j_\iota \\ t_k + \tau_{\iota_0 + \text{ mod } (k, \iota)} \quad \text{otherwise} \end{cases}
$$

$$
\sigma(t_{k+1}) = \quad \text{mod } (k + \iota_0, \iota) + 1 \quad k = 0, 1, \cdots. \tag{3.73}
$$

In the above switching strategy, the switching index sequence is cyclic. The observer-based switching mechanism is incorporated into the first and last phases. The strategy degenerates into (3.65) for the case $\iota = 2$.

Unlike in Theorem 3.33, we lack a guaranteed dwell-time for the observer-based switching mechanism in general. This is because Item 3) in the proof of Theorem 3.33 does not necessarily hold without Assumption 3.1. Nevertheless, the observer-based switching mechanism still works in a random manner, that is, when the switching is in the first or last phases and

$$
\hat{x}(t) \in \{x : x^T P x < -r_\sigma(t) x^T x\}
$$

the observer-based switching mechanism will work and decide the next switching time. See Example 3.41 for a simulation study.

The following theorem states the main result for the generalized class of systems.

Theorem 3.39. *For system (3.70), suppose that Assumptions 3.4 and 3.5 hold. Then, under the switching strategy (3.73), we have*

(i) *the system state and the estimator are bounded if the perturbations are bounded;*

(ii) *the system state and the estimator are bounded and convergent if the perturbations are bounded and convergent; and*

(iii) *the system state and the estimator are exponentially convergent if the perturbations are exponentially convergent.*

Proof. The theorem can be proven in the same way as in the proofs of Theorems 3.36 and 3.38. We hence omit the details. □

3.6 Numerical Examples

Example 3.40. Consider the switched linear system $\Sigma(A_i)_{\{1,2\}}$ with

$$
A_1 = \begin{bmatrix} -2.1 & 1.4 & 5.9 \\ -8.0 & -5.7 & -0.2 \\ 0.6 & 5.8 & 1.6 \end{bmatrix} \text{ and } A_2 = \begin{bmatrix} 1.0 & -0.5 & -2.8 \\ 4.8 & -5.0 & 1.1 \\ -1.0 & -6.6 & -2.1 \end{bmatrix}. \tag{3.74}
$$

Simple calculation shows that $wA_1+(1-w)A_2$ is Hurwitz for $0.33 \leq w \leq 0.75$. By letting $w_1 = 0.37$ and $w_2 = 0.73$, the average matrix is

$$A_0 = w_1 A_1 + w_2 A_2 = \begin{bmatrix} -0.1470 & 0.2030 & 0.4190 \\ 0.0640 & -5.2590 & 0.6190 \\ -0.4080 & -2.0120 & -0.7310 \end{bmatrix}$$

which possesses eigenvalues at $-0.5806 \pm 0.1148\sqrt{-1}$, and -4.9758. Solving the Lyapunov equation $A_0^T P + P A_0 = -I_3$ yields

$$P = \begin{bmatrix} 1.9737 & -0.0924 & 0.4999 \\ -0.0924 & 0.1800 & -0.2312 \\ 0.4999 & -0.2312 & 0.7748 \end{bmatrix}.$$

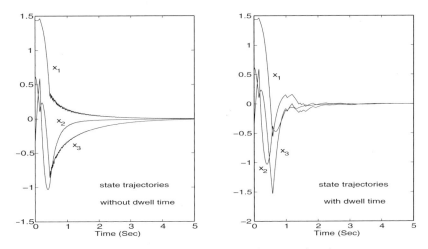

Fig. 3.5. State trajectories of system (3.74)

First, we fix the period $\tau = 0.08$. Hence, we have $\tau_1 = 0.0296$ and $\tau_2 = 0.0504$. In the following simulations, we assume that the system has the initial state

$$x_0 = [1.4435, -0.3510, 0.6232]^T.$$

Let $r_1 = r_2 = 0.1$. Figure 3.5 shows the state trajectories of the switched system under the state-feedback switching law (3.21) and the combined switching law (3.45), respectively. It can be seen that both trajectories converge with satisfactory rates. However, for the former case, the number of switches is 382, but for the latter, only 44. That is, by introducing the dwell time, we substantially reduce the switching frequency in this example. On the other hand, if we just exploit the time-driven periodic switching signal (3.17) at the cyclic

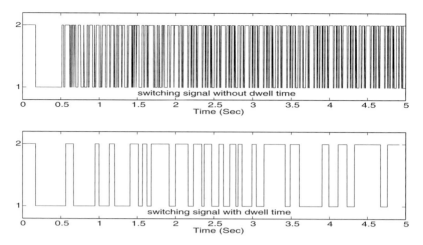

Fig. 3.6. Switching signals with and without dwell time

period τ, the number of switches is 124. That is, the combined switching strategy reduces more than half of the switches by introducing the state-feedback switching mechanism. The resultant switching signals are shown in Figure 3.6.

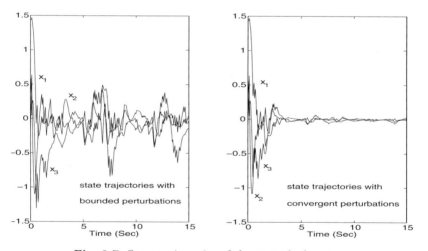

Fig. 3.7. State trajectories of the perturbed systems

Next, consider the system with perturbations. Let

$$f_1(t) = \begin{bmatrix} \sin(2t) \\ -0.5\,\mathrm{sat}(t-1) \\ \mathrm{sgn}(\cos(3t)) \end{bmatrix} \text{ and } f_2(t) = \begin{bmatrix} -te^{-t} \\ \ln\left(\frac{2+t^2}{1+t^2}\right) \\ -0.5 \end{bmatrix}$$

and

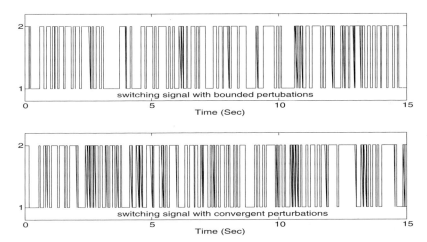

Fig. 3.8. Switching signals of the perturbed systems

$$\bar{f}_i(t) = \frac{1}{1+t} f_i(t) \quad i = 1, 2$$

where sat($\cdot$) is the saturation function with unit limits, and sgn($\cdot$) is the signum function. It is clear that f_1, f_2 are bounded and $\bar{f}_1, \bar{f}_2$ are convergent. Figure 3.7 shows the state trajectories of the systems with perturbations f_k and $\bar{f}_k$, respectively. The corresponding switching signals are given in Figure 3.8. It is clear that the switching frequencies are much higher than that of the nominal system.

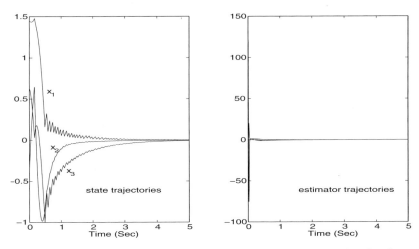

Fig. 3.9. State and estimator trajectories of the nominal system under the observer-based switching signal

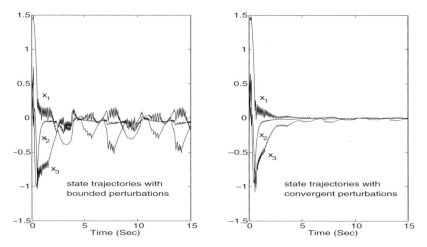

Fig. 3.10. State trajectories of the perturbed systems under the observer-based switching signal

Finally, suppose that the full state information is not available. Assume that x_1 is measurable for the first subsystem and x_2 for the other. That is,

$$C_1 = [1, 0, 0] \text{ and } C_2 = [0, 1, 0].$$

We assign the eigenvalues of $A_i - L_i C_i$ to be $\{\xi, \xi \pm \xi\sqrt{-1}\}$ where ξ is chosen such that (3.64) is satisfied. In this way, we have the high-gain matrices

$$L_1 = [746.80000, 1255903.1437, -255818.7443]^T$$
$$L_2 = [59560.0425, 356.9000, -207024.5944]^T.$$

Suppose that the estimator has the initial condition

$$\hat{x}_0 = [0, 0, 0]^T.$$

Figure 3.9 depicts the trajectories of the state and the estimator. Although the system state behaves properly, the estimator undergoes very high transient overshoot. State trajectories for the perturbed systems are shown in Figure 3.10.

Example 3.41. Consider the switched linear system $\Sigma(A_i)_{\{1,2\}}$ with

$$A_1 = \begin{bmatrix} 2.440 & -6.253 \\ 2.572 & -9.540 \end{bmatrix} \text{ and } A_2 = \begin{bmatrix} 1.053 & 17.578 \\ -0.345 & 1.947 \end{bmatrix}. \quad (3.75)$$

Numerical verification shows that Assumption 3.1 does not hold for this system. As a result, the system is not quadratically stabilizable by any switching signal. However, a simple search indicates that $e^{\tau_2 A_2} e^{\tau_1 A_1}$ is Schur when $\tau_1 = 0.60$ and $\tau_2 = 1.10$. Solving the discrete-time Lyapunov function

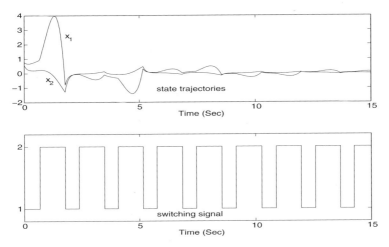

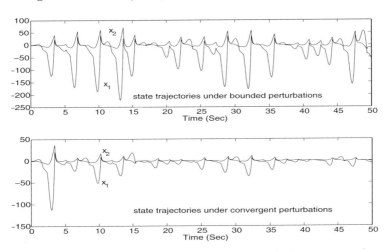

Fig. 3.11. State trajectory and switching signal of system (3.75)

Fig. 3.12. State trajectories of the perturbed systems

$$\left(e^{\tau_2 A_2} e^{\tau_1 A_1}\right)^T P e^{\tau_2 A_2} e^{\tau_1 A_1} - P + 0.5 I_2 = 0$$

we obtain

$$P = \begin{bmatrix} 6.5383 & -3.6203 \\ -3.6203 & 2.7962 \end{bmatrix}.$$

Let $r_1 = r_2 = 0.1$ and suppose that the system has the initial condition

$$x(0) = [0.7812, \ 0.5690]^T.$$

Figure 3.11 shows the state trajectory and switching signal under the switching strategy (3.73). It is clear that the switching path is nearly periodic, hence the dominant time of the state-feedback mechanism is far less than that of the time-driven mechanism, though detailed information clearly shows that the state-feedback mechanism does work in most switching periods. Figure 3.12 depicts the state trajectories of the perturbed system with bounded perturbations

$$f_1(t) = \begin{bmatrix} 0.3 \\ -\operatorname{sgn}(\sin(2t)) \end{bmatrix} \text{ and } f_2(t) = \begin{bmatrix} -0.5\operatorname{sat}(t^2 - 3t) \\ \ln(\cos(t)) \end{bmatrix}$$

and convergent perturbations

$$\bar{f}_i(t) = \frac{1}{\ln(2 + t^2)} f_i(t) \quad i = 1, 2$$

respectively. It can be seen that the perturbed trajectories have remarkably larger bounds than that of the nominal system, indicating that the system is quite sensitive to the perturbations. This is not surprising because the switching frequency is rather low so that the effect of the perturbations can accumulate during the constant (non-switching) process.

3.7 Discrete-time Switched Systems

In this section, we consider the discrete-time switched linear autonomous system given by

$$x_{k+1} = A_\sigma x_k \tag{3.76}$$

where $x \in \mathbf{R}^n$, and $A_i \in \mathbf{R}^{n \times n}$ for $i \in M$, and σ is the switching signal taking values from the index set M.

3.7.1 Contractive and Pre-contractive

In this subsection, we discuss some properties related to the convergence of matrix multiplications from a finite set of matrices. It is well-known that any trajectory of a linear time-invariant system

$$x_{k+1} = A x_k$$

is convergent if and only if A is Schur, *i.e.*, the spectral radius of A is less than one. For the switched linear system, we aim to find similar properties.

First, let us characterize what we mean by saying a set of matrices is convergent.

Definition 3.42. *A set* $\{H_1, \cdots, H_N\}$ *of* $n \times n$ *matrices is* switched convergent, *provided that, for each* $x \in \mathbf{R}^n$, *there is a sequence* $\{j_i(x)\}_{i=1}^\infty$ *with* $j_i(x) \in \{1, \cdots, N\}$, *such that the vector sequence*

$$x, H_{j_1(x)}x, H_{j_2(x)}H_{j_1(x)}x, \cdots$$

converges to the origin, 0.

When $N = 1$, the concept degenerates to the usual concept of matrix convergence.

Example 3.43. Consider the four symmetric matrices

$$H_1 = \begin{bmatrix} 0.5 & 0 \\ 0 & 2 \end{bmatrix} \quad H_2 = \begin{bmatrix} 1.25 & -0.75 \\ -0.75 & 1.25 \end{bmatrix}$$

and

$$H_3 = \begin{bmatrix} 2 & 0 \\ 0 & 0.5 \end{bmatrix} \quad H_4 = \begin{bmatrix} 1.25 & 0.75 \\ 0.75 & 1.25 \end{bmatrix}.$$

It can be routinely verified that each A_i has spectral radius 2. But set $\{H_i\}_{i=1}^4$ is switched convergent, as seen from the following facts:

(i) H_1 contracts each nonzero vector in the closed cone C_1, co-axial with the x_1-axis having vertex angle at the origin, and measuring $45°$. H_2, H_3 and H_4 act similarly on vectors in cones C_2, C_3 and C_4 which are counterclockwise rotations of C_1 through $45°$, $90°$, and $135°$, respectively.
(ii) There is a real number $\beta \in (0,1)$, such that, for each non-origin $x \in \mathbf{R}^2$, we have

$$\min_{i=1}^{4}\{\|H_i x\|\} < \beta \|x\|.$$

Thus, the set is switched convergent.

The example suggests the following definition.

Definition 3.44. *Let* $\| \cdot \|_v$ *be any norm on* $\mathbf{R}^n$. *The set* $\{H_1, \cdots, H_N\}$ *of* $n \times n$ *matrices is* contractive *w.r.t.* $\| \cdot \|_v$, *provided that, for any non-origin* $x \in \mathbf{R}^n$, *there is an* $i \in \{1, \cdots, N\}$, *such that* $\|H_i x\|_v < \|x\|_v$.

Usually, the above concept is norm-dependent. To avoid this situation, we propose the following concept.

Definition 3.45. *Let* $\| \cdot \|_v$ *be any norm on* $\mathbf{R}^n$. *The set* $\{H_1, \cdots, H_N\}$ *of* $n \times n$ *matrices is* pre-contractive *w.r.t.* $\| \cdot \|_v$, *provided that, for any non-origin* $x \in \mathbf{R}^n$, *there is a finite sequence* $\{j_i(x)\}_{i=1}^{n(x)}$, $j_i(x) \in \{1, \cdots, N\}$, *such that*

$$\left\| \left(\Pi_{i=n(x)}^{1} H_{j_i(x)} \right) x \right\|_v < \|x\|_v.$$

The following result establishes the equivalence between switched convergence and pre-contractiveness *w.r.t.* a norm.

Theorem 3.46. *Let $\| \cdot \|_v$ be any norm on $\mathbf{R}^n$, and $K = \{H_1, \cdots, H_N\}$ be a set of $n \times n$ matrices. Then, the matrix set is pre-contractive w.r.t. $\| \cdot \|_v$ if and only if it is switched convergent.*

Proof. It is clear that switched convergence of K implies pre-contractiveness of K to any norm.

Assume that K is pre-contractive w.r.t. $\| \cdot \|_v$. Then, for each $x \in \mathbf{R}^n$, $x \neq 0$, there is a finite sequence $\{j_i(x)\}_{i=1}^{n(x)}$, $j_i(x) \in \{1, \cdots, N\}$, such that

$$\left\| \left(\Pi_{i=n(x)}^{1} H_{j_i(x)} \right) x \right\|_v < \|x\|_v. \tag{3.77}$$

Note that, for any $y = \lambda x$ with $\lambda \neq 0$, to maintain Inequality (3.77), it suffices to choose $j_i(y) = j_i(x)$ and $n(y) = n(x)$. Note also that, once (3.77) holds for x, it also holds for nearby vectors with the same $n(x)$ and $j_i(x)$, *i.e.*,

$$\left\| \left(\Pi_{i=n(x)}^{1} H_{j_i(x)} \right) y \right\|_v < \|x\|_v \quad y - x \in \mathbf{B}_r$$

for sufficiently small r. Due to the Finite Covering Theorem, there is a positive integer l such that $n(x) \leq l$ for all $x \neq 0$. As a consequence, the augmented matrix set

$$\bar{K} \stackrel{def}{=} \left\{ \Pi_{i=k}^{1} H_{j_i} : k \leq l, 1 \leq j_i \leq N \right\}$$

is contractive w.r.t. $\| \cdot \|_v$. This means that, for each $x \neq 0$, there exist an $H \in \bar{K}$, and a $\beta(x) \in (0, 1)$, such that

$$\|H_i x\|_v < \beta(x) \|x\|_v.$$

Applying the Finite Covering Theorem once again, we can infer that there is a $\beta \in (0, 1)$ such that

$$\beta(x) \leq \beta \quad \forall\, x \neq 0.$$

The above reasonings show that, for any $x \neq 0$, there is a sequence $\{j_i(x)\}_{i=1}^{n(x)}$ such that $n(x) < l$, and

$$\left\| \left(\Pi_{i=n(x)}^{1} H_{j_i(x)} \right) x \right\|_v \leq \beta \|x\|_v.$$

To show that K is switched convergent, select $x \in \mathbf{R}^n$ and define recursively an index sequence and a vector sequence by

$$
\begin{aligned}
y_0 &= x \\
n_0 &= n(y_0) \\
p_i(x) &= j_i(y_0) \quad i = 1, \cdots, n_0 \\
y_{k+1} &= \left(\Pi_{i=n_k}^{n_{k-1}+1} H_{p_i(y_k)} \right) y_k \\
n_{k+1} &= n_k + n(y_{k+1}) \quad\quad\quad\quad\quad k = 0, 1, \cdots \\
p_i(x) &= j_{i-n_k}(x) \quad i = n_k + 1, \cdots, n_{k+1}
\end{aligned}
$$

where we set $n_{-1} = 0$. In this way, we have the sequence $\{p_i(x)\}_{i=1}^{\infty}$. It is clear from this construction that, for any $k \in \mathbf{N}_+$, we have

$$\| \left(\Pi_{i=n_k}^1 H_{p_i(x)} \right) x \|_v \leq \beta^{k+1} \|x\|_v. \tag{3.78}$$

Let

$$\gamma = \max\{\|H_i\|_v\}_{i \in M}.$$

For any $\delta \in \mathbf{N}^+$, there exist a nonnegative integer k, and an $s \in \{1, \cdots, l-1\}$, such that

$$\delta = n_k + s.$$

Accordingly, we have

$$\| \left(\Pi_{i=\delta}^1 H_{p_i(x)} \right) x \|_v \leq \|\Pi_{i=\delta}^{n_k+1} H_{p_i(x)}\|_v \| \left(\Pi_{i=n_k}^1 H_{p_i(x)} \right) x \|_v$$
$$\leq \gamma^s \beta^{k+1} \|x\|_v \leq \gamma^{(l-1)} \beta^{k+1} \|x\|_v. \tag{3.79}$$

As $\delta \to \infty$, $k \to \infty$ and the theorem follows. $\square$

Corollary 3.47. *Pre-contractiveness is norm-independent.*
Proof. As switched convergence is norm-independent, by Theorem 3.46, the corollary follows. $\square$

The estimation (3.79) is very important in the following derivations. In fact, as $n_{k+1} - n_k \leq l$ for all $k = 0, 1, \cdots$, (3.79) implies that

$$\| \left(\Pi_{i=\delta}^1 H_{p_i(x)} \right) x \|_v \leq \rho \alpha^{\delta} \|x\|_v \quad \forall x \in \mathbf{R}^n \quad \delta = 1, 2, \cdots \tag{3.80}$$

where $\rho = \gamma^{(l-1)}$, and $\alpha = \beta^{\frac{1}{l}}$. Note that $\alpha < 1$, and both α and ρ do not depend on x. This observation leads to the following result.

Corollary 3.48. *If $K = \{H_i\}_{i \in M}$ is contractive w.r.t. $\| \cdot \|_v$, then K is switched convergent.*
Proof. It is straightforward. $\square$

Although the pre-contractiveness is norm-independent, the contractiveness is norm-dependent in general. For example, the matrix $H = \begin{bmatrix} 0 & 1 \\ 0 & 0 \end{bmatrix}$ is contractive w.r.t. the norm $\|x\|_Q = \left(x^T Q x \right)^{\frac{1}{2}}$ with $Q = \begin{bmatrix} \frac{1}{2} & 0 \\ 0 & 1 \end{bmatrix}$, but is not contractive w.r.t. the Euclidean norm $\| \cdot \|_2$. However, for some matrix sets, the contractiveness w.r.t. one norm implies the contractiveness w.r.t. another related norm, as constructed in the next theorem.

Theorem 3.49. *Suppose that $K = \{H_i\}_{i \in M}$ is contractive to $\|\cdot\|$. Let $\{\omega_i\}_{i=1}^n$ be a basis of $\mathbf{R}^n$ such that*

$$H_j \omega_i = \begin{cases} \lambda_j \omega_i & i = j \\ 0 & i \neq j \end{cases}$$

for some $\lambda_i \in \mathbf{R}$, $i = 1, \cdots, n$. *For each* $x \in \mathbf{R}^n$ *with* $x = \sum_{i=1}^{n} x_i \omega_i$, *let*

$$\|x\|_* = \sum_{i=1}^{n} |x_i| \|\omega_i\|.$$

Then, $\| \cdot \|_*$ *is a norm on* $\mathbf{R}^n$ *and* K *is contractive w.r.t.* $\| \cdot \|_*$.
Proof. It is straightforward and is therefore omitted. $\square$

Finally, we point out with an example that pre-contractiveness may not imply contractiveness *w.r.t.* any norm.

Example 3.50. Let $K = \{H_1, H_2\}$ with

$$H_1 = \begin{bmatrix} 3 & 0 \\ 0 & 0 \end{bmatrix} \text{ and } H_2 = \begin{bmatrix} 0 & 0 \\ 0 & 4 \end{bmatrix}.$$

As $H_1 H_2 = 0$, K is pre-contractive. Suppose that there is a norm $\| \cdot \|$ on $\mathbf{R}^2$ *w.r.t.* which K is contractive. Let

$$\omega_1 = \begin{bmatrix} 1 \\ 0 \end{bmatrix} \text{ and } \omega_2 = \begin{bmatrix} 0 \\ 1 \end{bmatrix}.$$

Then, by Theorem 3.49,

$$\|x\|_* = |x_1| \|\omega_1\| + |x_2| \|\omega_2\|$$

defines a norm on $\mathbf{R}^2$ and K is contractive *w.r.t.* this norm. Taking any $x = \begin{bmatrix} y \\ 1 \end{bmatrix}$ with $y > 0$, we have either $\|H_1 x\|_* < \|x\|_*$ or $\|H_2 x\|_* < \|x\|_*$. This implies that either $y < \frac{\|\omega_2\|}{2\|\omega_1\|}$ or $y > \frac{3\|\omega_2\|}{\|\omega_1\|}$. This is not true for $y \in [\frac{\|\omega_2\|}{2\|\omega_1\|}, \frac{3\|\omega_2\|}{\|\omega_1\|}]$. This contradiction means that K is not contractive *w.r.t.* any norm.

3.7.2 Algebraic Criteria

For a discrete-time linear time-invariant system, it is well known that the system is stable when its poles are located in the open unit ball of the complex plane. For stabilizability of switched linear systems, we have similar criteria as follows.

Theorem 3.51. *Suppose that the switched linear system (3.76) is consistently stabilizable. Then, there is a* $k \in M$ *such that*

$$|\Pi_{i=1}^{n} \lambda_i(A_k)| \leq 1$$

where $\lambda_i(A)$, $1 \leq i \leq n$ *are the eigenvalues of matrix* A. *Furthermore, if the system is consistently asymptotically stabilizable, then the inequality is strict.*

Proof. The theorem can be proven in a similar manner as in the proof of its continuous-time counterpart (Theorem 3.4) and we shall only summarize the main points here.

According to Definition 3.2, by setting $\varepsilon = 1$, there exist a positive number δ, and switching signal σ, such that $\|x_0\| \leq \delta$ implies that $\|\phi(t; 0, x_0, \sigma)\| \leq 1$ for $t \geq t_0$. Hence, we have

$$\|A_{\sigma(s)} \cdots A_{\sigma(1)} A_{\sigma(0)} x_0\| \leq 1 \quad \forall \, x_0 \in \mathbf{B}_\delta \quad s = 0, 1, \cdots.$$

As a result, all entries of the matrices

$$A_{\sigma(0)}, A_{\sigma(1)} A_{\sigma(0)}, \cdots, A_{\sigma(s)} \cdots A_{\sigma(1)} A_{\sigma(0)}, \cdots$$

must be bounded by $\frac{1}{\delta}$. Suppose that

$$\varrho = \min_{k \in M} \{|\Pi_{i=1}^n \lambda_i(A_k)|\} > 1.$$

Then, we have

$$|\det A_k| = |\Pi_{i=1}^n \lambda_i(A_k)| \geq \varrho \quad k \in M.$$

As a result,

$$|\det A_{\sigma(s)} \cdots A_{\sigma(1)} A_{\sigma(0)}| \geq \varrho^s \to \infty \quad \text{as } s \to \infty.$$

This contradicts the boundedness of entries of the matrices. This establishes the former part of the theorem. The latter part can be proven in a similar way. $\square$

Theorem 3.52. *If a switched linear system is pointwise stabilizable, then there is a $k \in M$ such that*

$$\mathrm{sv}_{min}(A_k) \leq 1$$

where $\mathrm{sv}_{min}(\cdot)$ denotes the smallest singular value. Furthermore, if the system is pointwise asymptotically stabilizable, then the inequality is strict.

Proof. We proceed to prove by contradiction.

Suppose that the minimum singular value of each A_k is greater than 1. This implies that

$$A_k^T A_k > I_n \quad k \in M.$$

As the index M is finite, there is a positive real number ϵ such that

$$A_k^T A_k \geq (1 + \epsilon) I_n \quad k \in M.$$

Take $V(x) = x^T x$. It is easily seen that, for any state trajectory, we have

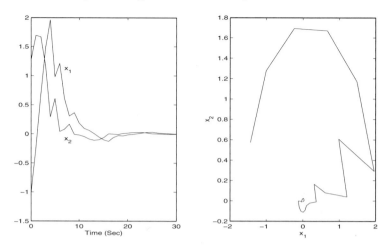

Fig. 3.13. State trajectory and phase portrait of system (3.81)

$$V(x_{k+1}) - V(x_k) > \epsilon V(x_k) \quad k \in M.$$

According to the Lyapunov Theorem, every non-trivial trajectory diverges to infinity *w.r.t.* any switching signal, hence the system is unstable.

The latter part can be proven in the same way. □

As an example, it can be easily verified that the system $\Sigma(A_i)_{\bar{2}}$ with

$$A_1 = \begin{bmatrix} \frac{1}{2} & 0 \\ 0 & \frac{21}{10} \end{bmatrix} \text{ and } A_2 = \frac{21}{20} \begin{bmatrix} \frac{\sqrt{3}}{2} & \frac{1}{2} \\ -\frac{1}{2} & \frac{\sqrt{3}}{2} \end{bmatrix} \tag{3.81}$$

does not satisfy Theorem 3.51, hence it is not consistently stabilizable. However, this system is pointwise asymptotically stabilizable. Indeed, applying A_2 on any state x once means that the state rotates clockwise through $30°$ with the norm increasing by one twentieth, *i.e.*

$$\|A_2 x\| = \frac{21}{20}\|x\|.$$

It can be seen that, by applying A_2 up to five times, any state can be steered into one of the two $30°$ cones centered at the x_1-axis. On the other hand, for any x in either cone, we have

$$\|A_1 x\| \le 0.7271\|x\|.$$

Note that

$$0.7271 \left(\frac{21}{20}\right)^5 \approx 0.93 < 1$$

which implies that

$$\|A_1 A_2^k x\| < \|x\| \quad \forall \, x \in \mathbf{R}^2 \quad k = 0, 1, \cdots, 5.$$

Based on this observation, we propose the following state-feedback switching strategy: If the state is in one of the two $30°$ cones centered at the x_1-axis, activate the first subsystem, otherwise activate the second subsystem. Under this switching law, the switched system is asymptotically stable. Figure 3.13 shows this with the state trajectory and phase portrait initialized at

$$x_0 = [-1.4409, 0.5711]^T.$$

3.7.3 Equivalence Among the Stabilizability Notions

By definition, asymptotic stabilizability implies switched convergence. A question naturally arises: Is the converse true? By means of (3.80), we can not only answer this question, but also go further to prove the following result.

Theorem 3.53. *For the discrete-time switched linear system, the following statements are equivalent:*

(i) the switched system is pointwise asymptotically stabilizable;
(ii) the switched system is pointwise exponentially stabilizable; and
(iii) the switched system is switched convergent.

Proof. It is clear that $(ii) \implies (i) \implies (iii)$. Thus we only need to prove $(iii) \implies (ii)$.

Suppose that the switched system is switched convergent. Following the proof of Theorem 3.46, for any given $x \in \mathbf{R}^n$, and any norm $\|\cdot\|_v$ in $\mathbf{R}^n$, there is an index sequence $p_1(x), p_2(x), \cdots$, such that

$$\left\| \left(\Pi_{i=k}^0 A_{p_i(x)} \right) x \right\|_v \leq \rho \alpha^i \|x\|_v \quad \forall \, x \in \mathbf{R}^n \quad k = 1, 2, \cdots$$

for some positive numbers α and ρ which are independent of x. This clearly implies that the switched system is exponentially stabilizable. $\square$

This theorem shows that switched convergence actually implies (hence is equivalent to) pointwise exponential stabilizability. This is a nice property to have and is quite useful in many situations.

Next, we present equivalent characteristics of consistent stabilizability.

Theorem 3.54. *For the discrete-time switched linear system, the following statements are equivalent:*

(i) the system is consistently asymptotically stabilizable;
(ii) the system is consistently exponentially stabilizable;
(iii) there exist a natural number k, and an index sequence $i_1, \cdots, i_k$, such that matrix $A_{i_k} \cdots A_{i_1}$ is Schur; and
(iv) for any real number $s \in (0, 1)$, there exist a natural number $l = l(s)$, and an index sequence $i_1, \cdots, i_l$, such that

$$\|A_{i_l} \cdots A_{i_1}\| \leq s.$$

Proof. It can be seen that (iii) is equivalent to switched convergence by means of consistent switching signals. By Theorem 3.53, we have $(iv) \implies (iii) \iff (ii) \iff (i)$. Hence we need only to prove that $(iii) \implies (iv)$.

Suppose that (iii) holds. This means that matrix $A_{i_k} \cdots A_{i_1}$ is Schur for some $i_1, \cdots, i_k$. Accordingly, we have

$$(A_{i_k} \cdots A_{i_1})^j \to 0 \quad \text{as } j \to \infty.$$

This clearly implies (iv). $\square$

3.7.4 Robustness Analysis

In this subsection, we address two robustness issues for stabilization of the discrete-time switched linear system. One is where the disturbance depends linearly on the state, and the other is where the disturbance is a nonlinear function of time.

First, suppose that the switched system undergoes small perturbations:

$$x_{k+1} = (A_\sigma + B_\sigma)x_k \tag{3.82}$$

where $B_i \in \mathbf{R}^{n \times n}$ is the structured perturbation of A_i for $i \in M$.

Theorem 3.55. *Suppose that nominal system (3.76) is pointwise asymptotically stabilizable. Then, there is a positive number κ, such that the perturbed system (3.82) is also pointwise asymptotically stabilizable if $\|B_i\| < \kappa$, $i \in M$.*
Proof. By Theorems 3.46 and 3.53, the nominal system is pre-contractive *w.r.t.* any norm on $\mathbf{R}^n$. Following from the proof of Theorem 3.46, we can divide the state space $\mathbf{R}^n$ into a finite number of cones

$$\cup_{i=1}^l \mathcal{W}_i = \mathbf{R}^n$$

where in each cone $\mathcal{W}_i$, there is an index sequence $j_1^i, \cdots, j_{s_i}^i$ such that

$$\|(\Pi_{k=s_i}^1 A_{j_k^i})x\| \le \beta\|x\| \quad \forall\, x \in \mathcal{W}_i \tag{3.83}$$

for some $\beta < 1$. Taking the perturbations into account, we have

$$\left(\Pi_{k=s_i}^1 (A_{j_k^i} + B_{j_k^i})\right) x = \left(\Pi_{k=s_i}^1 A_{j_k^i}\right) x + \vartheta(A_{j_1^i}, B_{j_1^i}, \cdots, A_{j_{s_i}^i}, B_{j_{s_i}^i})x$$

where $\vartheta(\cdot)$ is the summation of all multiplications that involve at least one B_i's. Note that, as $\|B_i\| \to 0$, we have $\|\vartheta(\cdot)\| \to 0$. In view of (3.83), together with the fact that ϑ depends continuously on the B_i's, for a sufficiently small positive real number κ, we have

$$\left\| \left(\Pi_{k=s_i}^1 (A_{j_k^i} + B_{j_k^i})\right) x \right\| < \|x\| \quad \forall\, x \in \mathcal{W}_i \quad \|B_i\| \le \kappa.$$

This means that the perturbed switched system is still pre-contractive w.r.t. $\|\cdot\|$ if $\|B_i\| \leq \kappa$. Again, by Theorems 3.46 and 3.53, the perturbed switched system is pointwise asymptotically stabilizable. □

Next, we turn to the robustness analysis for switched systems subject to nonlinear time-varying perturbations. For this, consider a perturbed switched system given by

$$x_{k+1} = A_\sigma x_k + f_\sigma(k) \tag{3.84}$$

where $f_i \colon \mathbf{N}_+ \mapsto \mathbf{R}^n$, $i \in M$ represent system perturbations or uncertainties.

Theorem 3.56. *For perturbed system (3.84), suppose that the nominal system is consistently asymptotically stabilizable. Then, under any stabilizing periodic switching path, we have*

(a) the system state is bounded if the perturbation is bounded;
(b) the system state is bounded and convergent if the perturbation is bounded and convergent; and
(c) the system state is exponentially convergent if the perturbation is exponentially convergent.

Proof. The proof is similar to the proof of Theorem 3.23 and here we shall only outline the main points.

Suppose that σ is a periodic switching path that asymptotically stabilizes the nominal system. Let T be its period. Then, it can be seen that, there exists a natural number k such that

$$\left\| \Pi_{i=kT-1}^0 A_{\sigma(i)} \right\| < 1.$$

From this we can find positive real numbers α and β with $\alpha < 1$, such that

$$\left\| \Pi_{i=k_2}^{k_1} A_{\sigma(i)} \right\| \leq \beta \alpha^{k_2 - k_1 + 1} \quad k_2 \geq k_1. \tag{3.85}$$

For each state trajectory of the perturbed switched system, we have

$$\|x_{k+1}\| \leq \left\| \Pi_{i=k}^0 A_{\sigma(i)} \right\| \|x_0\| + \sum_{j=0}^{k} \left(\left\| \Pi_{i=k}^{j+1} A_{\sigma(i)} \right\| \|f_{\sigma(j)}(j)\| \right)$$

$$\leq \beta \alpha^{k+1} \|x_0\| + \beta \sum_{j=0}^{k} \left(\alpha^{k-j} \|f_{\sigma(j)}(j)\| \right).$$

This inequality guarantees that the state is bounded for bounded perturbations, convergent for bounded and convergent perturbations, and exponentially convergent for exponentially convergent perturbations. This completes the proof of the theorem. □

The theorem asserts that each stabilizing periodic switching signal for the nominal system is in fact also robust to time-varying nonlinear perturbations.

3.8 Notes and References

As a switched system consists of a number of subsystems and a rule that orchestrates the switching among them, the study of stability for these systems mainly includes two major categories: the definite stability of switched stable systems where all the subsystems are stable, and the stabilizability of switched unstable systems where none of the subsystems are stable. In the literature, much effort has been devoted to establishing tools for stability analysis, such as the Lyapunov approach [17, 1, 100, 112, 123, 23, 61]. In contrast, relatively less attention has been paid to the design of stabilizing switching signals for switched unstable systems [158, 44, 159, 7] (an exception is the extensive study for second-order switched systems, see, e.g., [164, 67, 68, 69]), which is what we have discussed in this chapter.

In Section 3.2, we presented general results and criteria for pointwise stabilizability as well as consistent stabilizability. The main material was taken from [130], and Theorem 3.9 was adopted from [136].

The periodic switching design is based on the average technique. Under the periodic switching signal, the switched system is linear time-varying and hence the theory of time-varying systems is applicable. Accordingly, the robustness properties presented in Theorem 3.23 can be seen as a special case of [79, Lemma 5.2].

The state-feedback switching design is based on an appropriate state-space partition. Switching signal (3.21) was proposed in [158, 159]. Lemma 3.26 was also adopted from there. The proof of the well-posedness is based on the simple idea that 'bounded speed' plus 'positive distance to travel' imply guaranteed dwell time. This idea was also applied to the proofs of Theorems 3.28 and 3.33. A similar idea has been used for switching control of nonholonomic systems [64]. The modified switching strategies in Subsections 3.4.2 and 3.4.3 were proposed in [131]. The combined switching strategies in Section 3.5 were presented in [132].

For discrete-time switched systems, the stabilization and robustness issues can be addressed in an analogical manner. However, there are at least two significant differences between the continuous-time systems and the discrete-time systems. One is that ill-posed phenomenon does not exist in discrete time, thus, we do not need to worry about the well-posedness of discrete-time systems. The other is that there is no discrete-time version of the CBH formula, hence the average approach does not apply to discrete-time systems. Most of the results in Subsection 3.7.1 were taken from [126, 129]. The other parts of Section 3.7 were newly developed using similar ideas from the continuous-time systems.

Besides the switching signals presented in this chapter, there are also several other kinds of switching strategies in the literature. For example, switched stable systems can be stabilized by the (average) dwell-time switching signals [105, 64]. The robustness analysis of the switching signals can be found in [170, 171].

4

Controllability, Observability, and Normal Forms

4.1 Introduction

In this chapter, we address several fundamental issues which reveal intrinsic system properties and pave the way for tackling control synthesis problems.

A fundamental pre-requisite for the design of linear feedback control systems is a good understanding of the structural properties of the linear systems under consideration. These properties are closely related to the concepts of controllability, observability and stability of control systems which are of fundamental importance in the literature of systems and control. However, for controllability and observability analysis of switched linear control systems, a much more difficult situation arises since both the control input and the switching signal are design variables to be determined, and thus the interaction between them must be fully understood.

Here, we present several complete criteria for controllability/observability in the geometric notions. By means of the criteria, we are able to decompose a switched system into a controllable part and an uncontrollable part as well as an observable part and an unobservable part. More elegant normal forms can be obtained using equivalent coordinate and feedback transformations. These normal forms reveal clearly the system structures and pave the way for further investigation of synthesis problems such as feedback stabilization which will be addressed in the next chapter.

When a continuous-time system is connected to a digital device, the overall system can be represented by a single-rate or multi-rate sampled-data system. The sampled-data system sets up a bridge between the continuous-time switched linear system and its discrete-time counterpart. This scheme also enables us to address the continuous-time/discrete-time switched linear systems in a unified framework. Criteria are obtained for sampling without loss of controllability. Several combined digital control with regular switch schemes are developed. These results provide tractable strategies for the practical control of switched linear systems.

In this chapter, we also address several fundamental issues which are closely related to controllability and observability. In particular, we examine several controllability notions from the nonlinear control theory and establish the equivalence among them; we address the controllability with switching and/or input constraints; and we briefly discuss the local controllability and decidability issues.

4.2 Definitions and Preliminaries

4.2.1 Definitions

Consider the switched linear control system given by

$$\delta x(t) = A_\sigma x(t) + B_\sigma u(t)$$
$$y(t) = C_\sigma x(t) \tag{4.1}$$

where $x \in \mathbf{R}^n$ is the state, $u \in \mathbf{R}^p$ is the input, $y \in \mathbf{R}^q$ is the output, $\sigma \in M$ is the switching signal, and δ is the derivative operator in continuous time and the shift forward operator in discrete time.

In the system representation, we do not impose any full rank condition on the input matrices B_k, $k \in M$. This implicitly allows us to consider the case when the column numbers of B_i are not the same. Indeed, suppose that B_k is of $n \times p_k$ for $k \in M$. Let $p = \max\{p_k : k \in M\}$, and expand each B_k to $n \times p$ by adding zero columns when necessary. Then, the expanded system is of the form (4.1).

Let $\phi(t; t_0, x_0, u, \sigma)$ denote the state trajectory at time t of switched system (4.1) starting from $x(t_0) = x_0$ with input u and switching path σ.

In the following definitions of controllability, reachability, etc., we always set $t_0 = 0$. As stated in Section 1.3.5, this assumption does not lose any generality. In addition, the switching signal is assumed to be taken for $\mathcal{S}$, the set of well-defined switching paths over the defined time interval.

Definition 4.1. State $x \in \mathbf{R}^n$ is controllable, if there exist a time instant $t_f > t_0$, a switching path $\sigma: [t_0, t_f] \mapsto M$, and an input $u: [t_0, t_f] \mapsto \mathbf{R}^p$, such that $\phi(t_f; t_0, x, u, \sigma) = 0$. The controllable set of system (4.1), denoted by $\mathcal{C}(C_i, A_i, B_i)_M$ or $\mathcal{C}$ in short, is the set of states which are controllable.

Definition 4.2. System (4.1) is said to be (completely) controllable, if its controllable set is the total state space, $\mathbf{R}^n$.

The reachability counterparts can be defined in the same fashion as follows.

Definition 4.3. State $x \in \mathbf{R}^n$ is reachable, if there exist a time instant $t_f > t_0$, a switching path $\sigma: [t_0, t_f] \mapsto M$, and an input $u: [t_0, t_f] \mapsto \mathbf{R}^p$, such that $\phi(t_f; t_0, 0, u, \sigma) = x$. The reachable set of system (4.1), denoted by $\mathcal{R}(C_i, A_i, B_i)_M$ or $\mathcal{R}$ in short, is the set of states which are reachable.

Definition 4.4. *System (4.1) is said to be* (completely) reachable, *if its reachable set is* $\mathbf{R}^n$.

The concepts of observability and reconstructibility can be defined in a similar manner.

Definition 4.5. *State x is said to be* unobservable, *if for any switching path σ, there exists an input u such that*

$$C_\sigma \phi(t; t_0, x, u, \sigma) = C_\sigma \phi(t; t_0, 0, u, \sigma) \quad \forall \, t \geq t_0.$$

The unobservable set *of system (4.1), denoted by* $\mathcal{UO}(C_i, A_i, B_i)_M$ *or* $\mathcal{UO}$ *in short, is the set of states which are unobservable.*

In other words, the unobservable set includes the initial states which cannot be distinguished from the origin using knowledge of the future output and input.

Definition 4.6. *System (4.1) is said to be* (completely) observable, *if its unobservable set is null.*

Definition 4.7. *State x is said to be* unreconstructible, *if for any switching path σ, there exists an input u such that*

$$C_\sigma \phi(t; t_0, x, u, \sigma) = C_\sigma \phi(t; t_0, 0, u, \sigma) \quad \forall \, t \leq t_0.$$

The unreconstructible set *of system (4.1), denoted by* $\mathcal{UR}(C_i, A_i, B_i)_M$ *or* $\mathcal{UR}$ *in short, is the set of states which are unreconstructible.*

In other words, the unreconstructible set includes the initial states which cannot be identified from the origin using knowledge of the past output and input.

Definition 4.8. *System (4.1) is said to be* (completely) reconstructible, *if its unreconstructible set is null.*

The above controllable/reachable/unobservable/unreconstructible sets are defined on the set of switching paths. In fact, the concepts can be confined to a fixed switching path as follows.

Let $\sigma \colon \mathbf{R} \mapsto M$ be a given switching path.

Definition 4.9. *State $x \in \mathbf{R}^n$ is* controllable via σ, *if there exist a time instant $t_f > t_0$, and an input $u \colon [t_0, t_f] \mapsto \mathbf{R}^p$, such that $\phi(t_f; t_0, x, u, \sigma) = 0$. The* controllable set via σ, *denoted by* $\mathcal{C}_\sigma(C_i, A_i, B_i)_M$ *or* $\mathcal{C}_\sigma$ *in short, is the set of states which are controllable via σ.*

Definition 4.10. *State $x \in \mathbf{R}^n$ is* reachable via σ, *if there exist a time instant $t_f > t_0$, and an input $u \colon [t_0, t_f] \mapsto \mathbf{R}^p$, such that $\phi(t_f; t_0, 0, u, \sigma) = x$. The* reachable set via σ, *denoted by* $\mathcal{R}_\sigma(C_i, A_i, B_i)_M$ *or* $\mathcal{R}_\sigma$ *in short, is the set of states which are reachable via σ.*

Definition 4.11. *State $x \in \mathbf{R}^n$ is unobservable via σ, if there exists an input u, such that*

$$C_\sigma \phi(t; t_0, x, u, \sigma) = C_\sigma \phi(t; t_0, 0, u, \sigma) \quad \forall \ t \geq t_0.$$

The unobservable set *of system, denoted by $\mathcal{UO}_\sigma(C_i, A_i, B_i)_M$ or $\mathcal{UO}_\sigma$ in short, (4.1) is the set of states which are unobservable via σ.*

Definition 4.12. *State x is said to be* unreconstructible *via σ, if there exists an input u, such that*

$$C_\sigma \phi(t; t_0, x, u, \sigma) = C_\sigma \phi(t; t_0, 0, u, \sigma) \quad \forall \ t \leq t_0.$$

The unreconstructible set *via σ of system (4.1), denoted by $\mathcal{UR}_\sigma(C_i, A_i, B_i)_M$ or $\mathcal{UR}_\sigma$ in short, is the set of states which are unreconstructible via σ.*

The concepts of complete controllability via σ, complete reachability via σ, etc., can be defined accordingly in the standard sense.

It can be seen that

$$\mathcal{C}(C_i, A_i, B_i)_M = \cup_{\sigma \in \mathcal{S}} \mathcal{C}_\sigma(C_i, A_i, B_i)_M$$
$$\mathcal{R}(C_i, A_i, B_i)_M = \cup_{\sigma \in \mathcal{S}} \mathcal{R}_\sigma(C_i, A_i, B_i)_M$$

and

$$\mathcal{UO}(C_i, A_i, B_i)_M = \cap_{\sigma \in \mathcal{S}} \mathcal{UO}_\sigma(C_i, A_i, B_i)_M$$
$$\mathcal{UR}(C_i, A_i, B_i)_M = \cap_{\sigma \in \mathcal{S}} \mathcal{UR}_\sigma(C_i, A_i, B_i)_M.$$

4.2.2 Elementary Analysis

We consider the continuous-time switched linear system given by

$$\dot{x}(t) = A_\sigma x(t) + B_\sigma u(t)$$
$$y(t) = C_\sigma x(t). \tag{4.2}$$

Given an initial state $x(0) = x_0$, an input u, and a switching path $\sigma \colon [0, t_f] \mapsto M$, the solution of Equation (4.2) is given by

$$x(t) = \phi(t; t_0, x_0, u, \sigma) = e^{A_{i_k}(t - t_k)} e^{A_{i_{k-1}}(t_k - t_{k-1})} \cdots e^{A_{i_0}(t_1 - t_0)} x_0$$

$$+ e^{A_{i_k}(t - t_k)} \cdots e^{A_{i_1}(t_2 - t_1)} \int_0^{t_1} e^{A_{i_0}(t_1 - \tau)} B_{i_0} u(\tau) d\tau + \cdots$$

$$+ e^{A_{i_k}(t - t_k)} \int_{t_{k-1}}^{t_k} e^{A_{i_{k-1}}(t_k - \tau)} B_{i_{k-1}} u(\tau) d\tau + \int_{t_k}^{t} e^{A_{i_k}(t - \tau)} B_{i_k} u(\tau) d\tau$$

$$t_k < t \leq t_{k+1} \quad 1 \leq k \leq s \tag{4.3}$$

where $\{0, t_1, \cdots, t_s\}$ is the switching time sequence of σ in $[t_0, t_f)$, $\{i_0 = \sigma(0+), \cdots, i_s = \sigma(t_s+)\}$ is the switching index sequence of σ in $[t_0, t_f)$, and $t_{s+1} = t_f$.

It can be seen that the reachable set is

$$\mathcal{R}(C_i, A_i, B_i)_M = \{x = \phi(t; 0, 0, u, \sigma) \colon t \geq 0, u \in U^p, \sigma \in \mathcal{S}_{[0,t]}\}$$
$$= \{e^{A_{i_k} h_k} \cdots e^{A_{i_1} h_1} \int_0^{h_0} e^{A_{i_0} \tau} B_{i_0} u(\tau) d\tau + \cdots + \int_0^{h_k} e^{A_{i_k} \tau} B_{i_k} u(\tau) d\tau \colon$$
$$k \in \mathbf{N}_+, i_j \in M, h_j > 0, u \in U^p\}$$

where U^p is the set of pth-dimensional piecewise continuous vector functions. It can be seen that the set is independent of the output. For clarity, we denote by $\mathcal{R}$ the set $\mathcal{R}(C_i, A_i, B_i)_M$ for system $\Sigma(C_i, A_i, B_i)_M$ without ambiguity.

By Lemma 2.3, for any matrices $A \in \mathbf{R}^{n \times n}$, $B \in \mathbf{R}^{n \times p}$ and $t > 0$, we have

$$\left\{ \int_0^t e^{A\tau} Bu(\tau) d\tau \colon u \in U^p \right\} = \sum_{j=0}^{n-1} A^j \operatorname{Im} B \tag{4.4}$$

where $\operatorname{Im} B$ denotes the image space of B, i.e., $\operatorname{Im} B = \{Bz \colon z \in \mathbf{R}^p\}$.

Denote $\mathcal{B}_k = \operatorname{Im} B_k$, and $\mathcal{D}_k = \sum_{j=0}^{n-1} A_k^j \mathcal{B}_k$ for $k \in M$. It follows from (4.4) that the reachable set can be expressed as

$$\mathcal{R} = \cup_{k=1}^\infty \cup_{i_0, \cdots, i_k \in M} \cup_{h_1, \cdots, h_k > 0} (e^{A_{i_k} h_k} \cdots e^{A_{i_1} h_1} \mathcal{D}_{i_0} + \cdots + \mathcal{D}_{i_k}). \tag{4.5}$$

Similarly, the controllable set is

$$\mathcal{C}(C_i, A_i, B_i)_M = \cup_{k=1}^\infty \cup_{i_0, \cdots, i_k \in M} \cup_{h_0, \cdots, h_k > 0} (e^{-A_{i_0} h_0} \mathcal{D}_{i_0}$$
$$+ \cdots + e^{-A_{i_0} h_0} \cdots e^{-A_{i_k} h_k} \mathcal{D}_{i_k}). \tag{4.6}$$

The set is also independent of the output and is denoted as $\mathcal{C}$ when the context is clear.

Let $\mathbf{A} = \{A_i, i \in M\}$ and $\mathcal{B} = \sum_{i \in M} \operatorname{Im} B_i$. Denote by $\mathcal{V}$ the multiple controllable subspace of $(\mathbf{A}, \mathcal{B})$ (c.f. Section 2.4). That is, $\mathcal{V}$ is the smallest subspace of $\mathbf{R}^n$ that is invariant under each A_i and contains each $\operatorname{Im} B_i$ for $i \in M$. The expression of $\mathcal{V}$ in the system matrices is

$$\mathcal{V} = \sum_{i_0, \cdots, i_{n-1} \in M}^{j_1, \cdots, j_{n-1} \in \underline{\mathbf{n}}} A_{i_{n-1}}^{j_{n-1}} \cdots A_{i_1}^{j_1} \mathcal{B}_{i_0}. \tag{4.7}$$

Recall that for a matrix A and a subspace $\mathcal{W}$, $\Gamma_A \mathcal{W}$ denotes the smallest A-invariant subspace that contains $\mathcal{W}$. It is clear that $e^{At} \operatorname{Im} B \subset \Gamma_A \operatorname{Im} B$ for all $A \in \mathbf{R}^{n \times n}$, $B \in \mathbf{R}^{n \times p}$ and $t \in \mathbf{R}$. This gives

$$\mathcal{R} \subset \cup_{k=1}^\infty \cup_{i_0, \cdots, i_k \in M} (\Gamma_{A_{i_k}} \cdots \Gamma_{A_{i_1}} \mathcal{D}_{i_0} + \cdots + \mathcal{D}_{i_k}) \subset \mathcal{V} \tag{4.8}$$

and

$$\mathcal{C} \subset \cup_{k=1}^\infty \cup_{i_0, \cdots, i_k \in M} (\mathcal{D}_{i_0} + \cdots + \Gamma_{A_{i_0}} \cdots \Gamma_{A_{i_{k-1}}} \mathcal{D}_{i_k}) \subset \mathcal{V}. \tag{4.9}$$

As a simple consequence, we have the following proposition.

Proposition 4.13. *If switched linear system (4.2) is controllable or reachable, then*

$$\mathcal{V} = \mathbf{R}^n.$$

4.2.3 A Heuristic Example

For a switched linear system with two or more subsystems, if the number of switches is subject to certain restrictions, the reachable set and controllable set are not necessarily linear subspaces, and the sets may not be coincident with each other. To see this, let $\mathcal{R}_j$ and $\mathcal{C}_j$ denote the sets of states which are reachable from and controllable to the origin within j switches, respectively. $\mathcal{R}_j$ and $\mathcal{C}_j$ may differ from each other for certain js as illustrated in the following example.

Example 4.14. Consider system (4.2) with $n = 4$, $m = 2$, and

$$A_1 = \begin{bmatrix} 0\,0\,0\,0 \\ 0\,0\,0\,0 \\ 0\,1\,0\,0 \\ 0\,0\,0\,0 \end{bmatrix} \quad B_1 = \begin{bmatrix} 1 \\ 0 \\ 0 \\ 0 \end{bmatrix} \quad A_2 = \begin{bmatrix} 0\,0\,0\,0 \\ 1\,0\,0\,0 \\ 0\,0\,0\,0 \\ 0\,0\,0\,0 \end{bmatrix} \quad B_2 = \begin{bmatrix} 0 \\ 0 \\ 0 \\ 0 \end{bmatrix}. \quad (4.10)$$

It can be calculated that

$$\mathcal{V} = \{ \begin{bmatrix} 1 \\ 0 \\ 0 \\ 0 \end{bmatrix}, \begin{bmatrix} 0 \\ 1 \\ 0 \\ 0 \end{bmatrix}, \begin{bmatrix} 0 \\ 0 \\ 1 \\ 0 \end{bmatrix} \}.$$

Simple computation gives

$$\mathcal{R}_0 = \mathrm{span}\{ \begin{bmatrix} 1 \\ 0 \\ 0 \\ 0 \end{bmatrix} \} \text{ and } \mathcal{R}_1 = \{ \begin{bmatrix} a \\ at \\ 0 \\ 0 \end{bmatrix} : a \in \mathbf{R}, t \geq 0 \}.$$

Note that set $\mathcal{R}_1$ is neither a subspace nor a union of countable subspaces. Further calculation yields

$$\mathcal{R}_2 = \{ \begin{bmatrix} a \\ b \\ bt \\ 0 \end{bmatrix} : a, b \in \mathbf{R}, t \geq 0 \} \text{ and } \mathcal{R}_3 = \{ \begin{bmatrix} a \\ at_3 + b \\ bt_2 \\ 0 \end{bmatrix} : a, b \in \mathbf{R}, t_2, t_3 \geq 0 \}.$$

Sets $\mathcal{R}_2$ and $\mathcal{R}_3$ are strict subsets of $\mathcal{V}$, and $\mathcal{R}_3$ strictly includes $\mathcal{R}_2$ as a subset.

Repeating this process, we have

$$\mathcal{R}_4 = \mathrm{span}\{\begin{bmatrix} 1 \\ 0 \\ 0 \\ 0 \end{bmatrix}, \begin{bmatrix} 0 \\ 1 \\ 0 \\ 0 \end{bmatrix}, \begin{bmatrix} 0 \\ 0 \\ 1 \\ 0 \end{bmatrix}\} = \mathcal{V}.$$

By analogy, the controllable counterparts are given by

$$\mathcal{C}_0 = \mathrm{span}\{\begin{bmatrix} 1 \\ 0 \\ 0 \\ 0 \end{bmatrix}\} \text{ and } \mathcal{C}_1 = \{\begin{bmatrix} a \\ -at \\ 0 \\ 0 \end{bmatrix} : a \in \mathbf{R}, t \geq 0\}$$

$$\mathcal{C}_2 = \{\begin{bmatrix} a \\ b \\ -bt \\ 0 \end{bmatrix} : a, b \in \mathbf{R}, t \geq 0\}$$

$$\mathcal{C}_3 = \{\begin{bmatrix} a \\ -at_2 + b \\ -bt_1 \\ 0 \end{bmatrix} : a, b \in \mathbf{R}, t_1, t_2 \geq 0\}$$

$$\mathcal{C}_4 = \mathrm{span}\{\begin{bmatrix} 1 \\ 0 \\ 0 \\ 0 \end{bmatrix}, \begin{bmatrix} 0 \\ 1 \\ 0 \\ 0 \end{bmatrix}, \begin{bmatrix} 0 \\ 0 \\ 1 \\ 0 \end{bmatrix}\} = \mathcal{V}.$$

To summarize, for system (4.10), we have the following observations:

(i) $\mathcal{C} = \mathcal{R} = \mathcal{V}$;

(ii) not all $\mathcal{R}_j$ and $\mathcal{C}_j$ are subspaces, and $\mathcal{R}_j \neq \mathcal{C}_j$ for $j = 1, 2, 3$; and

(iii) the dimension of $\mathcal{V}$ is three, while it needs four switches to transfer an arbitrarily given state in $\mathcal{V}$ to the origin.

Item (i) reveals that both the controllable set and the reachable set are subspaces, and the two sets are exactly the multiple controllable subspace $\mathcal{V}$. Items (ii) and (iii), however, indicate that complex phenomena may arise when switching between different subsystems occurs. For (ii), the difference is due to incomplete switching which is a unique phenomenon of switched systems. For (iii), it is natural to raise the question:

For a given switched system, what is the least number of switches required to transfer any arbitrarily given state in the controllable set to the origin?

This question is still open for future investigation.

4.2.4 Two Supporting Lemmas

As expressed in (4.3), the state transition matrix for switched system (4.2) is a multiple multiplication of matrix functions of the form e^{At}. Accordingly, properties of exponential matrix functions play an important role in the structural analysis for switched linear systems. In this subsection, a couple of rank properties for exponential matrix functions are presented. These properties are crucial to the derivations of the controllability/observability criteria in Section 4.3.1.

Lemma 4.15. *For any given matrix $A \in \mathbf{R}^{n \times n}$ and subspace $\mathcal{B} \subseteq \mathbf{R}^n$, the following equation holds for almost all $t_1, t_2, \cdots, t_n \in \mathbf{R}$*

$$e^{At_1}\mathcal{B} + e^{At_2}\mathcal{B} + \cdots + e^{At_n}\mathcal{B} = \Gamma_A \mathcal{B}.$$

Proof. Let matrix B be such that $\operatorname{Im} B = \mathcal{B}$, and $\mathcal{Z}$ be the smallest subspace of $\mathbf{R}^n$ that contains the subspaces $e^{At}\mathcal{B}$ for all $t \in \mathbf{R}$. That is, $\mathcal{Z}$ is spanned by the set of vectors

$$\{e^{At}Bz \colon t \in \mathbf{R}, z \in \mathbf{R}^n\}.$$

By Lemma 2.3, $\mathcal{Z}$ is exactly the controllable subspace of matrix pair (A, B):

$$\mathcal{Z} = \operatorname{span}\{e^{At}Bz \colon t \in \mathbf{R}, z \in \mathbf{R}^n\} = \Gamma_A \mathcal{B}.$$

Suppose that $e^{At_j^0}Bz_j$, $j = 1, \cdots, n$, spans subspace $\mathcal{Z}$, i.e.,

$$\mathcal{Z} = \operatorname{span}\{e^{At_1^0}Bz_1, \cdots, e^{At_n^0}Bz_n\}.$$

This implies that

$$e^{At_1^0}\mathcal{B} + \cdots + e^{At_n^0}\mathcal{B} = \Gamma_A \mathcal{B}$$

or equivalently,

$$\operatorname{rank}[e^{At_1^0}B, \cdots, e^{At_n^0}B] = \dim(\Gamma_A \mathcal{B}).$$

Denote integer $r = \dim(\Gamma_A \mathcal{B})$, and define matrix function

$$L(t_1, \cdots, t_n) = [e^{At_1}B, \cdots, e^{At_n}B].$$

Choose a nonsingular submatrix M_0 with maximal rank in $L(t_1^0, \cdots, t_n^0)$. Therefore, M_0 is nonsingular and $\operatorname{rank}M_0 = \operatorname{rank}L(t_1^0, \cdots, t_n^0) = r$. Denote the corresponding submatrix of $L(t_1, \cdots, t_n)$ as $M(t_1, \cdots, t_n)$, and its determinant as $d(t_1, \cdots, t_n)$.

Since each entry in matrix $M(t_1, \cdots, t_n)$ is an analytic function of variables $t_1, \cdots, t_n$, $d(t_1, \cdots, t_n)$ is also an analytic function of its arguments. As

$$d(t_1^0, \cdots, t_n^0) \neq 0$$

its zeros form a proper variety of $\mathbf{R}^n$ (*c.f.* Section 2.6). Therefore, the non-regularity of matrix $M(t_1, \cdots, t_n)$ is a generic property. This implies that

$$\mathrm{rank} L(t_1, \cdots, t_n) \geq \mathrm{rank}\, M(t_1, \cdots, t_n) = r$$

for almost all $t_1, \cdots, t_n$. Together with the fact that $\mathcal{Z} \subseteq \Gamma_A \mathcal{B}$, we can conclude that

$$e^{At_1} \mathcal{B} + \cdots + e^{At_n} \mathcal{B} = \Gamma_A \mathcal{B}$$

for almost all $t_1, \cdots, t_n$. $\square$

Lemma 4.16. *For any given matrices $A_k \in \mathbf{R}^{n \times n}$ and $B_k \in \mathbf{R}^{n \times p}$, $k = 1, 2$, inequality*

$$\mathrm{rank}[A_1 e^{A_2 t} B_1, B_2] \geq \mathrm{rank}[A_1 B_1, B_2] \tag{4.11}$$

holds for almost all $t \in \mathbf{R}$.

Proof. Denote matrix function $\Omega(t) = [A_1 e^{A_2 t} B_1, B_2]$. Choose a nonsingular submatrix G with maximal rank in $\Omega(0) = [A_1 B_1, B_2]$. Denote the corresponding submatrix of $\Omega(t)$ as $\Delta(t)$, and its determinant as $\varsigma(t)$. It is standard that all elements of $\Delta(t)$ are linear combinations of the form $t^k e^{\lambda t}$, hence $\varsigma : \mathbf{R} \mapsto \mathbf{R}$ is an analytic function. Because $\varsigma(0) = \det G \neq 0$, the zeros of $\varsigma(t)$ form a proper variety in $\mathbf{R}$. As a result, the property $\varsigma(t) \neq 0$ is generic. Accordingly, for almost all t, $\Delta(t)$ is nonsingular. Therefore,

$$\mathrm{rank} \Omega(t) \geq \mathrm{rank} \Delta(t) = \mathrm{rank} G = \mathrm{rank}[A_1 B_1, B_2]$$

for almost all t. $\square$

Note that inequality (4.11) cannot be substituted by equality, as shown by the following example

$$A_1 = I_3 \quad A_2 = \begin{bmatrix} 0 & 1 & 0 \\ 0 & 0 & 1 \\ 0 & 0 & 0 \end{bmatrix} \text{ and } B_1 = B_2 = \begin{bmatrix} 0 \\ 0 \\ 1 \end{bmatrix} \quad t \neq 0.$$

For this example, it can be verified that

$$\mathrm{rank}[A_1 e^{A_2 t} B_1, B_2] = 2 > 1 = \mathrm{rank}[A_1 B_1, B_2] \quad t \neq 0.$$

4.3 Controllability and Observability in Continuous Time

4.3.1 Controllability and Reachability

In this subsection, we identify the controllable set and the reachable set for switched linear systems.

Theorem 4.17. *For switched linear system (4.2), the reachable set is*

$$\mathcal{R} = \mathcal{V}. \tag{4.12}$$

Proof. Define the nested subspaces

$$\mathcal{V}_1 = \mathcal{D}_1 + \cdots + \mathcal{D}_m$$
$$\mathcal{V}_{j+1} = \Gamma_{A_1}\mathcal{V}_j + \cdots + \Gamma_{A_m}\mathcal{V}_j \quad j = 1, \cdots, n-1.$$

Then, we have $\mathcal{V} = \mathcal{V}_n$.

We are to design a switching path σ such that each state in $\mathcal{V}$ can be reached from the origin via this switching path. That is, $\mathcal{R}_\sigma(C_i, A_i, B_i)_M = \mathbf{R}^n$.

Assume that the switching index sequence of σ is cyclic, *i.e.*,

$$i_0 = 1, i_1 = 2, \cdots, i_{m-1} = m, i_m = 1, i_{m+1} = 2, \cdots, i_{2m-1} = m,$$
$$\cdots, i_{l-m+1} = 1, i_{l-m+2} = 2, \cdots, i_l = m \tag{4.13}$$

where the number l and the switching time sequence $0, t_1, \cdots, t_l$ are to be determined later.

Let $t_f > t_l$. From (4.5), the reachable set at t_f is

$$\mathcal{R}_\sigma(t_f) = e^{A_{i_l}h_l} \cdots e^{A_{i_1}h_1}\mathcal{D}_{i_0} + \cdots + e^{A_{i_l}h_l}\mathcal{D}_{i_{l-1}} + \mathcal{D}_{i_l}$$

where $h_j = t_{j+1} - t_j$, $j = 0, 1, \cdots, l-1$ and $h_l = t_f - t_l$.

Since

$$e^{A_{i_l}h_l} \cdots e^{A_{i_1}h_1}\mathcal{D}_{i_0} + \cdots + e^{A_{i_l}h_l}\mathcal{D}_{i_{l-1}} + \mathcal{D}_{i_l}$$
$$= e^{A_{i_l}h_l}\left(e^{A_{i_{l-1}}h_{l-1}} \cdots e^{A_{i_1}h_1}\mathcal{D}_{i_0} + \cdots + e^{A_{i_{l-1}}h_{l-1}}\mathcal{D}_{i_{l-2}} + \mathcal{D}_{i_{l-1}}\right) + \mathcal{D}_{i_l}$$

it follows from Lemma 4.16 that

$$\dim\left(e^{A_{i_l}h_l} \cdots e^{A_{i_1}h_1}\mathcal{D}_{i_0} + \cdots + e^{A_{i_l}h_l}\mathcal{D}_{i_{l-1}} + \mathcal{D}_{i_l}\right)$$
$$\geq \dim\left(e^{A_{i_{l-1}}h_{l-1}} \cdots e^{A_{i_1}h_1}\mathcal{D}_{i_0} + \cdots + e^{A_{i_{l-1}}h_{l-1}}\mathcal{D}_{i_{l-2}} + \mathcal{D}_{i_{l-1}} + \mathcal{D}_{i_l}\right)$$

for almost all h_l.

By repeatedly applying Lemma 4.16, for almost all $h_l, \cdots, h_{l-m+1}$, we have

$$\dim\left(e^{A_{i_l}h_l} \cdots e^{A_{i_1}h_1}\mathcal{D}_{i_0} + \cdots + e^{A_{i_l}h_l}\mathcal{D}_{i_{l-1}} + \mathcal{D}_{i_l}\right)$$
$$\geq \dim\left(e^{A_{i_{l-1}}h_{l-1}} \cdots e^{A_{i_1}h_1}\mathcal{D}_{i_0} + \cdots + \mathcal{D}_{i_{l-1}} + \mathcal{D}_{i_l}\right)$$
$$\vdots$$
$$\geq \dim\left(e^{A_{i_{\tau_1}}h_{\tau_1}} \cdots e^{A_{i_1}h_1}\mathcal{D}_{i_0} + \cdots + e^{A_{i_{\tau_1}}h_{\tau_1}}\mathcal{D}_{i_{\tau_1-1}} + \mathcal{D}_{i_{\tau_1}} + \cdots + \mathcal{D}_{i_l}\right)$$
$$= \dim\left(e^{A_{i_{\tau_1}}h_{\tau_1}} \cdots e^{A_{i_1}h_1}\mathcal{D}_{i_0} + \cdots + \mathcal{D}_{i_{\tau_1}} + \mathcal{V}_1\right)$$

where $\tau_1 = l - m$.

It follows from Lemma 4.16 that

$$\dim\left(e^{A_{i_{\tau_1}}h_{\tau_1}}\cdots e^{A_{i_1}h_1}\mathcal{D}_{i_0} + \cdots + \mathcal{D}_{i_{\tau_1}} + \mathcal{V}_1\right)$$
$$= \dim\left(e^{A_{i_{\tau_1}}h_{\tau_1}}e^{A_{i_{\tau_1-1}}h_{\tau_1-1}}\left(e^{A_{i_{\tau_1-2}}h_{\tau_1-2}}\cdots e^{A_{i_1}h_1}\mathcal{D}_{i_0} + \cdots\right.\right.$$
$$\left.\left.+ e^{A_{i_{\tau_1-2}}h_{\tau_1-2}}\mathcal{D}_{i_{\tau_1-3}} + \mathcal{D}_{i_{\tau_1-2}}\right) + e^{A_{i_{\tau_1}}h_{\tau_1}}\mathcal{D}_{i_{\tau_1-1}} + \mathcal{D}_{i_{\tau_1}} + \mathcal{V}_1\right)$$
$$\geq \dim\left(e^{A_{i_{\tau_1}}h_{\tau_1}}\left(e^{A_{i_{\tau_1-2}}h_{\tau_1-2}}\cdots e^{A_{i_1}h_1}\mathcal{D}_{i_0} + \cdots + e^{A_{i_{\tau_1-2}}h_{\tau_1-2}}\mathcal{D}_{i_{\tau_1-3}}\right.\right.$$
$$\left.\left.+ \mathcal{D}_{i_{\tau_1-2}}\right) + e^{A_{i_{\tau_1}}h_{\tau_1}}\mathcal{D}_{i_{\tau_1-1}} + \mathcal{D}_{i_{\tau_1}} + \mathcal{V}_1\right)$$
$$= \dim\left(e^{A_{i_{\tau_1}}h_{\tau_1}}\left(e^{A_{i_{\tau_1-2}}h_{\tau_1-2}}\cdots e^{A_{i_1}h_1}\mathcal{D}_{i_0} + \cdots + e^{A_{i_{\tau_1-2}}h_{\tau_1-2}}\mathcal{D}_{i_{\tau_1-3}}\right.\right.$$
$$\left.\left.+ \mathcal{D}_{i_{\tau_1-2}} + \mathcal{D}_{i_{\tau_1-1}}\right) + \mathcal{D}_{i_{\tau_1}} + \mathcal{V}_1\right)$$

for almost all h_{τ_1-1}.

By the same reasoning, we have

$$\dim\left(e^{A_{i_{\tau_1}}h_{\tau_1}}\cdots e^{A_{i_1}h_1}\mathcal{D}_{i_0} + \cdots + \mathcal{D}_{i_{\tau_1}} + \mathcal{V}_1\right)$$
$$\geq \dim\left(e^{A_{i_{\tau_1}}h_{\tau_1}}\left(e^{A_{i_{\tau_1-2}}h_{\tau_1-2}}\cdots e^{A_{i_1}h_1}\mathcal{D}_{i_0} + \cdots + e^{A_{i_{\tau_1-2}}h_{\tau_1-2}}\mathcal{D}_{i_{\tau_1-3}}\right.\right.$$
$$\left.\left.+ \mathcal{D}_{i_{\tau_1-2}} + \mathcal{D}_{i_{\tau_1-1}}\right) + \mathcal{D}_{i_{\tau_1}} + \mathcal{V}_1\right)$$
$$\geq \dim\left(e^{A_{i_{\tau_1}}h_{\tau_1}}\left(e^{A_{i_{\tau_1-3}}h_{\tau_1-3}}\cdots e^{A_{i_1}h_1}\mathcal{D}_{i_0} + \cdots + e^{A_{i_{\tau_1-3}}h_{\tau_1-3}}\mathcal{D}_{i_{\tau_1-4}}\right.\right.$$
$$\left.\left.+ \mathcal{D}_{i_{\tau_1-3}} + \cdots + \mathcal{D}_{i_{\tau_1-1}}\right) + \mathcal{D}_{i_{\tau_1}} + \mathcal{V}_1\right)$$
$$\vdots$$
$$\geq \dim\left(e^{A_{i_{\tau_1}}h_{\tau_1}}\left(e^{A_{i_{\tau_1-m}}h_{\tau_1-m}}\cdots e^{A_{i_1}h_1}\mathcal{D}_{i_0} + \cdots + e^{A_{i_{\tau_1-m}}h_{\tau_1-m}}\mathcal{D}_{i_{\tau_1-m-1}}\right.\right.$$
$$\left.\left.+ \mathcal{D}_{i_{\tau_1-m}} + \cdots + \mathcal{D}_{i_{\tau_1-1}}\right) + \mathcal{D}_{i_{\tau_1}} + \mathcal{V}_1\right)$$
$$= \dim\left(e^{A_{i_{\tau_1}}h_{\tau_1}}e^{A_{i_{\tau_1-m}}h_{\tau_1-m}}\left(e^{A_{i_{\tau_1-m-1}}h_{\tau_1-m-1}}\cdots e^{A_{i_1}h_1}\mathcal{D}_{i_0} + \cdots\right.\right.$$
$$\left.\left.+ \mathcal{D}_{i_{\tau_1-m-1}}\right) + e^{A_{i_{\tau_1}}h_{\tau_1}}\mathcal{V}_1 + \mathcal{D}_{i_{\tau_1}} + \mathcal{V}_1\right)$$

for almost all $h_j, j = \tau_1 - 1, \cdots, \tau_1 - m + 1$.

Continuing with the above process gives

$$\dim\left(e^{A_{i_{\tau_1}}h_{\tau_1}}\cdots e^{A_{i_1}h_1}\mathcal{D}_{i_0} + \cdots + \mathcal{D}_{i_{\tau_1}} + \mathcal{V}_1\right)$$
$$\geq \dim\left(e^{A_{i_{\tau_1}}h_{\tau_1}}e^{A_{i_{\tau_1-m}}h_{\tau_1-m}}\left(e^{A_{i_{\tau_1-m-1}}h_{\tau_1-m-1}}\cdots e^{A_{i_1}h_1}\mathcal{D}_{i_0} + \cdots\right.\right.$$
$$\left.\left.+ \mathcal{D}_{i_{\tau_1-m-1}}\right) + e^{A_{i_{\tau_1}}h_{\tau_1}}\mathcal{V}_1 + \mathcal{D}_{i_{\tau_1}} + \mathcal{V}_1\right)$$
$$\geq \dim\left(e^{A_{i_{\tau_1}}h_{\tau_1}}e^{A_{i_{\tau_1-m}}h_{\tau_1-m}}e^{A_{i_{\tau_1-2m}}h_{\tau_1-2m}}\left(e^{A_{i_{\tau_1-2m-1}}h_{\tau_1-2m-1}}\cdots\right.\right.$$
$$\left.\left.\times e^{A_{i_{\tau_1-2m}}h_{\tau_1-2m}}e^{A_{i_1}h_1}\mathcal{D}_{i_0} + \cdots + \mathcal{D}_{i_{\tau_1-2m-1}}\right)\right.$$
$$\left.+ e^{A_{i_{\tau_1}}h_{\tau_1}}e^{A_{i_{\tau_1-m}}h_{\tau_1-m}}\mathcal{V}_1 + e^{A_{i_{\tau_1}}h_{\tau_1}}\mathcal{V}_1 + \mathcal{D}_{i_{\tau_1}} + \mathcal{V}_1\right)$$
$$\vdots$$
$$\geq \dim\left(e^{A_{i_{\tau_1}}h_{\tau_1}}e^{A_{i_{\tau_1-m}}h_{\tau_1-m}}\cdots e^{A_{i_{\tau_1-nm}}h_{\tau_1-nm}}\left(e^{A_{i_{\tau_1-nm-1}}h_{\tau_1-nm-1}}\cdots\right.\right.$$
$$\left.\left.\times e^{A_{i_1}h_1}\mathcal{D}_{i_0} + \cdots + \mathcal{D}_{i_{\tau_1-nm-1}}\right) + e^{A_{i_{\tau_1}}h_{\tau_1}}\cdots e^{A_{i_{\tau_1-nm+m}}h_{\tau_1-nm+m}}\mathcal{V}_1\right.$$

$$+ \cdots + e^{A_{i_{\tau_1}} h_{\tau_1}} \mathcal{V}_1 + \mathcal{D}_{i_{\tau_1}} + \mathcal{V}_1)$$

$$= \dim(e^{A_{i_{\tau_1}} h_{\tau_1}} e^{A_{i_{\tau_1-m}} h_{\tau_1-m}} \cdots e^{A_{i_{\tau_1-nm}} h_{\tau_1-nm}} (e^{A_{i_{\tau_1-nm-1}} h_{\tau_1-nm-1}} \cdots$$

$$\times e^{A_{i_1} h_1} \mathcal{D}_{i_0} + \cdots + \mathcal{D}_{i_{\tau_1-nm-1}}) + e^{A_{i_{\tau_1}} (h_{\tau_1} + \cdots + h_{\tau_1-nm+m})} \mathcal{V}_1$$

$$+ \cdots + e^{A_{i_{\tau_1}} h_{\tau_1}} \mathcal{V}_1 + \mathcal{D}_{i_{\tau_1}} + \mathcal{V}_1) \tag{4.14}$$

for almost all $h_j, j = \tau_1 - mn + 1, \cdots, \tau_1 - mn + m - 1, \tau_1 - mn + m + 1, \cdots, \tau_1 - mn + 2m - 1, \cdots, \tau_1 - m + 1, \cdots, \tau_1 - 1$. The relationships, $i_j = i_{j+m}, j = 1, 2, \cdots$, are used in the last equation of (4.14).

From Lemma 4.15, we have

$$\Gamma_{A_{i_{\tau_1}}} \mathcal{V}_1 = e^{A_{i_{\tau_1}} h_{\tau_1}} \mathcal{V}_1 + \cdots + e^{A_{i_{\tau_1}} (h_{\tau_1} + h_{\tau_1-m} + \cdots + h_{\tau_1-mn+m})} \mathcal{V}_1 \tag{4.15}$$

for almost all $h_j, j = \tau_1, \tau_1 - m, \cdots, \tau_1 - mn$. Accordingly, we can rewrite (4.14) as

$$\dim (e^{A_{i_{\tau_1}} h_{\tau_1}} \cdots e^{A_{i_1} h_1} \mathcal{D}_{i_0} + \cdots + \mathcal{D}_{i_{\tau_1}} + \mathcal{V}_1)$$

$$\geq \dim(e^{A_{i_{\tau_1}} h_{\tau_1}} \cdots e^{A_{i_{\tau_1-nm}} h_{\tau_1-nm}} (e^{A_{i_{\tau_1-nm-1}} h_{\tau_1-nm-1}} \cdots e^{A_{i_1} h_1} \mathcal{D}_{i_0}$$

$$+ \cdots + \mathcal{D}_{i_{\tau_1-nm-1}}) + \Gamma_{A_{i_{\tau_1}}} \mathcal{V}_1 + \mathcal{D}_{i_{\tau_1}}).$$

Applying Lemma 4.16 once again, for almost all $h_j, j = \tau_1, \tau_1 - m, \cdots, \tau_1 - mn$, we have

$$\dim (e^{A_{i_{\tau_1}} h_{\tau_1}} \cdots e^{A_{i_1} h_1} \mathcal{D}_{i_0} + \cdots + \mathcal{D}_{i_{\tau_1}} + \mathcal{V}_1)$$

$$\geq \dim(e^{A_{i_{\tau_1}} h_{\tau_1}} \cdots e^{A_{i_{\tau_1-nm}} h_{\tau_1-nm}} (e^{A_{i_{\tau_1-nm-1}} h_{\tau_1-nm-1}} \cdots e^{A_{i_1} h_1} \mathcal{D}_{i_0}$$

$$+ \cdots + \mathcal{D}_{i_{\tau_1-nm-1}}) + \Gamma_{A_{i_{\tau_1}}} \mathcal{V}_1 + \mathcal{D}_{i_{\tau_1}})$$

$$\geq \dim(e^{A_{i_{\tau_1-m}} h_{\tau_1-m}} \cdots e^{A_{i_{\tau_1-nm}} h_{\tau_1-nm}} (e^{A_{i_{\tau_1-nm-1}} h_{\tau_1-nm-1}} \cdots e^{A_{i_1} h_1} \mathcal{D}_{i_0}$$

$$+ \cdots + \mathcal{D}_{i_{\tau_1-nm-1}}) + \Gamma_{A_{i_{\tau_1}}} \mathcal{V}_1 + \mathcal{D}_{i_{\tau_1}})$$

$$\vdots \tag{4.16}$$

$$\geq \dim(e^{A_{i_{\tau_1-nm}} h_{\tau_1-nm}} (e^{A_{i_{\tau_1-nm-1}} h_{\tau_1-nm-1}} \cdots e^{A_{i_1} h_1} \mathcal{D}_{i_0} + \cdots$$

$$+ \mathcal{D}_{i_{\tau_1-nm-1}}) + \Gamma_{A_{i_{\tau_1}}} \mathcal{V}_1 + \mathcal{D}_{i_{\tau_1}})$$

$$= \dim(e^{A_{i_{\tau_1-nm}} h_{\tau_1-nm}} e^{A_{i_{\tau_1-nm-1}} h_{\tau_1-nm-1}} \cdots e^{A_{i_1} h_1} \mathcal{D}_{i_0} + \cdots$$

$$+ e^{A_{i_{\tau_1-nm}} h_{\tau_1-nm}} \mathcal{D}_{i_{\tau_1-nm-1}} + \mathcal{D}_{i_{\tau_1-nm}} + \Gamma_{A_{i_{\tau_1}}} \mathcal{V}_1)$$

where the relationship $\mathcal{D}_{i_{\tau_1}} = \mathcal{D}_{i_{\tau_1-mn}}$ is used.

Since each of Equations (4.15) and (4.16) holds for almost all $h_j, j = \tau_1, \tau_1 - m, \cdots, \tau_1 - mn$, almost all choices of $h_j, j = \tau_1, \tau_1 - m, \cdots, \tau_1 - mn$ satisfy (4.15) and (4.16) simultaneously.

Continuing with this process, we can prove that, for almost all $h_j, j = \tau_1 - mn, \cdots, \tau_1 - m^2 n + 1$, we have

$$\dim \left(e^{A_{i_{\tau_1}} h_{\tau_1}} \cdots e^{A_{i_1} h_1} \mathcal{D}_{i_0} + \cdots + \mathcal{D}_{i_{\tau_1}} + \mathcal{V}_1\right)$$

$$\geq \dim(e^{A_{i_{\tau_1 - mn}} h_{\tau_1 - mn}} \cdots e^{A_{i_1} h_1} \mathcal{D}_{i_0} + \cdots + e^{A_{i_{\tau_1 - mn}} h_{\tau_1 - mn}} \mathcal{D}_{i_{\tau_1 - mn - 1}}$$

$$+ \mathcal{D}_{i_{\tau_1 - mn}} + \Gamma_{A_{i_{\tau_1}}} \mathcal{V}_1)$$

$$\vdots$$

$$\geq \dim(e^{A_{i_{\tau_2}} h_{\tau_2}} \cdots e^{A_{i_1} h_1} \mathcal{D}_{i_0} + \cdots + \mathcal{D}_{i_{\tau_2}} + \Gamma_{A_{i_{\tau_1}}} \mathcal{V}_1$$

$$+ \cdots + \Gamma_{A_{i_{\tau_1 - m + 1}}} \mathcal{V}_1)$$

$$= \dim(e^{A_{i_{\tau_2}} h_{\tau_2}} \cdots e^{A_{i_1} h_1} \mathcal{D}_{i_0} + \cdots + \mathcal{D}_{i_{\tau_2}} + \mathcal{V}_2)$$

where $\tau_2 = \tau_1 - m^2 n$.
Proceeding with the above reasonings, we finally have

$$\dim \left(e^{A_{i_l} h_l} \cdots e^{A_{i_1} h_1} \mathcal{D}_{i_0} + \cdots + e^{A_{i_l} h_l} \mathcal{D}_{i_{l-1}} + \mathcal{D}_{i_l}\right)$$

$$\geq \dim(e^{A_{i_{\tau_n}} h_{\tau_n}} \cdots e^{A_{i_1} h_1} \mathcal{D}_{i_0} + \cdots + \mathcal{D}_{i_{\tau_n}} + \mathcal{V}_n) \geq \dim \mathcal{V} \quad (4.17)$$

where $\tau_n = l - \sum_{k=0}^{n-1} m(mn)^k$.
Let $l \geq \sum_{k=0}^{n-1} m(mn)^k - 1$, then from (4.8) and (4.17), it follows that

$$\mathcal{R}_\sigma(t_f) = \mathcal{V}$$

which implies (4.12). $\square$

Theorem 4.18. *For switched linear system (4.2), the controllable set is*

$$\mathcal{C} = \mathcal{V}. \tag{4.18}$$

Proof. The proof is completely parallel to that of Theorem 4.17 and is hence omitted. $\square$

From the above theorems, the controllable set and the reachable set are always identical. Moreover, the set forms a subspace of the state space, which is exactly the smallest A_i-invariant subspace that contains $\sum_{k \in M} \operatorname{Im} B_k$, $\mathcal{V}$. For a linear time-invariant system, the subspace is the controllable subspace of (A, B). This explains the reason why $\mathcal{V}$ is termed the controllable subspace of the switched system.

Corollary 4.19. *For switched linear system (4.2), the following statements are equivalent:*

(i) the system is completely controllable;
(ii) the system is completely reachable; and
(iii) $\mathcal{V} = \mathbf{R}^n$.

Remark 4.20. The geometric criterion *(iii)* is equivalent to the algebraic criterion

$$\text{rank}[B_1, \cdots, B_m, A_1 B_1, \cdots, A_m B_1, \cdots, A_1 B_m, \cdots, A_m B_m,$$
$$A_1^2 B_1, \cdots, A_m A_1 B_1, \cdots, A_1^2 B_m, \cdots, A_m A_1 B_m, \cdots,$$
$$A_1^{n-1} B_1, \cdots, A_m A_1^{n-2} B_1, \cdots, A_1 A_m^{n-2} B_m, \cdots, A_m^{n-1} B_m] = n.$$

These criteria generalize the well-known controllability criteria for linear time-invariant systems (see, *e.g.*, [21] and [77]).

Remark 4.21. Due to Corollary 4.19, we can give an equivalent definition of controllability as follows.

Definition 4.22. *System (4.2) is said to be* (completely) *controllable, if for any states x_0 and x_f, there exist a time instant $t_f > 0$, a switching path $\sigma \colon [0, t_f] \mapsto M$, and an input $u \colon [0, t_f] \mapsto \mathbf{R}^p$, such that $x(t_f; 0, x_0, u, \sigma) = x_f$.*

Remark 4.23. From the proof of Theorem 4.17, it can be seen that reachability can be achieved through one switching path in any finite time. That is, there exists a switching path σ, such that for and any time $T > 0$, and any states x_0 and x_f in $\mathcal{V}$, there exist a time instant $t_f \leq T$, and an input $u \colon [0, t_f] \mapsto \mathbf{R}^p$, such that $x(t_f; 0, x_0, u, \sigma) = x_f$. In particular, we have

$$\mathcal{C}_\sigma(A_i, B_i)_M = \mathcal{R}_\sigma(A_i, B_i)_M = \mathcal{V}$$

which shows that controllability (reachability) can be achieved via a fixed path.

Remark 4.24. The controllable and reachable sets are invariant under different permutations of A_k and B_k for $k \in M$. That is, suppose that both $j_1, \cdots, j_m$ and $l_1, \cdots, l_m$ are permutations of $1, \cdots, m$, then the controllable (reachable) set of system (4.2) coincide with that of the system given by

$$\dot{x}(t) = \bar{A}_\sigma x(t) + \bar{B}_\sigma u(t) \tag{4.19}$$

where $\bar{A}_k = A_{j_k}$, and $\bar{B}_k = B_{l_k}$ for $k \in M$.

Remark 4.25. As discussed in Section 2.4, a basis for $\mathcal{V}$ is of the form

$$\{b_1, A_{i_{1,1}} b_1, A_{i_{k_1,1}} \cdots A_{i_{1,1}} b_1, \cdots,$$
$$b_{n_0}, A_{i_{1,n_0}} b_{n_0}, A_{i_{k_{n_0},n_0}} \cdots A_{i_{1,n_0}} b_{n_0}\} \tag{4.20}$$

where $b_j \in \cup_{j \in M} \text{Im} \, B_j$, $k_j \geq 0$, and $1 \leq i_{l,j} \leq m$ for $l = 1, \cdots, k_j$ and $j = 1, \cdots, n_0$. Since the number of vectors in (4.20) is not more than n, there are at most n different subsystems whose parameters appear in (4.20). That is to say, at most n subsystems contribute to the controllability and reachability. By removing the redundant subsystems from the switched system, we assume $m \leq n$ without loss of generality.

4.3.2 Observability and Reconstructibility

By Definition 4.5, state x is said to be unobservable, if for any switching path σ, there exists an input u such that

$$C_\sigma \phi(t; 0, x, u, \sigma) = C_\sigma \phi(t; 0, 0, u, \sigma) \quad \forall \, t \geq 0.$$

From the expression of the state solution in (4.3), this implies that

$$C_{i_k} e^{A_{i_k}(t-t_k)} \cdots e^{A_{i_0} t_1} x = 0 \quad \forall \, t_k < t \leq t_{k+1}$$

where $\{(0, i_0), (t_1, i_1), \cdots\}$ is the switching sequence of σ. By the arbitrariness of σ, we have

$$C_{i_k} e^{A_{i_k} h_k} \cdots e^{A_{i_0} h_0} x = 0 \quad \forall \, k \in \mathbf{N}_+ \quad h_j > 0 \quad i_j \in M.$$

This is equivalent to

$$C_{i_k} A_{i_k}^{l_k} \cdots A_{i_0}^{l_0} x = 0 \quad \forall \, k \in \mathbf{N}_+ \quad l_j \in \mathbf{N}_+ \quad i_j \in M. \tag{4.21}$$

Define a nested sequence of subspaces by

$$\mathcal{O}_1 = \operatorname{Im} C_1^T + \cdots + \operatorname{Im} C_m^T$$
$$\mathcal{O}_{j+1} = \Gamma_{A_1^T} \mathcal{O}_j + \cdots + \Gamma_{A_m^T} \mathcal{O}_j \quad j = 1, 2, \cdots.$$

Let

$$\mathcal{O} = \mathcal{O}_n = \sum_{i=1}^{\infty} \mathcal{O}_i$$

and

$$\mathcal{U} = \mathcal{O}^{\perp} = \{x' : \, <x, x'> = 0 \quad \forall \, x \in \mathcal{O}\}$$

where $< \cdot, \cdot >$ denotes the standard inner product in $\mathbf{R}^n$.

Theorem 4.26. *For switched linear system (4.2), the unobservable set is the subspace $\mathcal{U}$.*

Proof. From (4.21), we have

$$x \in \mathcal{UO} \iff x \in \cap_{k \in \mathbf{N}_+} \cap_{i_1, \cdots, i_k \in M}^{l_1, \cdots l_k \in \mathbf{N}_+} \operatorname{Ker}(C_{i_k} A_{i_k}^{l_k} \cdots A_{i_0}^{l_0}).$$

This means that

$$x \in \mathcal{UO} \iff x \in \left(\sum_{k \in \mathbf{N}_+} \sum_{i_1, \cdots, i_k \in M}^{l_1, \cdots l_k \in \mathbf{N}_+} \operatorname{Im}(C_{i_k} A_{i_k}^{l_k} \cdots A_{i_0}^{l_0})^T \right)^{\perp} \iff x \in \mathcal{O}^{\perp}.$$

Therefore, we have

$$\mathcal{UO} = \mathcal{O}^{\perp} = \mathcal{U}. \quad \square$$

Theorem 4.27. *For switched linear system (4.2), the unreconstructible set is subspace $\mathcal{U}$.*

Proof. The theorem can be proven in the same way as for Theorem 4.26 and is hence omitted. $\square$

Corollary 4.28. *For switched linear system (4.2), the following statements are equivalent:*

(i)   the system is completely observable;
(ii) the system is completely reconstructible;
(iii) system $\Sigma(A_i^T, C_i^T)_M$ is completely controllable;
(iv) system $\Sigma(A_i^T, C_i^T)_M$ is completely reachable; and
(v) $\mathcal{O} = \mathbf{R}^n$.

The corollary establishes the principle of duality, namely, if we term the system $\Sigma(A_i^T, C_i^T)_M$ as the *dual system* of $\Sigma(C_i, A_i)_M$, then, the complete observability (reconstructibility) of a switched system is equal to the complete reachability (controllability) of its dual system. The principle of duality plays an important role in the analysis and control of switched linear systems.

4.3.3 Path Planning for Controllability

In this subsection, we study the following switching control design problem for switched system (4.2).

Switching Control Design Problem Given any two states x_0 and x_f in the controllable subspace $\mathcal{V}$, find a switching path σ and a control input u to steer the system from x_0 to x_f in a finite time.

Combining the proof of Theorem 4.17 and the geometric approach of linear systems [160], we can formulate a procedure to address this problem as follows.

From the proof of Theorem 4.17, we can find a natural number l, positive real numbers $h_1, \cdots, h_l$, and an index sequence $i_0, \cdots, i_l$, such that Equation (4.17) holds. This, together with (4.8), implies that

$$\mathcal{V} = e^{A_{i_l} h_l} \cdots e^{A_{i_1} h_1} \mathcal{D}_{i_0} + \cdots + e^{A_{i_l} h_l} \mathcal{D}_{i_{l-1}} + \mathcal{D}_{i_l}. \tag{4.22}$$

Fix a positive real number h_0. Define the switching time sequence as

$$t_0 = 0 \quad t_k = t_{k-1} + h_{k-1} \quad k = 1, \cdots, l+1.$$

From Lemma 2.3, for any $k \in M$ and $t > 0$, we have

$$\mathcal{D}_k = \operatorname{Im} W_t^k \tag{4.23}$$

where

$$W_t^k = \int_0^t e^{A_k(t-\tau)} B_k B_k^T e^{A_k^T(t-\tau)} d\tau.$$

Combining (4.22) with (4.23) leads to

$$\mathcal{V} = e^{A_{i_l}h_l} \cdots e^{A_{i_1}h_1} \operatorname{Im} W_{h_0}^{i_0} + \cdots + e^{A_{i_l}h_l} \operatorname{Im} W_{h_{l-1}}^{i_{l-1}} + \operatorname{Im} W_{h_l}^{i_l}. \quad (4.24)$$

The path planning problem is to find, for any initial state x_0 and target state x_f, both from the controllable subspace, a control input u such that

$$x_f = x(t_{l+1}) = e^{A_{i_l}h_l} \cdots e^{A_{i_0}h_0} x_0 + e^{A_{i_l}h_l} \cdots e^{A_{i_1}h_1} \int_0^{t_1} e^{A_{i_0}(t_1-\tau)} B_{i_0} u(\tau)d\tau$$

$$+ \cdots + \int_{t_l}^{t_{l+1}} e^{A_{i_l}(t_{l+1}-\tau)} B_{i_l} u(\tau)d\tau. \quad (4.25)$$

To this end, consider the piecewise continuous control strategy given by

$$u(t) = B_{i_k}^T e^{A_{i_k}^T(t_{k+1}-t)} a_{k+1} \quad t_k \le t < t_{k+1} \quad k = 0, 1, \cdots, l \quad (4.26)$$

where $a_k \in \mathbf{R}^n$, $k = 1, \cdots, l+1$ are constant vector variables to be determined.

Combining (4.25) with (4.26) gives

$$x_f - e^{A_{i_l}h_l} \cdots e^{A_{i_1}h_1} e^{A_{i_0}h_0} x_0$$

$$= e^{A_{i_l}h_l} \cdots e^{A_{i_1}h_1} \int_{t_0}^{t_1} e^{A_1(t_1-\tau)} B_1 B_1^T e^{A_1^T(t_1-\tau)} d\tau a_1$$

$$+ \cdots + \int_{t_l}^{t_{l+1}} e^{A_{i_l}(t_{l+1}-\tau)} B_{i_l} B_{i_l}^T e^{A_{i_l}^T(t_{l+1}-\tau)} d\tau a_{l+1}.$$

This is equivalent to

$$x_f - e^{A_{i_l}h_l} \cdots e^{A_{i_0}h_0} x_0 = [e^{A_{i_l}h_l} \cdots e^{A_{i_1}h_1} W_{h_0}^{i_0}, \cdots, W_{h_l}^{i_l}] a \quad (4.27)$$

where $a = [a_1^T, \cdots, a_{l+1}^T]^T$.

Note that

$$x_f - e^{A_{i_l}h_l} \cdots e^{A_{i_0}h_0} x_0 \in \mathcal{V}.$$

It follows from (4.24) that linear equation (4.27) with unknown a has at least one solution. The solution(s) of Equation (4.27) can be computed by symbolic or numerical softwares.

Suppose that $a_0 = [a_{0,1}^T, \cdots, a_{0,l+1}^T]^T$ is a solution of Equation (4.27). Define the control input as

$$u(t) = B_{i_k}^T e^{A_{i_k}^T(t_{k+1}-t)} a_{0,k+1} \quad t_k \le t < t_{k+1} \quad k = 0, 1, \cdots, l \quad (4.28)$$

and the switching path as

$$\sigma(t) = i_k \quad \text{for } t \in [t_k, t_{k+1}) \quad k = 0, 1, \cdots, l. \quad (4.29)$$

Following the above reasonings, we have

$$x_f = x(t_{l+1}; 0, x_0, u, \sigma).$$

That is, the piecewise continuous control input (4.28) and the switching path (4.29) constitute a solution for the switching control problem of switched system (4.2).

Example 4.29. Consider the switched systems given by

$$A_1 = 0 \quad B_1 = e_1 \quad A_j = e_j e_{j-1}^T \quad B_j = 0 \quad j = 2, \cdots, m \quad m \leq n \quad (4.30)$$

where e_j, $1 \leq j \leq n$ is the unit column vector with the jth entry equal to one.

To compute the controllable subspace $\mathcal{V}$, we follow the procedure presented in Section 2.4.

It can be readily seen that

$$\mathcal{W}_0 = \text{span}\{e_1\}.$$

By searching the independent vectors in

$$\mathcal{W}_1 = \text{span}\{e_1, A_j e_1, j = 1, \cdots, m\}$$

we obtain

$$\mathcal{W}_1 = \text{span}\{e_1, A_2 e_1\} = \text{span}\{e_1, e_2\}.$$

Continuing with this process, we have

$$\mathcal{W}_k = \text{span}\{e_1, \cdots, e_k, A_j e_k \quad j = 1, \cdots, m\} = \text{span}\{e_1, \cdots, e_{k+1}\}$$

for $k = 2, \cdots, m - 1$, and

$$\mathcal{W}_m = \text{span}\{e_1, \cdots, e_m, A_j e_m \quad j = 1, \cdots, m\}$$
$$= \text{span}\{e_1, \cdots, e_m\} = \mathcal{W}_{m-1}.$$

Thus, $\mathcal{V} = \mathcal{W} = \mathcal{W}_{m-1}$. According to Theorems 4.17 and 4.18 , the controllable (reachable) set is

$$\mathcal{R} = \mathcal{C} = \text{span}\{e_1, \cdots, e_m\}$$

which is an m-dimensional subspace. If $m = n$, then the switched system is controllable and reachable.

Next, we address the switching control problem for system (4.30). Following the path planning procedure, we consider the periodic switching index sequence and piecewise continuous input.

Let us choose the switching time sequence to be

$$t_0 = 0 \quad t_1 = 1 \quad t_2 = 2 \quad \cdots .$$

Accordingly, $h_k = h = 1$ for $k = 0, 1, \cdots$. Simple calculation gives

$$e^{A_1 h} = I_n \text{ and } e^{A_j h} = I_n + A_j \quad j = 2, \cdots, m.$$

Let $l = m l_0$ with l_0 to be determined. Under the periodic switching index sequence (4.13), we can compute

$$\dim \left(e^{A_{i_l} h} \cdots e^{A_{i_1} h} \mathcal{D}_{i_0} + \cdots + e^{A_{i_l} h} \mathcal{D}_{i_{l-1}} + \mathcal{D}_{i_l} \right)$$
$$= \dim(e^{A_{i_l} h} \cdots e^{A_2 h} \mathcal{D}_1 + e^{A_{i_l} h} e^{A_{i_{l-m+1}} h} \cdots e^{A_{i_{l-m}} h} \mathcal{D}_1 + \cdots + \mathcal{D}_1)$$
$$= \dim(Q^{l_0} \mathcal{B}_1 + Q^{l_0 - 1} \mathcal{B}_1 + \cdots + \mathcal{B}_1)$$

where

$$Q = e^{A_m h} e^{A_{m-1} h} \cdots e^{A_2 h} = I_n + A_2 + \cdots + A_m.$$

It can be verified that vectors $B_1, QB_1, \cdots, Q^{m-1} B_1$ are linearly independent, and

$$\mathcal{V} = \text{span}\{B_1, QB_1, \cdots, Q^{m-1} B_1\}.$$

Accordingly, we choose $l_0 = m - 1$.

Simple calculation gives

$$W_h^1 = e_1 e_1^T \text{ and } W_t^k = 0 \quad k = 2, \cdots, m.$$

For any given states x_0 and x_f in $\mathcal{V}$, consider equation

$$[Q^{m-1} W_h^1, \cdots, Q W_h^1, W_h^1] a = x_f - Q^{m-1} x_0. \tag{4.31}$$

Let P denote the submatrix of $[Q^{m-1} B_1, \cdots, Q B_1, B_1]$ consisting of the first m rows. It is clear that P is nonsingular. Denote

$$a_0 = [P^{-1}, 0](x_f - Q^{m-1} x_0).$$

A solution of Equation (4.31) is given by

$$a = [a_0(1), 0, \cdots, 0, a_0(2), 0, \cdots, 0, \cdots, a_0(m), 0, \cdots, 0]^T$$

where $a_0(j)$ denotes the jth entry of vector a_0.

4.4 Controllability and Observability in Discrete Time

Consider a discrete-time switched linear control system given by

$$x_{k+1} = A_\sigma x_k + B_\sigma u_k$$
$$y_k = C_k x_k \tag{4.32}$$

where $x_k \in \mathbf{R}^n$ and $u_k \in \mathbf{R}^p$ are the state and the input, respectively, $\sigma \colon \mathbf{N}_+ \mapsto M$ is the switching path to be designed.

It can be calculated that

$$
\begin{aligned}
x_k &= A_{i_{k-1}} \cdots A_{i_0} x_0 + A_{i_{k-1}} \cdots A_{i_1} B_{i_0} u_0 \\
&\quad + \cdots + A_{i_{k-1}} B_{i_{k-2}} u_{k-2} + B_{i_{k-1}} u_{k-1}
\end{aligned}
\tag{4.33}
$$

where $i_j = \sigma(j)$ for $j \in \underline{k}$.

Define

$$
\mathcal{C}(i_0, \cdots, i_k) = (A_{i_k} \cdots A_{i_0})^{-1} (A_{i_k} \cdots A_{i_1} B_{i_0} + \cdots + B_{i_k}).
\tag{4.34}
$$

Let $\mathcal{C}_k$ denote the set of states that can be transferred to the origin within k steps. It can be readily seen that

$$
\mathcal{C}_k = \cup_{i_0, \cdots, i_{k-1} \in M} \mathcal{C}(i_0, \cdots, i_{k-1})
\tag{4.35}
$$

and

$$
\mathcal{C} = \cup_{k=1}^{\infty} \mathcal{C}_k
\tag{4.36}
$$

where $\mathcal{C}$ is the controllable set of system (4.32).

Define

$$
\mathcal{R}(i_0, \cdots, i_k) = A_{i_k} \cdots A_{i_1} B_{i_0} + \cdots + A_{i_k} B_{i_{k-1}} + B_{i_k}.
$$

Let $\mathcal{R}_k$ denote the set of states that are reachable from the origin within k steps. It can be readily seen that

$$
\mathcal{R}_k = \cup_{i_0, \cdots, i_{k-1} \in M} \mathcal{R}(i_0, \cdots, i_{k-1})
$$

and

$$
\mathcal{R} = \cup_{k=1}^{\infty} \mathcal{R}_k = \cup_{k=1}^{\infty} \cup_{i_0, \cdots, i_{k-1} \in M} \mathcal{R}(i_0, \cdots, i_{k-1})
\tag{4.37}
$$

where $\mathcal{R}$ is the reachable set of system (4.32).

4.4.1 General Results

Theorem 4.30. *The switched linear system (4.32) is controllable if and only if there exist an integer $k < \infty$, and $i_0, \cdots, i_k$, such that*

$$
\mathrm{Im}(A_{i_k} \cdots A_{i_1} A_{i_0}) \subseteq \mathcal{R}(i_0, \cdots, i_k).
\tag{4.38}
$$

Proof. From (4.34), (4.35) and (4.36), the controllable set of system (4.32) is given by

$$\mathcal{C} = \cup_{k=1}^{\infty} \cup_{i_0,\cdots,i_{k-1} \in M} ((A_{i_{k-1}} \cdots A_{i_0})^{-1}(A_{i_{k-1}} \cdots A_{i_1}\mathcal{B}_{i_0} + \cdots$$
$$+ A_{i_{k-1}}\mathcal{B}_{i_{k-2}} + \mathcal{B}_{i_{k-1}})).$$

That is, the controllable set can be expressed as a union of countable subspaces of $\mathbf{R}^n$. Because $\mathbf{R}^n$ cannot be expressed as a countable union of lower-dimensional subspaces (Baire's Category Theorem, see Section 2.2), to ensure controllability of system (4.32), we have

$$\mathcal{C}(i_0,\cdots,i_k) = (A_{i_k} \cdots A_{i_0})^{-1}(A_{i_k} \cdots A_{i_1}\mathcal{B}_{i_0} + \cdots + A_{i_k}\mathcal{B}_{i_{k-1}}$$
$$+ \mathcal{B}_{i_k}) = \mathbf{R}^n \tag{4.39}$$

for some $k < \infty$ and $i_0,\cdots,i_k \in M$. That is

$$(A_{i_k} \cdots A_{i_1}\mathcal{B}_{i_0} + \cdots + A_{i_k}\mathcal{B}_{i_{k-1}} + \mathcal{B}_{i_k}) \supseteq \mathrm{Im}(A_{i_k} \cdots A_{i_0}). \quad \square$$

For the reachability of switched linear systems, a similar criterion can be obtained as follows.

Theorem 4.31. *The switched linear system (4.32) is reachable if and only if there exist an integer $k < \infty$, and $i_0,\cdots,i_k \in M$, such that*

$$\mathcal{R}(i_0,\cdots,i_k) = \mathbf{R}^n. \tag{4.40}$$

Proof. From (4.37), the reachable set is the countable union of subspaces $\mathcal{R}(i_0,\cdots,i_{k-1})$. The theorem follows easily from the Baire's Category Theorem. $\square$

In view of the above theorems on reachability and controllability, the following criteria are readily obtained for observability and reconstructibility by using the principle of duality.

Theorem 4.32. *The switched linear system (4.32) is observable if and only if there exist an integer $k < \infty$, and $i_0,\cdots,i_k$, such that*

$$\mathcal{E}_{i_0} + A_{i_0}^T\mathcal{E}_{i_1} + \cdots + A_{i_0}^T \cdots A_{i_{k-1}}^T\mathcal{E}_{i_k} = \mathbf{R}^n \tag{4.41}$$

where $\mathcal{E}_i = \mathrm{Im}\, E_i^T$ for $i = 1,\cdots,m$.

Theorem 4.33. *The switched linear system (4.32) is reconstructible if and only if there exist an integer $k < \infty$, and $i_0,\cdots,i_k$, such that*

$$\mathcal{E}_{i_0} + A_{i_0}^T\mathcal{E}_{i_1} + \cdots + A_{i_0}^T \cdots A_{i_{k-1}}^T\mathcal{E}_{i_k} \supseteq \mathrm{Im}(A_{i_0}^T \cdots A_{i_k}^T). \tag{4.42}$$

Note that the conditions of theorems are not verifiable in general. The proofs do not provide any information on how to find switching paths for controllability, reachability, *etc.* As a result, the scheme is not constructive.

In the remainder of this subsection, we focus on the structure of the controllable and reachable sets. As given in (4.36) and (4.37), both sets are unions of countable subspaces. Unlike in continuous time, the controllable/reachable sets are not subspaces anymore, and the two sets are not identical in general.

Example 4.34. Consider system (4.32) with $n = 4, m = 2$, and

$$
A_1 = \begin{bmatrix} 0\,0\,0\,0 \\ 0\,0\,0\,0 \\ 0\,0\,1\,0 \\ 0\,0\,1\,0 \end{bmatrix} \quad B_1 = \begin{bmatrix} 1 \\ 0 \\ 0 \\ 0 \end{bmatrix} \quad A_2 = \begin{bmatrix} 0\,0\,0\,0 \\ 1\,0\,0\,0 \\ 0\,0\,0\,1 \\ 0\,0\,0\,1 \end{bmatrix} \quad B_2 = \begin{bmatrix} 0 \\ 0 \\ 0 \\ 0 \end{bmatrix}.
$$

Simple calculation gives

$$
\begin{aligned}
\mathcal{V} &= \operatorname{span}\{e_1, e_2\} \\
\mathcal{R} &= \operatorname{span}\{e_1\} \cup \operatorname{span}\{e_2\} \\
\mathcal{C} &= \operatorname{span}\{e_1, e_2, e_3\} \cup \operatorname{span}\{e_1, e_2, e_4\}.
\end{aligned}
$$

Note that neither the controllable set nor the reachable set is a subspace of the total space. Furthermore, $\mathcal{R} \subset \mathcal{V} \subset \mathcal{C}$, where the subset relationships are strict. □

By (4.37), the reachable set is the union of countable subspace

$$
\mathcal{R} = \cup_{k \in \mathbf{N}^+} \cup_{i_0, \cdots, i_{k-1} \in M} \mathcal{R}(i_0, \cdots, i_{k-1}).
$$

Each such subspace is said to be a *component* of the reachable set. A component $\mathcal{W}$ is said to be *maximal*, if there is no other component which strictly contains $\mathcal{W}$ as a subset.

It can be seen that the reachable set is the union of its maximal components. For a switched system $\Sigma(A_i, B_i)_M$, we denote by $c(A_i, B_i)_M$ the integer k such that the reachable set is the union of exactly k maximal components, if such an integer exists. Otherwise, let $c(A_i, B_i)_M = \infty$.

The following example shows that for any given integer $k \leq \infty$, there always exists a switched system such that the reachable set is the union of exactly k maximal components.

Example 4.35. Let $n \geq 3$, k is a given natural number. For each $\theta \in \mathbf{R}$, let

$$
R_2(\theta) = \begin{bmatrix} \cos\theta & -\sin\theta \\ \sin\theta & \cos\theta \end{bmatrix}.
$$

Let J_l denote the $l \times l$ Jordan block

$$
J_l = \begin{bmatrix} 0\,1\,\cdots\,0 \\ \ddots \\ 0\,0\,\cdots\,1 \\ 0\,0\,\cdots\,0 \end{bmatrix}.
$$

Define

$$
A_1 = \begin{cases} \operatorname{diag}(R_2(\frac{2\pi}{k}), J_{n-2}) & \text{if } k \text{ is odd} \\ \operatorname{diag}(R_2(\frac{\pi}{k}), J_{n-2}) & \text{if } k \text{ is even.} \end{cases}
$$

Let

$$A_2 = \mathrm{diag}(0_2, J_{n-2}) \quad B_1 = e_3 \quad B_2 = e_2.$$

Then, we have

$$\mathcal{R} = \begin{cases} \cup_{i=1}^{k} \mathrm{span}\left\{e_3, \begin{bmatrix} R_2(\frac{2i\pi}{k})\hat{e}_2 \\ 0 \end{bmatrix}\right\} & \text{if } k \text{ is } \text{odd} \\[4mm] \cup_{i=1}^{k} \mathrm{span}\left\{e_3, \begin{bmatrix} R_2(\frac{i\pi}{k})\hat{e}_2 \\ 0 \end{bmatrix}\right\} & \text{if } k \text{ is } \text{even} \end{cases}$$

where $\hat{e}_2 = \begin{bmatrix} 0 \\ 1 \end{bmatrix}$.

As each of the k components is maximal, we have $c(A_i, B_i)_{\bar{2}} = k$.

If we let $A_1 = \mathrm{diag}(R_2(1), J_{n-2})$ and keep others unchanged, then, it can be seen that the reachable set is the union of countable maximal components, that is, $c(A_i, B_i)_{\bar{2}} = \infty$.

The component of controllable/unobservable/unreconstructible sets can be discussed in the same manner, and we leave this to the reader as an exercise.

4.4.2 Reversible Systems

System (4.32) is said to be *reversible*, if all matrices A_i, $i = 1, \cdots, m$ are non-singular. As had been proven in [46], any causal discrete-time (input-output) system can be realized by means of a reversible state variable representation. Accordingly, reversible system representation is very general and applicable to a large class of systems.

Let $\mathcal{V}$ denote the minimal subspace which is invariant under A_i for $i \in M$ and contains $\sum_{j=1}^{m} \mathcal{B}_j$.

Since $A^i \mathrm{Im} B \subseteq \Gamma_A \mathrm{Im} B$ for all $A \in \mathbf{R}^{n \times n}$, $B \in \mathbf{R}^{n \times p}$ and $i \geq 0$, we know that the reachable set satisfies

$$\mathcal{R} \subseteq \cup_{k=0}^{\infty} \cup_{i_0, \cdots, i_{k-1} \in M} \left(\Gamma_{A_{i_{k-1}}} \cdots \Gamma_{A_{i_1}} \mathcal{B}_{i_0} + \cdots + \mathcal{B}_{i_{k-1}} \right) \subseteq \mathcal{V}. \quad (4.43)$$

Similarly, the controllable set satisfies

$$\mathcal{C} \subseteq \cup_{k=0}^{\infty} \cup_{i_0, \cdots, i_{k-1} \in M} \left(\Gamma_{A_{i_0}^{-1}} \mathcal{B}_{i_0} + \cdots + \Gamma_{A_{i_0}^{-1}} \cdots \Gamma_{A_{i_{k-1}}^{-1}} \mathcal{B}_{i_{k-1}} \right) \subseteq \mathcal{V}.$$

In what follows, we present verifiable criteria of controllability and reach-ability for reversible switched linear systems. As in continuous time, we prove that the reachable and controllable sets are nothing but subspace $\mathcal{V}$.

Theorem 4.36. *Suppose that the switched linear system (4.32) is reversible. Then, its reachable set is*

$$\mathcal{R} = \mathcal{V}. \quad (4.44)$$

Proof. Let us proceed by contradiction. Suppose that

$$\dim \mathcal{R}(i_0, \cdots, i_k) = \max\{\dim \mathcal{R}(l_0, \cdots, l_j) \colon l_0, \cdots, l_j \in M, j = 0, 1, \cdots\}$$
$$< \dim \mathcal{V}.$$

It follows that, for any arbitrary given integers $l_0, \cdots, l_j$, we have

$$\mathcal{R}(l_0, \cdots, l_j, i_0, \cdots, i_k) = \mathcal{R}(i_0, \cdots, i_k)$$

which implies that

$$(A_{i_k} \cdots A_{i_0})(A_{l_j} \cdots A_{l_1} \mathcal{B}_{l_0}) \subseteq \mathcal{R}(i_0, \cdots, i_k). \tag{4.45}$$

On the other hand, we have

$$\mathcal{V} = \sum_{\substack{i_1, \cdots, i_n \in M}}^{j_1, \cdots, j_n \in \underline{\mathbf{n}}} A_{i_n}^{j_n} \cdots A_{i_1}^{j_1} \mathcal{B}_{i_1}. \tag{4.46}$$

Since j and $l_0, \cdots, l_j$ in (4.45) can arbitrarily take any values, we obtain

$$(A_{i_k} \cdots A_{i_0})\mathcal{V} \subseteq \mathcal{R}(i_0, \cdots, i_k)$$

which is a contradiction because

$$\dim[(A_{i_k} \cdots A_{i_0})\mathcal{V}] = \dim \mathcal{V} > \dim \mathcal{R}(i_0, \cdots, i_k)$$

where the equality follows from the identity $\dim A\mathcal{V} = \dim \mathcal{V}$ for any nonsingular matrix $A \in \mathbf{R}^{n \times n}$ and subspace $\mathcal{V} \subseteq \mathbf{R}^n$.

Accordingly, we have

$$\dim \mathcal{R}(i_0, \cdots, i_k) = \dim \mathcal{V}. \tag{4.47}$$

It follows from (4.43) that

$$\mathcal{R} = \mathcal{V}. \quad \square$$

Theorem 4.37. *Suppose that the switched linear system (4.32) is reversible. Then, its controllable set is*

$$\mathcal{C} = \mathcal{V}. \tag{4.48}$$

Proof. This theorem can be proven following the same argument as in the proof of Theorem 4.36, the details are hence omitted. $\square$

Corollary 4.38. *For a reversible switched linear system, the following statements are equivalent:*

(i) the system is completely controllable;
(ii) the system is completely reachable; and
(iii) $\mathcal{V} = \mathbf{R}^n$.

Proof. The corollary follows directly from Theorems 4.36 and 4.37. $\Box$

Remark 4.39. The criteria are in the same form as those in the continuous-time case. However, it should be noted that the proofs are quite different. In particular, the proof of Theorem 4.36 is proceeded by contradiction and does not provide a constructive procedure to plan a path for controllability.

The observability and reconstructibility criteria can be obtained by the principle of duality.

Theorem 4.40. *For a reversible switched linear system, the following statements are equivalent:*

(i) the system is completely observable;
(ii) the system is completely reconstructible; and
(iii) $\mathcal{O} = \mathbf{R}^n$.

Example 4.41. (Controllability of a multi-rate sampled-data system)
Consider the linear continuous time-invariant system given by

$$
\dot{x} = Ax + Bu(t) = \begin{bmatrix} 0 & -100\pi & 0 & 0 \\ 100\pi & 0 & 0 & 0 \\ 0 & 0 & \frac{3}{2}\pi & 0 \\ 0 & 0 & 0 & \frac{3}{2}\pi \end{bmatrix} x + \begin{bmatrix} 10 & 0 \\ 10 & 0 \\ 0 & 10 \\ 0 & 10 \end{bmatrix} u(t) \quad (4.49)
$$

which can be verified to be controllable.
The corresponding sampled-data system is given by

$$
x_{k+1} = A_T x_k + B_T u_k
$$

where T is the sampling interval, and

$$
x_k = x(kT) \quad u_k = u(kT) \quad A_T = e^{AT} \quad B_T = \int_0^T e^{\tau A} d\tau B.
$$

Suppose that the sampling intervals are chosen to be $T_1 = 0.01$ and $T_2 = 0.015$, respectively. Then, the corresponding matrix pairs (A_{T_1}, B_{T_1}) and (A_{T_2}, B_{T_2}) are

$$
A_{T_1} = \begin{bmatrix} -1 & 0 & 0 & 0 \\ 0 & -1 & 0 & 0 \\ 0 & 0 & -0.5 & -\sin\frac{3}{2}\pi \\ 0 & 0 & \sin\frac{3}{2}\pi & -0.5 \end{bmatrix} \text{ and } B_{T_1} = \begin{bmatrix} 0.1 & 0 \\ 0.1 & 0 \\ 0 & -0.05 - 0.1\sin\frac{3}{2}\pi \\ 0 & -0.05 - 0.1\sin\frac{3}{2}\pi \end{bmatrix}
$$

and

$$A_{T_2} = \begin{bmatrix} 0 & 1 & 0 & 0 \\ -1 & 0 & 0 & 0 \\ 0 & 0 & -1 & 0 \\ 0 & 0 & 0 & -1 \end{bmatrix} \text{ and } B_{T_2} = \begin{bmatrix} 0.15 & 0 \\ -0.15 & 0 \\ 0 & -0.15 \\ 0 & -0.15 \end{bmatrix}$$

respectively. It can be verified that

$$\text{rank}[B_{T_i}, A_{T_i} B_{T_i}, \cdots, A_{T_i}^3 B_{T_i}] = 3 \quad i = 1, 2$$

which show that the corresponding sampled-data systems are not controllable.

Now, we consider the multi-rate sampling of system (4.49) with sampling rate of either T_1 or T_2. A question naturally arises: Does there exist a sampling strategy such that the resultant switched system is controllable? That is, is the switched system (4.32) with $A_i = A_{T_i}$, $B_i = B_{T_i}$, $i = 1, 2$ controllable or not?

Simple computation gives

$$\mathcal{V} \supseteq \text{span}\{B_1, B_2, A_1 B_1, A_2 B_2\} = \mathbf{R}^4.$$

From Corollary 4.38, the controllability follows. In addition, it can be verified that

$$\mathcal{C}(2, 1) = \mathcal{R}(2, 1) = \mathbf{R}^4.$$

Accordingly, the switched system is controllable from (and reachable to) any point within 2 steps by choosing subsystem (A_2, B_2) at the first step and then switching to subsystem (A_1, B_1) at the second step.

This example shows that switching among different sampling rates may avoid singularity caused by inappropriate choice of sampling rates.

4.5 Canonical Decompositions

In this section, we investigate various normal forms of switched linear systems based on the controllability/observability criteria presented in the previous sections. Due to the similarity between the criteria for continuous-time systems and the counterparts for reversible discrete-time systems, we treat the two cases in a unifying framework.

Consider a switched linear control system described by

$$\delta x(t) = A_\sigma x(t) + B_\sigma u(t)$$
$$y(t) = C_\sigma x(t) \tag{4.50}$$

where δ denotes the derivative operator in continuous time and the shift forward operator in discrete time. In the discrete-time case, we assume that the system is reversible, that is, A_i for $i \in M$ are nonsingular matrices.

Suppose that T is a nonsingular $n \times n$ real matrix. By letting $\bar{x} = Tx$, it follows from (4.50) that

$$\delta\bar{x}(t) = TA_\sigma T^{-1}\bar{x}(t) + TB_\sigma u(t)$$
$$y(t) = C_\sigma T^{-1}\bar{x}(t). \tag{4.51}$$

This equation describes the same system dynamics in different bases of the state space. Thus the two systems are equivalent under the coordinate transformation $\bar{x} = Tx$.

For systems which are equivalent, their controllable/observable sets are also connected by the equivalence transformation in a clear manner.

Proposition 4.42. *Denote $\bar{\mathcal{V}}$ and $\bar{\mathcal{U}}$ the controllable set and unobservable set of system (4.51), respectively. Then, we have*

$$\bar{\mathcal{V}} = T\mathcal{V} \text{ and } \bar{\mathcal{U}} = T\mathcal{U}\mathcal{O}.$$

Proof. Simple calculation gives

$$\bar{\mathcal{V}} = \sum_{i_0,\cdots,i_{n-1}\in M}^{j_1,\cdots,j_{n-1}\in\underline{\mathbf{n}}} (TA_{i_{n-1}}T^{-1})^{j_{n-1}}\cdots(TA_{i_1}T^{-1})^{j_1}\operatorname{Im}(TB_{i_0})$$

$$= T\sum_{i_0,\cdots,i_{n-1}\in M}^{j_1,\cdots,j_{n-1}\in\underline{\mathbf{n}}} A_{i_{n-1}}^{j_{n-1}}\cdots A_{i_1}^{j_1}\operatorname{Im}B_{i_0} = T\mathcal{V}.$$

Similarly, suppose that $x \in \bar{\mathcal{U}}$. Then, we have $C_i T^{-1}(TA_j^k T^{-1})^k x = 0$ for all $i, j \in M$ and $k \in \underline{\mathbf{n}}$. This means that $T^{-1}x \in \mathcal{U}\mathcal{O}$. Therefore, $\bar{\mathcal{U}} = T\mathcal{U}\mathcal{O}$. $\square$

As a simple implication, we have the following corollary.

Corollary 4.43. *The properties of complete controllability/observability are invariant under any equivalence transformation.*

4.5.1 General Canonical Forms

By Theorems 4.18 and 4.37, the controllable set $\mathcal{C}$ of system (4.50) is a subspace of $\mathbf{R}^n$. Denote $n_1 = \dim \mathcal{C}$. Let $\gamma_1, \cdots, \gamma_{n_1}$ be a basis set of $\mathcal{C}$. Extend the basis to basis $\{\gamma_i\}_{i=1}^n$ of $\mathbf{R}^n$, and let T be the matrix of transition from the standard basis of $\mathbf{R}^n$ to $[\gamma_1, \cdots, \gamma_n]$. Let

$$\bar{A}_k = TA_k T^{-1} \text{ and } \bar{B}_k = TB_k \quad k \in M.$$

From the fact that $\mathcal{C}$ is invariant under A_k and contains $\operatorname{Im}B_k$ for each $k \in M$, we can prove that the matrices are in the following block form

$$\bar{A}_k = \begin{bmatrix} \bar{A}_{k,1} & \bar{A}_{k,2} \\ 0 & \bar{A}_{k,3} \end{bmatrix} \text{ and } \bar{B}_k = \begin{bmatrix} \bar{B}_{k,1} \\ 0 \end{bmatrix} \quad k \in M \tag{4.52}$$

where $\bar{A}_{k,1} \in \mathbf{R}^{n_1 \times n_1}$ and $\bar{B}_{k,1} \in \mathbf{R}^{n_1 \times p}$.

The above analysis is summarized in the following theorem.

Theorem 4.44. *Switched system* $\Sigma(A_k, B_k)_M$ *is equivalent to* $\Sigma(\bar{A}_k, \bar{B}_k)_M$.
Moreover, switched system $\Sigma(\bar{A}_{k,1}, \bar{B}_{k,1})_M$ *is completely controllable.*

System $\Sigma(\bar{A}_k, \bar{B}_k)_M$ in triangular form (4.52) is said to be in *controllability canonical form.*

By duality, we have the following observability canonical decomposition.

Theorem 4.45. *Switched system* $\Sigma(C_k, A_k)_M$ *is equivalent to* $\Sigma(\tilde{C}_k, \tilde{A}_k)_M$ *in the form*

$$\tilde{A}_k = \begin{bmatrix} \tilde{A}_{k,1} & 0 \\ \tilde{A}_{k,2} & \tilde{A}_{k,3} \end{bmatrix} \text{ and } \tilde{C}_k = \begin{bmatrix} \tilde{C}_{k,1}, 0 \end{bmatrix} \quad k \in M. \tag{4.53}$$

Moreover, switched system $\Sigma(\tilde{C}_{k,1}, \tilde{A}_{k,1})_M$ *is completely observable.*

A system in the above form is said to be in *observability canonical form.*

We can also decompose the system based on both controllability and observability as follows.

Theorem 4.46. *Switched system (4.50) is equivalent to the following system*

$$\sum \left([0 \ \bar{C}_{i2} \ 0 \ \bar{C}_{i4}], \begin{bmatrix} \bar{A}_{i11} & \bar{A}_{i12} & \bar{A}_{i13} & \bar{A}_{i14} \\ 0 & \bar{A}_{i22} & 0 & \bar{A}_{i24} \\ 0 & 0 & \bar{A}_{i33} & \bar{A}_{i34} \\ 0 & 0 & 0 & \bar{A}_{44} \end{bmatrix}, \begin{bmatrix} \bar{B}_{i1} \\ \bar{B}_{i2} \\ 0 \\ 0 \end{bmatrix} \right)_M. \tag{4.54}$$

In addition, switched system $\sum \left(\begin{bmatrix} \bar{A}_{i11} & \bar{A}_{i12} \\ 0 & \bar{A}_{i22} \end{bmatrix}, \begin{bmatrix} \bar{B}_{i1} \\ \bar{B}_{i2} \end{bmatrix} \right)_M$ *is completely controllable, and switched system* $\sum \left([\bar{C}_{i2} \ \bar{C}_{i4}], \begin{bmatrix} \bar{A}_{i22} & \bar{A}_{i24} \\ 0 & \bar{A}_{44} \end{bmatrix} \right)_M$ *is completely observable.*

Proof. Suppose that subspaces $\mathcal{Y}_i$, $i = 1, 2, 3, 4$ are chosen such that

$$\mathcal{Y}_1 = \mathcal{C} \cap \mathcal{UO} \quad \mathcal{C} = \mathcal{Y}_1 \oplus \mathcal{Y}_2 \quad \mathcal{UO} = \mathcal{Y}_1 \oplus \mathcal{Y}_3 \quad \oplus_{i=1}^4 \mathcal{Y}_i = \mathbf{R}^n.$$

Select a basis $\beta_{i1}, \cdots, \beta_{ir_i}$ for each $\mathcal{Y}_i$. Let

$$L = [\beta_{11}, \cdots, \beta_{1r_1}, \beta_{21}, \cdots, \beta_{2r_2}, \beta_{31}, \cdots, \beta_{3r_3}, \beta_{41}, \cdots, \beta_{4r_4}].$$

Let T be the matrix that transforms the standard basis of $\mathbf{R}^n$ to L.

Note that both subspaces $\mathcal{C}$ and $\mathcal{UO}$ are A_k-invariant for all $k \in M$. From this, it can be seen that the subspace $\mathcal{Y}_1$ is also A_k-invariant for all $k \in M$. This, together with Proposition 4.42, implies that

$$T\mathcal{C} = \text{span} \left\{ x = \begin{bmatrix} x^1 \\ 0 \end{bmatrix} : x^1 \in \mathbf{R}^{r_1+r_2} \right\}.$$

Similarly, we can prove that

$$TU\mathcal{O} = \text{span} \left\{ \begin{bmatrix} x^1 \\ 0 \\ x^3 \\ 0 \end{bmatrix} : x^1 \in \mathbf{R}^{r_1}, x^3 \in \mathbf{R}^{r_3} \right\}$$

and the theorem follows. $\square$

The form in (4.54) is said to be the *standard canonical form* of the switched system. It divides the state variables into four parts. The first is controllable but unobservable, the second is both controllable and observable, the third is both uncontrollable and unobservable, and the fourth is observable but uncontrollable. The partition is the same as in the linear time-invariant case (see, *e.g.*, [160]).

4.5.2 Controllable Systems: Single-input Case

In this subsection, we present normal forms for switched linear systems via both state and feedback transformations.

By introducing state feedback

$$u(t) = F_\sigma x(t) + G_\sigma v(t) \quad v \in \mathbf{R}^p \tag{4.55}$$

where G_i is nonsingular for $i \in M$, and v is the new input, the switched system $\Sigma(A_i, B_i)_M$ is transformed to $\Sigma(A_i + B_i F_i, B_i G_i)_M$.

Proposition 4.47. *The controllable subspace is invariant under any state feedback.*

Proof. Let $\mathcal{V}$ and $\bar{\mathcal{V}}$ denote the controllable subspaces of $\Sigma(A_i, B_i)_M$ and $\Sigma(A_i + B_i F_i, B_i G_i)_M$, respectively. As

$$(A_i + B_i F_i G_i)\mathcal{V} \subseteq A_i \mathcal{V} + \text{Im}\, B_i \subseteq \mathcal{V}$$

$\mathcal{V}$ is $(A_i + B_i F_i)$-invariant. This means that

$$\bar{\mathcal{V}} \subseteq \mathcal{V}.$$

On the other hand, system $\Sigma(A_i, B_i)_M$ can be seen as the transformed system from $\Sigma(A_i + B_i F_i, B_i G_i)_M$ via state feedback

$$v(t) = -G_{\sigma(t)}^{-1} F_{\sigma(t)} x(t) + G_{\sigma(t)}^{-1} u(t) \quad \forall\, t \geq t_0.$$

Hence $\mathcal{V} \subseteq \bar{\mathcal{V}}$, and the proposition follows. $\square$

If we implement both coordinate and feedback transformations, the structure of the controllable part can be made simpler than the canonical form presented in Section 4.5.1. To see this, we first focus on the controllable single-input systems as described below.

Definition 4.48. *Switched system* $\Sigma(A_i, B_i)_M$ *is said to be of* multi-input, *if* rank$[B_1, \cdots, B_m] \geq 1$. *The system is said to be of* single-input, *if* rank $B_k = 1$ *for some* $k \in M$ *and* $B_j = 0$ *for* $j \neq k$.

For a single-input system, by possibly re-indexing the subsystems, we can always assume that $B_1 \neq 0$ while $B_j = 0$ for $j \geq 2$.

Suppose that the single-input system $\Sigma(A_i, B_i)_M$ is completely controllable. The system is said to be *reducible*, if some subsystem can be discarded to produce a completely controllable system. Otherwise, the system is *irreducible*. From Remark 4.25, an irreducible system has at most n subsystems, *i.e.*, $m \leq n$. Note that the reducibility concept is consistent with the one defined in Section 2.4.

Let $\mathcal{Z}_0 = \Gamma_{A_1} \operatorname{Im} B_1$. Define recursively the following

$$\mathcal{Z}_j = \sum_{i \in M} \Gamma_{A_i} \mathcal{Z}_{j-1} \quad j = 1, 2, \cdots .$$

Denote $n_k = \dim \mathcal{Z}_k$ for $k \in \mathbf{N}_+$, and let $\rho = \min\{k \colon \mathcal{Z}_k = \mathbf{R}^n\} \leq n - n_0$. The following procedure resembles the one presented in Section 2.4 and can be used to find a basis of $\mathbf{R}^n$.

First, let $\gamma_i = A_1^{i-1} B_1$ for $i = 1, \cdots, n_0$.

Second, we can find a basis $\gamma_1, \cdots, \gamma_{n_1}$ of $\mathcal{Z}_1$ by searching the set

$$\{\gamma_1, \cdots, \gamma_{n_0}, A_k^l \gamma_j, k \in M, l \in \{1, \cdots, n-1\}, j = 1, \cdots, n_0\}$$

from left to right.

Continuing with the process, suppose that we have found a basis

$$\gamma_1, \cdots, \gamma_{n_0}, \cdots \gamma_{n_{l-1}+1}, \cdots, \gamma_{n_i}$$

for $\mathcal{Z}_i$. Then, by searching the set

$$\{\gamma_1, \cdots, \gamma_{n_l}, A_j^l \gamma_k, j \in M, l \in \{1, \cdots, n-1\}, k = n_{i-1}+1, \cdots, n_i\}$$

from left to right for linearly independent column vectors, we can find a basis

$$\gamma_1, \cdots, \gamma_{n_0}, \cdots, \gamma_{n_{i-1}+1}, \cdots, \gamma_{n_i}, \gamma_{n_i+1}, \cdots, \gamma_{n_{i+1}}$$

for $\mathcal{Z}_{i+1}$.

Finally, we can find a basis

$$\{\gamma_1, \cdots, \gamma_{n_0}, \cdots, \gamma_{n_{\rho-1}+1}, \cdots, \gamma_n\}$$

for $\mathbf{R}^n$.

With this procedure, for all $j \geq 2$, we can always express γ_j by $A_{i_j} \gamma_{k_j}$ with unique i_j and k_j. For $l \in \bar{n}$, let $\mathcal{E}_l = \operatorname{span}\{e_1, \cdots, e_l\}$. Denote $Q_1 = [\gamma_1, \cdots, \gamma_n]$.

As $\mathcal{Z}_0$ is A_1-invariant, we denote by A_{11} the restriction of A_1 in $\mathcal{Z}_0$. Suppose that the characteristic polynomial of A_{11} is

$$\det(sI - A_{11}) = s^{n_0} + \alpha_1 s_{n_0-1} + \cdots + \alpha_{n_0-1} s + \alpha_{n_0}.$$

Denote

$$Q_2 = \begin{bmatrix} 1 & \alpha_1 & \cdots & \alpha_{n_0-2} & \alpha_{n_0-1} \\ 0 & 1 & \cdots & \alpha_{n_0-3} & \alpha_{n_0-2} \\ & & \ddots & & \\ 0 & 0 & \cdots & 1 & \alpha_1 \\ 0 & 0 & \cdots & 0 & 1 \end{bmatrix} \quad \text{and } T = Q_1 \operatorname{diag}(Q_2, I_{n-n_0}).$$

Let F_1 be the first row of $T^{-1}A_1 T$. Introduce coordinate transformation $\bar{x} = T^{-1}x$ and state feedback $u(t) = -F_1\bar{x} + v(t)$ when $\sigma(t) = 1$, and denote by $\Sigma(\bar{A}_i, \bar{B}_i)_M$ the transformed system. It is clear that $\bar{B}_1 = e_1$ and the first row of $\bar{A}_1$ is zero. It follows from $\gamma_j = A_{i_j}\gamma_{k_j}$ that

$$T^{-1}A_{i_j}TT^{-1}\gamma_{k_j} = T^{-1}\gamma_j.$$

As $T^{-1}\gamma_j = e_j$, the above equation exactly states that the k_jth column of $\bar{A}_{i_j}$ is e_j. Similarly, from the fact

$$A_i\mathcal{Z}_l \subseteq \mathcal{Z}_{l+1} \quad \forall\, i \in M \;\; l \in \mathbf{N}_+$$

we know that the jth column of $\bar{A}_i$ is in $\mathcal{E}_{n_{l+1}}$. Hence, we arrive at the following conclusion.

Theorem 4.49. *The controllable single-input system $\Sigma(A_i, B_i)_M$ is equivalent, via suitable coordinate and feedback transformations, to normal system $\Sigma(\bar{A}_i, \bar{B}_i)_M$ with*

(i) $\bar{B}_1 = e_1$ and the first row of $\bar{A}_1$ is zero;
(ii) for all $j \geq n_l$ and $i \in M$, the jth column of $\bar{A}_i$ is in $\mathcal{E}_{n_{l+1}}$; and
(iii) for all $j \geq 2$, the k_jth column of $\bar{A}_{i_j}$ is e_j.

The normal system in the theorem is said to be the *single-input controllable normal form* of the controllable system. In particular, when the system degenerates into a linear time-invariant system, the normal form becomes

$$\left(\begin{bmatrix} 0 & 0 & \cdots & 0 & 0 \\ 1 & 0 & \cdots & 0 & 0 \\ & & \ddots & & \\ 0 & 0 & \cdots & 0 & 0 \\ 0 & 0 & \cdots & 1 & 0 \end{bmatrix}, \begin{bmatrix} 1 \\ 0 \\ \vdots \\ 0 \\ 0 \end{bmatrix} \right)$$

which is the standard normal form for controllable single-input systems.

Unlike the linear time-invariant case, normal forms for the controllable single-input switched system are usually not unique. In the following, we list the possible norm forms for second- and third-order switched systems by appropriate classification.

For $n = 2$, a controllable system satisfies either of the two cases:

(i) rank$[B_1, A_1 B_1] = 2$; and
(ii) rank$[B_1, A_1 B_1] = 1$, and rank$[B_1, A_2 B_1] = 2$.

In the former case, the normal form is

$$\bar{A}_1 = \begin{bmatrix} 0 & 0 \\ 1 & 0 \end{bmatrix} \text{ and } \bar{A}_2 = \begin{bmatrix} * & * \\ * & * \end{bmatrix}$$

where '*' stands for a real-valued number whose value cannot be determined by the controllability property. It is clear that subsystem $(\bar{A}_1, \bar{B}_1)$ is in the controllable normal form, and $\bar{A}_2$ is not necessarily in any specific form. In the latter case, the normal form is

$$\bar{A}_1 = \begin{bmatrix} 0 & 0 \\ 0 & * \end{bmatrix} \text{ and } \bar{A}_2 = \begin{bmatrix} 0 & * \\ 1 & * \end{bmatrix}.$$

Similarly, a third-order controllable single-input system with two subsystems falls into one of the five cases:

(a) rank$[B_1, A_1 B_1, A_1^2 B_1] = 3$;
(b) rank$[B_1, A_1 B_1, A_1^2 B_1] = 2$, rank$[B_1, A_1 B_1, A_2 B_1] = 3$;
(c) rank$[B_1, A_1 B_1, A_1^2 B_1] = $ rank$[B_1, A_1 B_1, A_2 B_1] = 2$, and
    rank$[B_1, A_1 B_1, A_2 A_1 B_1] = 3$;
(d) rank$[B_1, A_1 B_1] = 1$, rank$[B_1, A_2 B_1, A_1 A_2 B_1] = 3$; and
(e) rank$[B_1, A_1 B_1] = 1$, rank$[B_1, A_2 B_1, A_1 A_2 B_1] = 2$, and
    rank$[B_1, A_2 B_1, A_2^2 B_1] = 3$.

For case (a), the normal form is

$$\bar{A}_1 = \begin{bmatrix} 0 & 0 & 0 \\ 1 & 0 & 0 \\ 0 & 1 & 0 \end{bmatrix} \text{ and } \bar{A}_2 = \begin{bmatrix} * & * & * \\ * & * & * \\ * & * & * \end{bmatrix}.$$

For case (b), the first two columns of $\bar{A}_1$ and the first column of $\bar{A}_2$ are fixed, the other column of $\bar{A}_1$ is constrained but the other columns of $\bar{A}_2$ are totally unspecified. Hence, the matrices are in form

$$\bar{A}_1 = \begin{bmatrix} 0 & 0 & 0 \\ 1 & 0 & * \\ 0 & 0 & * \end{bmatrix} \text{ and } \bar{A}_2 = \begin{bmatrix} 0 & * & * \\ 0 & * & * \\ 1 & * & * \end{bmatrix}.$$

Cases (c)-(e) can be discussed in the same way, and the normal forms are

$$\bar{A}_1 = \begin{bmatrix} 0 & 0 & 0 \\ 1 & 0 & * \\ 0 & 0 & * \end{bmatrix} \text{ and } \bar{A}_2 = \begin{bmatrix} * & 0 & * \\ * & 0 & * \\ 0 & 1 & * \end{bmatrix}$$

$$\bar{A}_1 = \begin{bmatrix} 0 & 0 & 0 \\ 0 & 0 & * \\ 0 & 1 & * \end{bmatrix} \text{ and } \bar{A}_2 = \begin{bmatrix} 0 & * & * \\ 1 & * & * \\ 0 & * & * \end{bmatrix}$$

$$\bar{A}_1 = \begin{bmatrix} 0 & 0 & 0 \\ 0 & * & * \\ 0 & 0 & * \end{bmatrix} \text{ and } \bar{A}_2 = \begin{bmatrix} 0 & 0 & * \\ 1 & 0 & * \\ 0 & 1 & * \end{bmatrix}$$

respectively.

4.5.3 Feedback Reduction: Multi-input Case

For controllable multi-input switched linear systems, normal forms under co-ordinate and feedback transformations can be obtained using the same method as in the previous subsection. In particular, Theorem 4.49 can be extended to the multi-input case. However, the normal form may look more complex and its system structure may be less clear. Since the system decomposition is mainly used for addressing synthesis problems such as feedback stabilization and regulation, a better way is to change a multi-input problem into a single-input problem, just as in the standard linear system theory.

To turn a multi-input system into a single-input one, we need a non-regular linear state feedback of the form

$$u(t) = Fx + Gv \quad G \in \mathbf{R}^{m \times 1}$$

where the gain matrix G is a column vector. The idea of using non-regular state feedbacks in control system design can be traced back to the work of [65] which showed that a multi–input controllable linear system can always be brought to a single–input controllable linear system via a non-regular static state feedback, thus enabling an easy proof of the pole assignment theorem for the multi–input case. This idea was generalized to address nonlinear systems in [154, 155]. Other implementations of non-regular state feedbacks could be found in the well-known Morgan's problem [103] and feedback linearization [144, 53, 140].

The following lemma is an extension of [59, Lemma 2] from linear systems to switched linear systems.

Lemma 4.50. *Suppose that switched linear system $\Sigma(A_i, B_i)_M$ is controllable. Then, for any nonzero vector $b \in \sum_{k \in M} \mathrm{Im}\, B_k$, there exist index sequences $i_1, \cdots, i_{n-1}$ and $l_1, \cdots, l_{n-1}$ with $l_j \leq j$, and a vector sequence $u_{i_1}, \cdots, u_{i_{n-1}}$ in $\mathbf{R}^p$, such that the vectors defined by*

$$\eta_1 = b \text{ and } \eta_{k+1} = A_{i_k}\eta_{l_k} + B_{i_k}u_{i_k} \quad k = 1, \cdots, n-1 \qquad (4.56)$$

are independent.

Proof. We proceed by induction. As $\eta_1 \neq 0$, it is independent. Suppose that that $k < n - 1$ and $\eta_1, \cdots, \eta_k$ have been constructed according to (4.56) and are independent. Denote by $\mathcal{W}_k$ the linear subspace spanned by $\eta_1, \cdots, \eta_k$. We have to choose i_k, l_k and u_{i_k} such that

$$\eta_{k+1} = A_{i_k}\eta_{l_k} + B_{i_k}u_{i_k} \notin \mathcal{W}_k.$$

If this is not possible, then

$$A_{i_k}\eta_{l_k} + B_{i_k}u \in \mathcal{W}_k \quad \forall\ i_k \in M \quad l_k \in \bar{k} \quad u \in \mathbf{R}^p.$$

Let $u = 0$, we have

$$A_{i_k} \eta_{l_k} \in \mathcal{W}_k \quad \forall\, i_k \in M \ \ l_k \in \bar{k}.$$

In other words, $\mathcal{W}_k$ is A_i-invariant for all $i \in M$. At the same time, $B_{i_k} u \in \mathcal{W}_k$ for all $u \in \mathbf{R}^p$. This means that $\sum_{i \in M} \operatorname{Im} B_i \subseteq \mathcal{W}_k$. Now that $\mathcal{W}_k$ is A_i-invariant and contains $\sum_{i \in M} \operatorname{Im} B_i$, it contains the controllable subspace of system $\Sigma(A_i, B_i)_M$ as a subspace. This is a contradiction because system $\Sigma(A_i, B_i)_M$ is controllable. $\square$

Theorem 4.51. *Any controllable multi-input system can be changed into a controllable single-input system via suitable non-regular state feedback.*
Proof. Choose a nonzero vector b from $\cup_{i \in M} \operatorname{Im} B_i$. By Lemma 4.50, we can construct a basis of $\mathbf{R}^n$ according to (4.56). Let gain matrices F_i satisfy $F_{i_k} \eta_{l_k} = u_{i_k}$ for $k = 1, \cdots, n-1$. Note that the choice of such F_i is always possible since $\{\eta_k\}_{k=1}^n$ are independent. Each η_k can be expressed by

$$\eta_k = (A_{\kappa_j} + B_{\kappa_j} F_{\kappa_j}) \cdots (A_{\kappa_1} + B_{\kappa_1} F_{\kappa_1}) b$$

for some j and $\kappa_l \in M$. This implies that each η_k is in the controllable subspace of system $\Sigma(A_i + B_i F_i, b)_M$. Consequently, system $\Sigma(A_i + B_i F_i, b)_M$ is controllable.

Suppose that $b \in \operatorname{Im} B_j$. Let $b_j = b$ and $b_l = 0$, $l \neq j$. It can be seen that the single-input system $\Sigma(A_i + B_i F_i, b_i)_M$ is controllable. By introducing non-regular state feedback

$$u(t) = \begin{cases} F_j x(t) + G_j v(t) & \text{if } \sigma(t) = j \\ F_{\sigma(t)} x(t) & \text{otherwise} \end{cases}$$

where G_j satisfies $B_j G_j = b$, the original multi-input system $\Sigma(A_i, B_i)_M$ is changed into the single-input system $\Sigma(A_i + B_i F_i, b_i)_M$ which is controllable. $\square$

Example 4.52. Suppose that we have a multi-input system $\Sigma(A_i, B_i)_{\bar{2}}$ with

$$A_1 = \begin{bmatrix} 1 & -3 & 0 & 0 & -1 \\ -2 & 0 & 0 & 0 & 2 \\ 2 & -1 & 0 & 1 & -2 \\ 0 & 0 & 1 & 0 & 0 \\ 2 & -1 & 0 & 0 & -2 \end{bmatrix} \quad B_1 = \begin{bmatrix} 1 & 1 \\ 0 & 2 \\ 0 & 0 \\ 0 & 0 \\ 1 & 0 \end{bmatrix} \quad A_2 = 0 \quad B_2 = \begin{bmatrix} 0 \\ 0 \\ 1 \\ 0 \\ 0 \end{bmatrix}.$$

Let

$$F_1 = \begin{bmatrix} 0 & 0 & 0 & 0 & 0 \\ 1 & 0 & 0 & 0 & 0 \end{bmatrix} \quad G_1 = \begin{bmatrix} 1 \\ 0 \end{bmatrix} \quad F_2 = [1\ 0\ 0\ 0\ 0] \quad G_2 = 0.$$

It can be verified that the single-input system $\Sigma(A_i + B_i F_i, B_i G_i)_{\bar{2}}$ is controllable. By applying the searching procedure as in Section 4.5.2, we have

$$Q_1 = \begin{bmatrix} 1 & 1 & -4 & 0 & 0 \\ 0 & 2 & 0 & 0 & 0 \\ 0 & 0 & 0 & 1 & 0 \\ 0 & 0 & 0 & 0 & 1 \\ 1 & 0 & 0 & 0 & 0 \end{bmatrix} \quad \text{and } Q_2 = I_3.$$

Let $T = (Q_1 \operatorname{diag}(Q_2, I_2))^{-1}$. Let F_1' be the first row of matrix $T(A_1 + B_1 F_1)T^{-1}$, and $F_2' = [0, 0, 0]$. Denote

$$\bar{A}_1 = T(A_1 + B_1 F_1)T^{-1} - T B_1 G_1 F_1' = T A_1 T^{-1} + T B_1 (F_1 T^{-1} - G_1 F_1')$$

and

$$\bar{B}_1 = T B_1 G_1 \quad \bar{A}_2 = T(A_2 + B_2 F_2)T^{-1} \quad \bar{B}_2 = T B_2 G_2.$$

The system $\Sigma(\bar{A}_i, \bar{B}_i)_{\bar{2}}$ is in the normal form (c.f. Theorem 4.49) with

$$\bar{A}_1 = \begin{bmatrix} 0 & 0 & 0 & 0 & 0 \\ 1 & 0 & 0 & 0 & 0 \\ 0 & 1 & 0 & 0 & 0 \\ 0 & 0 & -8 & 0 & 1 \\ 0 & 0 & 0 & 1 & 0 \end{bmatrix} \quad \bar{B}_1 = \begin{bmatrix} 1 \\ 0 \\ 0 \\ 0 \\ 0 \end{bmatrix} \quad \bar{A}_2 = \begin{bmatrix} 0 & 0 & 0 & 0 & 0 \\ 0 & 0 & 0 & 0 & 0 \\ 0 & 0 & 0 & 0 & 0 \\ 1 & 1 & -4 & 0 & 0 \\ 0 & 0 & 0 & 0 & 0 \end{bmatrix} \quad \bar{B}_2 = 0.$$

This normal form is the reduced system from the original system via coordinate and feedback transformations

$$\bar{x} = T x \text{ and } u = (F_\sigma T^{-1} - G_\sigma F_\sigma')x + G_\sigma v.$$

4.6 Sampling and Digital Control

4.6.1 Sampling Without Loss of Controllability

Consider a linear time-invariant system given by

$$\dot{x}(t) = A x(t) + B u(t). \tag{4.57}$$

Under equidistant sampling and piecewise constant control, the corresponding sampled-data system is

$$x_{k+1} = C^\tau x_k + D^\tau u_k \tag{4.58}$$

where τ is the sampling period, and

$$x_k = x(k\tau) \quad u_k = u(k\tau) \quad C^\tau = e^{A\tau} \quad D^\tau = \int_0^\tau \exp(tA)dt B.$$

The problem of sampling without loss of controllability has been addressed extensively during the past thirty years (see, for example, [153] and the references therein). In particular, it is well-known that, the controllability property can be preserved under almost any sampling rate, except for possibly isolated points of the real axis.

Lemma 4.53. *Suppose that system (4.57) is controllable. If $s_i = \delta_i + \omega_i\sqrt{-1}$, $i = 1, \cdots, n$ are the eigenvalues of matrix A, then under the condition*

$$\exp(s_i\tau) \neq \exp(s_l\tau) \quad \forall \; s_i \neq s_l \tag{4.59}$$

or equivalently,

$$\tau \neq \frac{k\pi}{\omega_i - \omega_l} \quad \forall \; s_i \neq s_l \; s.t. \; \delta_i = \delta_l \quad k = \pm 1, \pm 2, \cdots \tag{4.60}$$

the sampled-data system (4.58) is also controllable.

For an uncontrollable system, the above result still holds in the sense of preserving the controllable subspace.

Lemma 4.54. *Let $\mathcal{C}_1$ and $\mathcal{C}_2$ denote the controllable subspaces of systems (4.57) and (4.58), respectively. Then, under the same condition as in Lemma 4.53, we have*

$$\mathcal{C}_1 = \mathcal{C}_2.$$

Proof. It follows from standard linear system theory that, via a proper nonsingular state transformation $z = Px$, system (4.57) can be transformed into

$$\dot{z}(t) = \bar{A}z(t) + \bar{B}u(t)$$

where

$$\bar{A} = PAP^{-1} = \begin{bmatrix} \bar{A}_c & \bar{A}_{12} \\ 0 & \bar{A}_{\bar{c}} \end{bmatrix} \quad \bar{B} = PB = \begin{bmatrix} \bar{B}_c \\ 0 \end{bmatrix}$$

and $(\bar{A}_c, \bar{B}_c)$ is controllable.

By the same state transformation $z_k = Px_k$, $k = 0, 1, \cdots$, the sampled-data system (4.58) is transformed into

$$z_{k+1} = \bar{C}z_k + \bar{D}u_k$$

where

$$\bar{C} = PC^\tau P^{-1} = \begin{bmatrix} \bar{C}_c & \bar{C}_{12} \\ 0 & \bar{C}_{\bar{c}} \end{bmatrix} \text{ and } \bar{D} = PD^\tau = \begin{bmatrix} \bar{D}_c \\ 0 \end{bmatrix}.$$

It is easy to verify that

$$\bar{C}_c = \exp(\bar{A}_c\tau) \text{ and } \bar{D}_c = \int_0^\tau \exp(\bar{A}_c t)dt\bar{B}_c.$$

Therefore, $(\bar{C}_c, \bar{D}_c)$ is the sampled-data system of $(\bar{A}_c, \bar{B}_c)$ with sampling period τ. Since the set of eigenvalues of $\bar{A}_c$ is a subset of the eigenvalue set of A, it follows from Lemma 4.53 that, under the assumption of the lemma, $(\bar{C}_c, \bar{D}_c)$ is also controllable. As a result, we have

$$\dim \Gamma_{C^\tau} \operatorname{Im} D^\tau = \dim \Gamma_{\bar{C}_c} \operatorname{Im} \bar{D}_c = \dim \Gamma_{\bar{A}_c} \operatorname{Im} \bar{B}_c = \dim \Gamma_A \operatorname{Im} B.$$

In view of the relationship that $\Gamma_{C^\tau} \operatorname{Im} D^\tau \subseteq \Gamma_A \operatorname{Im} B$, we have

$$\Gamma_{C^\tau} \operatorname{Im} D^\tau = \Gamma_A \operatorname{Im} B.$$

This establishes the lemma. $\square$

To prove the main result of this subsection, we also need the following supporting lemma.

Lemma 4.55. *Each A-invariant subspace is also $\exp(A\tau)$-invariant for all $\tau \in \mathbf{R}$. Conversely, if τ is selected such that Condition (4.60) holds, then each $\exp(A\tau)$-invariant subspace is also A-invariant.*
Proof. Suppose that $\mathcal{V}$ is an A-invariant subspace, it is readily seen that

$$A^k \mathcal{V} \subseteq \mathcal{V} \quad \forall \ k = 0, 1, \cdots.$$

In view of the power expansion of $\exp(A\tau)$, we have

$$\exp(A\tau)\mathcal{V} \subseteq \sum_{k=0}^{\infty} A^k \mathcal{V} = \mathcal{V}$$

which establishes the former part of the lemma.

To prove the latter part, let $\mathcal{V}$ be $\exp(A\tau)$-invariant. Choose a matrix D with $\mathcal{V} = \operatorname{Im} D$. By definition, $\Gamma_{\exp(A\tau)} \operatorname{Im} D = \mathcal{V}$. Let

$$C = \exp(A\tau) \text{ and } B = \left(\int_0^\tau \exp(tA)dt \right)^{-1} D.$$

It can be seen that (C, D) is the sampled-data system of (A, B) with sampling rate τ. By Lemma 4.54, it follows that

$$\Gamma_A \operatorname{Im} B = \Gamma_C \operatorname{Im} D = \mathcal{V}.$$

Therefore, $\mathcal{V}$ is A-invariant. $\square$

Next, we turn to the continuous-time switched linear system $\Sigma(A_k, B_k)_M$. Sampling all the subsystems (A_k, B_k), $k \in M$, with a (unified) period τ, we have m sampled-data systems (C_k^τ, D_k^τ), $k \in M$. Denote the discrete-time switched linear system

$$x_{k+1} = C_\sigma^\tau x_k + D_\sigma^\tau u_k \tag{4.61}$$

as the sampled-data system of the continuous-time system $\Sigma(A_k, B_k)_M$. It can be readily seen that the sampled-data system corresponds to the original system with synchronous switching path and piecewise constant control input. As a result, the controllable subspace of the original system includes that of the sampled-data system as a subset. The following theorem establishes the fact that the two subspaces are coincident for almost any sampling rate.

Theorem 4.56. *Let $\mathcal{C}_c$ and $\mathcal{C}_s$ denote the controllable subspaces of systems (4.2) and (4.61), respectively. Suppose that $s_{k,l} = \delta_{k,l} + \omega_{k,l}\sqrt{-1}$, $l = 1, \cdots, n$ are the eigenvalues of matrix A_k, then under the condition*

$$\exp(s_{k,l}\tau) \neq \exp(s_{k,i}\tau) \quad \forall\, l, i = 1, \cdots, n \quad k \in M \quad s_{k,l} \neq s_{k,i}$$

or equivalently,

$$\tau \neq \frac{\iota \pi}{\omega_{k,l} - \omega_{k,i}} \quad \forall\, s_{k,l} \neq s_{k,i} \quad s.t. \quad \delta_{k,l} = \delta_{k,i} \quad \iota = \pm 1, \pm 2, \cdots \quad (4.62)$$

we have

$$\mathcal{C}_c = \mathcal{C}_s.$$

Proof. It follows from Lemma 4.54 that

$$\Gamma_{C_k^\tau} \operatorname{Im} D_k^\tau = \Gamma_{A_k} \operatorname{Im} B_k \quad k \in M.$$

Since $\mathcal{C}_s$ is C_k^τ-invariant and contains $\operatorname{Im} D_k^\tau$, we have

$$\mathcal{C}_s \supseteq \sum_{k \in M} \Gamma_{C_k^\tau} \operatorname{Im} D_k^\tau \supseteq \sum_{k \in M} \operatorname{Im} B_k.$$

By Lemma 4.55, $\mathcal{C}_s$ is A_k-invariant for $k \in M$. Combining these facts leads to

$$\mathcal{C}_s \supseteq \Gamma_\Lambda \mathcal{B} = \mathcal{C}_c$$

where $\Lambda = \{A_1, \cdots, A_m\}$ and $\mathcal{B} = \operatorname{Im}[B_1, \cdots, B_m]$. On the other hand, it is obvious that

$$\mathcal{C}_s \subseteq \mathcal{C}_c.$$

Thus, the two controllable subspaces are identical. □

Remark 4.57. Theorem 4.56 sets up a bridge between a continuous-time switched linear system and its discrete-time counterpart. Indeed, any switching/control strategy for the latter can also be applicable to the former in terms of controllability. An advantage of this scheme is that we can achieve controllability via synchronous switching and piecewise constant control, which are more attractive than an 'arbitrary' switching/control law from the implementation point of view. This scheme also enables us to address the continuous-time/discrete-time switched linear systems in a unified framework.

4.6.2 Regular Switching and Digital Control

In this subsection, we discuss the possibility of controlling a continuous-time switched system by means of the cyclic and synchronous switching signals and the piecewise constant control inputs. As discussed in Remark 4.57, this problem can be addressed in the sampled-data framework.

Definition 4.58. *Switching path σ is said to be* cyclic, *if there is a subset* $\{j_1, \cdots, j_s\}$ *of M, such that the switching index sequence is*

$$\{j_1, \cdots, j_s, j_1, \cdots, j_s, \cdots\}.$$

Definition 4.59. *Switching path σ is said to be of* single-rate, *if there is a base rate ω, such that the switching time sequence is*

$$\{\omega, 2\omega, 3\omega, \cdots\}.$$

Note that any single-rate switching path is also a synchronous switching path (*c.f.* Section 3.2.3).

Roughly speaking, a cyclic switching path has a cyclic switching index sequence, and a single-rate switching path has an equidistant switching time sequence. Cyclic and/or single-rate switching paths are interesting from the viewpoint of implementation.

For a discrete-time switched system

$$x_{k+1} = C_\sigma x_k + D_\sigma u_k \tag{4.63}$$

suppose that the system is reversible, that is, C_k is nonsingular for all $k \in M$. Let Ψ denote the ordered set $\{1, 2, \cdots, m\}$, and define

$$\Psi^2 = \Psi \wedge \Psi = \{1, \cdots, m, 1, \cdots, m\}$$
$$\Psi^{i+1} = \Psi^i \wedge \Psi \quad i = 2, 3, \cdots.$$

Let $\mathcal{R}(h_1, \cdots, h_{km}; \Psi^k)$ denote the reachable set from the origin via the switching path with the switching time sequence $\{0, h_1, h_1 + h_2, \cdots, \sum_{j=1}^{km} h_j\}$ and the cyclic index sequence Ψ^k. Simple calculation gives

$$\mathcal{R}(h_1, \cdots, h_{km}; \Psi^k) = C_m^{h_{km}} \cdots C_2^{h_2} \operatorname{Im}[D_1, C_1 D_1, \cdots, C_1^{h_1 - 1} D_1]$$
$$+ C_m^{h_{km}} \cdots C_3^{h_3} \operatorname{Im}[D_2, C_2 D_2, \cdots, C_2^{h_2 - 1} D_2]$$
$$+ \cdots + C_m^{h_{km}} \operatorname{Im}[D_{m-1}, \cdots, C_{m-1}^{h_{k_{m-1}} - 1} D_{m-1}]$$
$$+ \operatorname{Im}[D_m, \cdots, C_m^{h_{km} - 1} D_m].$$

Suppose that

$$\dim \mathcal{R}(\eta_1, \cdots, \eta_{k^* m}; \Psi^{k^*})$$
$$= \max\{\dim \mathcal{R}(h_1, \cdots, h_{km}; \Psi^k): \ k \geq 0, h_j \in \mathbf{N}^+\}. \tag{4.64}$$

The existence of the maximum comes from the fact that all the dimensions are equal to or less than n, a finite number. Denote $\mathcal{R}^* = \mathcal{R}(\eta_1, \cdots, \eta_{k^* m}; \Psi^{k^*})$ and it is obvious that $\mathcal{R}^*$ is a subspace of $\mathcal{C}_d$, the controllable subspace of system (4.63).

Theorem 4.60. $\mathcal{R}^*$ *is exactly the controllable subspace of system (4.63).*
Proof. Consider the subspace

$$\mathcal{R}(\eta_1, \cdots, \eta_{k^*m}, \eta_1, \cdots, \eta_{k^*m}; \Psi^{2k^*}) = (C_m^{\eta_{k^*m}} \cdots C_1^{\eta_1})\mathcal{R}^* + \mathcal{R}^*.$$

It follows from (4.64) that

$$\dim[(C_m^{\eta_{k^*m}} \cdots C_1^{\eta_1})\mathcal{R}^* + \mathcal{R}^*] \leq \dim \mathcal{R}^*$$

which implies that

$$(C_m^{\eta_{k^*m}} \cdots C_1^{\eta_1})\mathcal{R}^* \subseteq \mathcal{R}^*. \tag{4.65}$$

Note that matrices C_k, $k \in M$ are nonsingular. Thus, we have

$$\dim[(C_m^{\eta_{k^*m}} \cdots C_1^{\eta_1})\mathcal{R}^*] = \dim \mathcal{R}^*.$$

This, together with (4.65), yields

$$(C_m^{\eta_{k^*m}} \cdots C_1^{\eta_1})\mathcal{R}^* = \mathcal{R}^*.$$

As a result, for any vector $\nu \notin \mathcal{R}^*$, we have $(C_m^{\eta_{k^*m}} \cdots C_1^{\eta_1})\nu \notin \mathcal{R}^*$.
Let $s_1, \cdots, s_m$ be any positive integers. Simple calculation gives

$$\mathcal{R}(\eta_1, \cdots, \eta_{k^*m}, s_1, \cdots, s_m, \eta_1, \cdots, \eta_{k^*m}; \Psi^{2k^*+m}) =$$
$$(C_m^{\eta_{k^*m}} \cdots C_1^{\eta_1})(C_m^{s_m} \cdots C_1^{s_1})\mathcal{R}^* + \cdots + \mathcal{R}^*.$$

It is not difficult to prove that

$$(C_m^{s_m} \cdots C_1^{s_1})\mathcal{R}^* \subseteq \mathcal{R}^*.$$

Indeed, suppose that

$$(C_m^{s_m} \cdots C_1^{s_1})\mathcal{R}^* \nsubseteq \mathcal{R}^*$$

then, from the above rationale, we have

$$(C_m^{\eta_{k^*m}} \cdots C_1^{\eta_1})(C_m^{s_m} \cdots C_1^{s_1})\mathcal{R}^* \nsubseteq \mathcal{R}^*$$

which contradicts (4.64). By the non-singularity of matrices C_k, we have

$$(C_m^{s_m} \cdots C_1^{s_1})\mathcal{R}^* = \mathcal{R}^* \tag{4.66}$$

for any $s_1, \cdots, s_m$. Accordingly, we have

$$C_m\mathcal{R}^* = C_m(C_m^{s_m} \cdots C_1^{s_1})\mathcal{R}^* = (C_m^{s_m+1} \cdots C_1^{s_1})\mathcal{R}^* = \mathcal{R}^*$$

which implies that $\mathcal{R}^*$ is C_m-invariant. In the same way, we can prove that $\mathcal{R}^*$ is C_k-invariant for all $k = m-1, \cdots, 1$.
    On the other hand, note that

$$\text{Im}\, D_m \subseteq \mathcal{R}^* \quad C_m^{\eta_k * m}\, \text{Im}\, D_{m-1} \subseteq \mathcal{R}^* \quad \cdots \quad C_m^{\eta_k * m} \cdots C_2^{\eta_2}\, \text{Im}\, D_1 \subseteq \mathcal{R}^*$$

which, together with the invariant property of $\mathcal{R}^*$, implies that

$$\text{Im}\, D_k \subseteq \mathcal{R}^* \quad k \in M.$$

The above analysis shows that subspace $\mathcal{R}^*$ is C_k-invariant for $k \in M$ and contains $\sum_{k \in M} \text{Im}\, D_k$ as a subspace. Therefore,

$$\mathcal{R}^* \supseteq \mathcal{C}_d.$$

This establishes the theorem. □

Remark 4.61. In this theorem, the switching index sequence is of the order $\Psi = (1, \cdots, m)$. It is obvious that the theorem still holds for any other order, that is, any permutation of Ψ.

Remark 4.62. Though Theorem 4.60 guarantees that controllability can be achieved via cyclic switching paths, it does not provide any constructive way of finding such a switching signal. A straightforward way is to check one by one, that is, to compute $\mathcal{R}(h_1, \cdots, h_{km}; \Psi^k)$ for any possible $k, h_1, \cdots, h_{km}$ and Ψ. To ensure effectiveness, it is desirable to have a finite number of candidates, or equivalently, that each state can be driven to the origin in finite steps via a cyclic switching signal. This leads naturally to the following question:

For any fixed n and m, is there a natural number η such that for any switched system (4.63) of order n and with m subsystems, it has

$$\cup_{\sum_{j=1}^{km} h_j \leq \eta} \mathcal{R}(h_1, \cdots, h_{km}; \Psi^k) = \mathcal{C}_d \ ?$$

If the answer is positive, then we can find a controllable switching path by exhaustively searching for the finite candidates. However, it seems that the above problem is quite difficult and we leave it open for further investigation.

The above theorem demonstrates that it suffices to exploit cyclic switching for the purpose of controllability. A question naturally arises: Is it possible to use more regular switching, for example, cyclic and single-rate switching, to achieve controllability? The next example gives a negative answer to this question.

Example 4.63. (Discrete-time system loses controllability under cyclic and single-rate switching)
Consider the switched system (4.63) with $n = 4, m = 2$, and

$$C_1 = \begin{bmatrix} 1\,0\,0\,0 \\ 0\,0\,0\,1 \\ 0\,1\,0\,0 \\ 0\,0\,1\,0 \end{bmatrix} \quad D_1 = \begin{bmatrix} 1 \\ 0 \\ 0 \\ 0 \end{bmatrix} \quad C_2 = \begin{bmatrix} 0\,1\,0\,0 \\ 1\,0\,0\,0 \\ 0\,0\,0\,1 \\ 0\,0\,1\,0 \end{bmatrix} \quad D_2 = \begin{bmatrix} 0 \\ 0 \\ 0 \\ 0 \end{bmatrix}.$$

It can be verified that

$$\text{rank}\left[D_1, C_2 D_1, C_1 C_2 D_1, C_1^2 C_2 D_1\right] = 4.$$

Therefore, the switched system is controllable.

On the other hand, none of the subsystems are controllable. In addition, if we denote by $\mathcal{R}(h)$ the reachable set via any cyclic switching path with single-rate h, then, we can verify that none of the sets is the total space. Indeed, it can be seen that

$$\mathcal{R}(h) = \sum_{i=0}^{\infty} (C_1^h C_2^h)^i \operatorname{Im} D_1 \cup \sum_{i=0}^{\infty} C_1^h (C_2^h C_1^h)^i \operatorname{Im} D_1$$

$$= \Gamma_{(C_1^h C_2^h)} \operatorname{Im} D_1 \cup C_1^h (\Gamma_{(C_2^h C_1^h)} \operatorname{Im} D_1).$$

Note that

$$C_1^3 = C_2^2 = I_4.$$

Therefore, $\Gamma_{(C_1^h C_2^h)} \operatorname{Im} D_1$ falls into one of the six cases:

$$\Gamma_{(C_1^0 C_2^0)} \operatorname{Im} D_1 \quad \Gamma_{(C_1^0 C_2^1)} \operatorname{Im} D_1 \quad \Gamma_{(C_1^1 C_2^0)} \operatorname{Im} D_1$$
$$\Gamma_{(C_1^2 C_2^0)} \operatorname{Im} D_1 \quad \Gamma_{(C_1^1 C_2^1)} \operatorname{Im} D_1 \quad \Gamma_{(C_1^2 C_2^1)} \operatorname{Im} D_1.$$

Simple computation shows that none of the above subspaces is the total space. Similarly, none of the subspaces $\Gamma_{(C_2^h C_1^h)} \operatorname{Im} D_1$ is the total space. As a result, for any natural number h, the set $\mathcal{R}(h)$ is not the total space and hence the controllability is lost.

The above example also excludes the possibility of achieving controllability via any cyclic switching path with a periodic time sequence. That is,

$$\mathcal{R}(h_1, h_2, h_1, h_2, \cdots ; \Psi^l) \neq \mathbf{R}^4$$

for any $h_1, h_2, l \in \mathbf{N}_+$ and Ψ either $\{1, 2\}$ or $\{2, 1\}$.

Next, we turn to the continuous-time case. By combining Theorem 4.56 with Theorem 4.60, we obtain the following theorem.

Theorem 4.64. *For the continuous-time switched linear system, any two states in the controllable subspace can be steered to each other via a cyclic and synchronous switching path and a piecewise constant control law.*

In general, controllability cannot be achieved by an cyclic and single-rate switching path, as illustrated in the following example.

Example 4.65. (Continuous-time system loses controllability under cyclic and single-rate switching)
Consider the continuous-time switched system $\Sigma(A_i, B_i)_{\bar{2}}$ with

$$A_1 = \begin{bmatrix} 0 & 0 & 0 \\ 1 & 0 & 0 \\ 0 & 0 & 0 \end{bmatrix} \text{ and } B_1 = \begin{bmatrix} 1 \\ 0 \\ 0 \end{bmatrix}$$

and

$$A_2 = \begin{bmatrix} 0 & 0 & 0 \\ 0 & 0 & 1 \\ 1 & 0 & 1 \end{bmatrix} \text{ and } B_2 = \begin{bmatrix} 0 \\ 0 \\ 0 \end{bmatrix}.$$

It can be verified that this system is controllable.

On the other hand, it can be calculated that, under a cyclic switching sequence with single duration rate τ, the reachable set is

$$\mathcal{R}(\tau) = \text{span}\{ \begin{bmatrix} 1 \\ 0 \\ 0 \end{bmatrix}, \begin{bmatrix} 0 \\ 1 \\ 1 \end{bmatrix} \}$$

which is a strictly proper subspace of the controllable subspace.

4.7 Further Issues

In this section, we further discuss some issues closely related to the controllability and observability of switched linear systems. These include the introduction of other controllability concepts, mainly from the context of nonlinear control systems, and their relationships with the concepts given in Section 4.2.1. We also address the controllability for constrained switched linear systems where the switching/input are subject to certain constraints. Finally, we briefly discuss the decidability of various controllability/observability notions.

4.7.1 Equivalence Among Different Controllability Notions

The controllability notions defined in Section 4.2.1 correspond to their counterparts in standard linear systems. As a switched linear system is essentially a nonlinear system, it is necessary to introduce the 'nonlinear' notions and make a comparison between them.

First, we introduce some global controllability notions. The notions are standard and can be found, for example, in the textbooks [74, 111].

Consider a nonlinear control system given by

$$\dot{x} = f(x, u) \quad x \in \mathbf{R}^n \quad u \in \mathbf{R}^p. \tag{4.67}$$

Definition 4.66. *For nonlinear system (4.67), a point x_2 is said to be:*

- *reachable from x_1, denoted by $x_2 \in R(x_1)$, if there exist a measurable input function, and a time $T > 0$, such that the trajectory of the controlled system satisfies $x(0) = x_1$ and $x(T) = x_2$;*
- *reachable from x_1 at a given time $T > 0$, denoted by $x_2 \in R_T(x_1)$, if x_2 is reachable from x_1 at the pre-assigned time instant T;*

- weakly reachable from x_1, denoted by $x_2 \in WR(x_1)$, if there exist a natural number l, and a state sequence $z_0 = x_1, z_1, \cdots, z_l = x_2$, such that either $z_i \in R(z_{i-1})$ or $z_{i-1} \in R(z_i)$ for each $i \in \bar{l}$; and
- weakly reachable from x_1 at a given time $T > 0$, denoted by $x_2 \in WR_T(x_1)$, if there exist a natural number l, a state sequence $z_0 = x_1, z_1, \cdots, z_l = x_2$, and a positive real number sequence $T_1, \cdots, T_l$, such that either $z_i \in R_{T_i}(z_{i-1})$ or $z_{i-1} \in R_{T_i}(z_i)$ for each $i \in \bar{l}$, and $\sum_{i \in \bar{l}} T_i = T$.

Note that WR is an equivalence relationship. That is,

(a) $x \in WR(x)$;
(b) $x \in WR(y) \implies y \in WR(x)$; and
(c) $x \in WR(y)$ and $y \in WR(z)$ implies that $x \in WR(z)$.

The relationship R, however, is not an equivalence in general. For example, for the non-affine one-dimensional system

$$\dot{x} = u^2$$

$x_2 \in R(x_1)$ if and only if $x_2 \geq x_1$, hence $1 \in R(0)$ but $0 \notin R(1)$. The non-symmetry comes from the non-reversibility of the time space. In view of this, we introduce the notion of symmetric switched systems as follows.

A switched linear system $\Sigma(A_i, B_i)_M$ is said to be *symmetric*, if for each A_i, $i \in M$, there is a $j \in M$ such that $A_j = -A_i$. For any symmetric switched linear system, R is an equivalence relationship.

Given a switched linear system $\Sigma(A_i, B_i)_M$, let its *symmetric closure system* be the symmetric switched linear system $\Sigma(C_j, D_j)_N$, where each (C_j, D_j), $j \in N$, is either (A_i, B_i) or $(-A_i, B_i)$ for some $i \in M$, and vice versa.

Definition 4.67. *Nonlinear system (4.67) is said to be:*

- controllable, *if* $R(x) = \mathbf{R}^n$ *for each* $x \in \mathbf{R}^n$;
- small-time controllable, *if* $\cup_{0 < T \leq t} R_T(x) = \mathbf{R}^n$ *for any* $t > 0$ *and* $x \in \mathbf{R}^n$;
- T-controllable, *if* $R_T(x) = \mathbf{R}^n$ *for any* $T > 0$ *and* $x \in \mathbf{R}^n$;
- weakly controllable, *if* $WR(x) = \mathbf{R}^n$ *for any* $x \in \mathbf{R}^n$;
- weakly small-time controllable, *if* $\cup_{0 < T \leq t} WR_T(x) = \mathbf{R}^n$ *for any* $t > 0$, *and* $x \in \mathbf{R}^n$; *and*
- weakly T-controllable, *if* $WR_T(x) = \mathbf{R}^n$ *for any* $T > 0$, *and* $x \in \mathbf{R}^n$.

It can be seen that

$$T - \text{controllable} \implies$$
$$\left\{ \begin{array}{l} \text{small-time controllable} \implies \text{controllable} \\ \text{weakly } T - \text{controllable} \implies \text{weakly small-time controllable} \end{array} \right\}$$
$$\implies \text{weakly controllable.} \tag{4.68}$$

It is worthy to notice that the above notions are global and bi-directional in nature. That is, the controllabilities require that any given initial state can be steered to any target state in the state space. Thus, the controllabilities are in fact imply both the controllability and reachability defined in Section 4.2.1.

Let us compare the notions with the one given in Definition 4.2. Note that the equivalence notion in Definition 4.22 is exactly the same as the first controllability notion in Definition 4.67. As a result, we only need to compare the notions in Definition 4.67.

Theorem 4.68. *For the switched linear system, the following statements are equivalent:*

(i)   the system is controllable;
(ii) the system is small-time controllable;
(iii) the system is T-controllable;
(iv) the system is weakly controllable;
(v)   the system is weakly small-time controllable; and
(vi) the system is weakly T-controllable.

Proof. In view of the relationship (4.68), we only need to prove the T-controllability from the weak controllability. To make use of the controllability criteria presented in the previous sections, we proceed by showing the following implications:

weak controllability $\Longrightarrow$ controllability $\Longrightarrow$ T-controllability.

To establish the former part, note that, for any switched linear system, weak controllability implies controllability of its symmetric closure system. In addition, the controllability of a switched linear system is equivalent to that of its symmetric closure system. To see this, recall that the controllable set of a switched system $\Sigma(A_i, B_i)_M$ is the smallest $\mathbf{A}$-invariant subspace containing $\sum_{i \in M} \mathrm{Im}\, B_i$, where $\mathbf{A} = \{A_1, \cdots, A_m\}$. It is clear that this subspace is also the smallest $\bar{\mathbf{A}}$-invariant and containing $\sum_{i \in M} \mathrm{Im}\, B_i$, where $\bar{\mathbf{A}} = \{A_1, \cdots, A_m, -A_1, \cdots, -A_m\}$. The controllability of a switched linear system hence coincides with that of its symmetric closure system.

The latter part can be derived from the path planning algorithm in Section 4.3.3. Indeed, note that Equation (4.24) holds for a fixed and finite l. As $\mathrm{Im}\, W_{t_1}^i = \mathrm{Im}\, W_{t_2}^i$ for any positive t_1 and t_2, it is always possible to choose h_i in (4.24) such that $\sum h_i \le T/2$ for any pre-assigned $T > 0$. That is to say, for any given states x_0 and x_f in the controllable subspace, the system can be steered from x_0 to the origin within $T/2$ time, and from the origin to x_f within $T/2$ time. These, together with the fact the origin is an equilibrium of the unforced system, imply that x_f can be attainable from x_0 using exactly T time for any $T > 0$. $\square$

The theorem shows that the different (global) controllability notions are in fact equivalent to each other for continuous-time switched linear systems.

4.7.2 Reachability Under Constrained Switching and Input

In the aforementioned concepts of controllability and observability, the switching path and the control input are assumed to be designed in an arbitrary manner. That is, we do not impose any restriction on the possible ways of switching and control. In many practical situations, this is not the case. For example, in workshops, the order of the activated subsystems is pre-assigned rather than arbitrarily assigned. In this case, for instance, we must first activate subsystem 1, then switch to subsystem 2, then subsystem 3, etc. This fixed sequence imposes a restriction on the switching signal. Another example is the control input undergoing certain saturations which imposes a restriction on the control input.

In this subsection, we discuss various switching/input constraints and their possible influences on controllability. We focus on the continuous-time switched systems.

For switched linear system $\Sigma(A_i, B_i)_M$, let $\phi(t; t_0, x_0, u, \sigma)$ denote the state trajectory at time t of the switched system starting from $x(t_0) = x_0$ with input u and switching signal σ.

Let $\mathcal{S}_0$ be the allowed set of switching paths, and $\mathcal{U}_0$ be the allowed set of input functions. The reachable set of the system under $\mathcal{S}_0$ and $\mathcal{U}_0$ is the set of states which are attainable in a finite time by appropriate choices of the inputs and switching paths in the allowed sets.

Definition 4.69. *The* reachable set *of the switched linear system at time $T >$ 0 starting from x under $\mathcal{S}_0$ and $\mathcal{U}_0$, denoted by $R(x, T, \mathcal{U}_0, \mathcal{S}_0)$, is*

$$R(x, T, \mathcal{U}_0, \mathcal{S}_0) = \{\phi(T; 0, x, u, \sigma) \colon u \in \mathcal{U}_0, \sigma \in \mathcal{S}_0\}.$$

The system is said to be (completely) *T-reachable under $\mathcal{S}_0$ and $\mathcal{U}_0$, if*

$$R(x, T, \mathcal{U}_0, \mathcal{S}_0) = \mathbf{R}^n \quad \forall\, x \in \mathbf{R}^n \quad T > 0.$$

First, we consider the case that the switching sequence is subject to certain restrictions.

4.7.2.1 Reachability Under Restricted Switching Signal

Suppose that G is a directed graph (digraph) composed of set M as the set of points and a set of arcs N, where $N \subseteq M \times M$. Let $G = (M, N)$ govern the allowed switchings from one subsystem to another. That is, for any $k \in M$, the (possibly empty) set $N_k = \{i \in M \colon (k, i) \in N\}$ defines the allowed subsystem indices following the kth subsystem. In other words, if $(k, i) \notin N$, then, any switching from the kth subsystem to the ith subsystem is prohibited. Accordingly, if N is a strict subset of $M \times M$, then the directed graph imposes a nontrivial restriction on the switching index sequence. Let $\mathcal{S}_G$ denote the set of switching paths obeying the restriction. On the other hand,

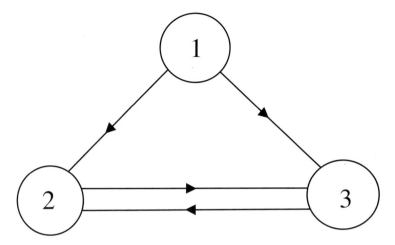

Fig. 4.1. The directed graph G in (4.69)

we assume temporarily that there is no constraint imposed on the input, and let $\mathcal{U}_G$ denote the set of piecewise continuous input functions.

In general, the set $R(0, T, \mathcal{U}_G, \mathcal{S}_G)$ under the digraph is a strict subset of the controllable subspace of the unconstrained system. For example, suppose that

$$G = (\{1, 2, 3\}, \{(1, 2), (1, 3), (2, 3), (3, 2)\}) \tag{4.69}$$

(the graph G is shown in Figure 4.1) with

$$A_1 = \begin{bmatrix} 0 & 1 & 1 \\ 0 & 0 & 0 \\ 0 & 0 & 0 \end{bmatrix} \quad A_2 = \begin{bmatrix} 0 & 0 & 0 \\ 0 & 0 & 1 \\ 0 & 0 & 0 \end{bmatrix} \quad A_3 = 0$$

$$B_1 = 0 \quad B_2 = 0 \quad B_3 = \begin{bmatrix} 0 \\ 0 \\ 1 \end{bmatrix}.$$

Then, the reachable set under the graph can be calculated to be

$$R(0, T, \mathcal{U}_G, \mathcal{S}_G) = \text{span}\{ \begin{bmatrix} 0 \\ 1 \\ 0 \end{bmatrix}, \begin{bmatrix} 0 \\ 0 \\ 1 \end{bmatrix} \}$$

which is a strict subset of the controllable subspace, $\mathcal{R}(A_i, B_i)_{\bar{3}} = \mathbf{R}^3$.

Given any sequence $i_1, \cdots, i_l$, we say that the sequence generates set L, if $i_j \in L$ for any $j = 1, \cdots, l$, and each element in L appears at least once in the sequence.

Recall that a closed walk in a digraph is an alternating sequence of points and arcs

$$v_0, x_1, v_1, \cdots, x_k, v_k$$

in which each arc x_i is $v_{i-1} v_i$ and $v_0 = v_k$. Each walk generates a subset of M which contains each point in the walk.

Theorem 4.70. *Suppose that L is a subset of M and directed graph G permits a closed walk which generates set L. If switched system $\Sigma(A_i, B_i)_L$ is completely controllable, then, we have*

$$R(x, T, \mathcal{U}_G, \mathcal{S}_G) = \mathbf{R}^n \quad \forall\, x \in \mathbf{R}^n \quad T > 0 \tag{4.70}$$

which means that the switched system $\Sigma(A_i, B_i)_M$ is completely T-reachable under graph G.

Proof. Suppose that the closed walk that generates set L is

$$k_1 \sim k_2 \sim \cdots \sim k_s \sim k_1$$

where '$\sim$' denote the corresponding arcs in the walk.

We consider a switching path with the cyclic switching index sequence

$$k_1, \cdots, k_s, k_1, \cdots, k_s, \cdots . \tag{4.71}$$

The number of switches l and the switching times $t_1, \cdots, t_l$ are to be determined later.

Note that this switching path is in the allowed switching set $\mathcal{S}_G$.

Let $t_f > t_l$. From (4.5), the reachable set at t_f is

$$\mathcal{R}(t_f) = e^{A_{k_s} h_l} \cdots e^{A_{k_2} h_1} \mathcal{D}_{k_1} + \cdots + e^{A_{k_s} h_l} \mathcal{D}_{k_{s-1}} + \mathcal{D}_{k_s}$$

where $h_j = t_{j+1} - t_j$, $j = 0, 1, \cdots, l-1$ and $h_l = t_f - t_l$.

Arrange a permutation of $L = \{i_1, \cdots, i_j\}$ such that $i_1, \cdots, i_j$ is a subsequence of $k_1, \cdots, k_s$. It is clear that the cyclic index sequence

$$i_1, \cdots, i_j, i_1, \cdots, i_j, \cdots$$

is a subsequence of index sequence (4.71). Denote the corresponding subsequence of $h_0, \cdots, h_l$ to be $\tau_1, \cdots, \tau_\mu$. Applying Lemma 4.16 repeatedly, we have

$$\dim \left(e^{A_{k_s} h_l} \cdots e^{A_{k_2} h_1} \mathcal{D}_{k_1} + \cdots + e^{A_{k_s} h_l} \mathcal{D}_{k_{s-1}} + \mathcal{D}_{k_s} \right)$$

$$\geq \dim \left(e^{A_{i_j} \tau_\mu} \cdots e^{A_{i_2} \tau_2} \mathcal{D}_{i_1} + \cdots + e^{A_{i_j} \tau_\mu} \mathcal{D}_{i_{j-1}} + \mathcal{D}_{i_j} \right)$$

for almost all $h_0, h_1, \cdots$. Choose l to be sufficiently large. This means that μ is also sufficiently large. By the proof of Theorem 4.17, for almost all $\tau_1, \cdots, \tau_\mu$, we have

$$e^{A_{i_j} \tau_\mu} \cdots e^{A_{i_2} \tau_2} \mathcal{D}_{i_1} + \cdots + e^{A_{i_j} \tau_\mu} \mathcal{D}_{i_{j-1}} + \mathcal{D}_{i_j} = \mathcal{R}(A_i, B_i)_L = \mathbf{R}^n.$$

As a result, $\mathcal{R}(t_f) = \mathbf{R}^n$. From the fact that

$$\mathcal{R}(t_f) \subseteq R(x, T, \mathcal{U}_G, \mathcal{S}_G)$$

the theorem follows. $\square$

Corollary 4.71. *Suppose that directed graph G permits a spanning and closed walk. Then, we have*

$$R(x, T, \mathcal{U}_G, \mathcal{S}_G) = \mathcal{R}(A_i, B_i)_M \quad \forall\, x \in \mathbf{R}^n \quad T > 0.$$

Proof. It is clear that the set $R(x, T, \mathcal{U}_G, \mathcal{S}_G)$ is a subset of $\mathcal{R}(A_i, B_i)_M$. On the other hand, from the proof of Theorem 4.70, we have

$$\dim \mathcal{R}_G(*) \geq \dim \mathcal{R}(A_i, B_i)_M$$

where $\mathcal{R}_G(*) \subseteq R(x, T, \mathcal{U}_G, \mathcal{S}_G)$ is the reachable set of switched system along some allowed switching path under graph G. This means that the two sets coincide with each other. $\square$

This corollary provides an important piece of information for the reachability of the switched linear system under the restricted switching mechanism. Indeed, according to the corollary, assume that the directed graph permits a spanning and closed walk, then, the switched system possesses the complete reachability under the graph. The assumption is very mild and can be met in many practical situations. For example, in a workshop, m working procedures are required to produce a product. Suppose that the procedure sequence is cyclic among $1, \cdots, m - 1$, but between any two procedures the mth procedure applies. This corresponds to the directed graph (with $m = 5$) depicted in Figure 4.2. It can be seen that a spanning and closed walk is

$$1 \sim m \sim 2 \sim m \sim \cdots \sim m - 1 \sim m \sim 1$$

which generates the set $M = \{1, \cdots, m\}$. According to Corollary 4.71, the reachable set under the graph is exactly the controllable subspace of the unconstrained switched system.

A special but very interesting case is that the subsystems are divided into different groups and any transition within a group is forbidden. This means that only transitions among different groups are allowed. A typical example is a production workshop with a set of procedures each of which can be implemented in several alternative ways. In this case, the switching is severely restricted as only transitions between groups (the procedures) are allowed. However, as a spanning and closed walk always exists in this case, the reachable set coincides with the controllable subspace of the unconstrained system.

Figure 4.3 shows an example where the subsystems are divided into three groups. A sample spanning and closed walk is

$$1 \sim 4 \sim 2 \sim 3 \sim 4 \sim 5 \sim 8 \sim 4 \sim 6 \sim 1 \sim 4 \sim 7 \sim 1.$$

Corollary 4.71 can be further extended to more general cases. For example, for a switched system with a restriction on the switching signal which is

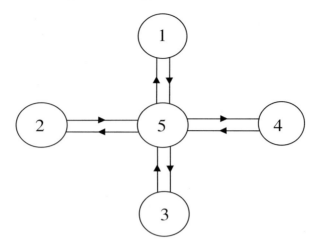

Fig. 4.2. The schematic of the directed graph with $m = 5$

not necessarily described by a directed graph, we can prove that, the reachable set under the restriction coincides with the controllable subspace of the unconstrained system, provided that, there exists an allowed switching index sequence where each individual index appears sufficiently many times. This assumption is very mild and holds for many interesting cases.

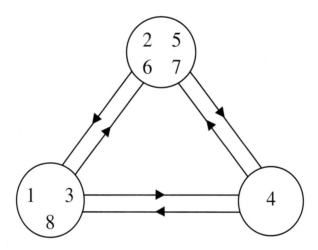

Fig. 4.3. The schematic of the grouped directed graph with $m = 8$

4.7.2.2 Reachability Under Restricted Control Input

In many practical situations, the control input is subject to hard constraints, such as saturation and/or non-symmetry (*e.g.*, v is allowed but $-v$ not). Under these situations, the controllability analysis may be more difficult.

Consider the local controllability of the origin of the switched linear system with input saturation

$$\dot{x}(t) = A_\sigma x(t) + B_\sigma(t)u(t) \quad u(t) \in U \tag{4.72}$$

where U is a neighborhood of the origin.

Note that the system model reflects many practical situations where the control input is subject to hard constraints such as saturation or force/enegy limit. A typical example for U is that the set is a ball in $\mathbf{R}^n$ centered at the origin. However, here we do not require the set to be convex or symmetric. The only intrinsic assumption is that it contains the origin as an interior point.

Let $\mathcal{U}$ denote the set of piecewise continuous input functions taking values from U.

By means of the reachability criteria presented in Section 4.3, we are able to obtain the following criterion.

Theorem 4.72. *The constrained T-reachable set $R(0,T,\mathcal{U},\mathcal{S})$ contains the origin as an interior for any $T > 0$ if and only if the unconstrained switched system is completely controllable, that is*

$$\mathcal{C}(A_i, B_i)_M = \mathbf{R}^n. \tag{4.73}$$

Proof. Let $\mathcal{V}$ denote $\mathcal{C}(A_i, B_i)_M$ for briefness. The necessity (only if) part can be easily proven by contradiction. Indeed, the violation of (4.73) means that the controllable subspace of the unconstrained switched system is a strict subspace of the total space. Accordingly, there is an $x \in \mathbf{R}^n/\mathcal{V}$. As

$$\lambda x \notin \mathcal{V} \quad \forall \, \lambda \neq 0$$

the unconstrained switched system is not locally reachable at the origin by definition. This implies that the constrained switched system is not locally reachable at the origin.

To prove the other part, we need to recall some formulas in Section 4.3.3. In (4.24), we proved that there exist a natural number l, and a cyclic index sequence $i_0, \cdots, i_l$, such that

$$\mathcal{V} = e^{A_{i_l}h_l} \cdots e^{A_{i_1}h_1} \operatorname{Im} W_{h_0}^{i_0} + \cdots + e^{A_{i_l}h_l} \operatorname{Im} W_{h_{l-1}}^{i_{l-1}} + \operatorname{Im} W_{h_l}^{i_l} \tag{4.74}$$

for almost any positive real number sequence $h_0, \cdots, h_l$, where

$$W_t^k = \int_0^t e^{A_k(t-\tau)} B_k B_k^T e^{A_k^T(t-\tau)} d\tau.$$

Fix such a sequence $h_0, \cdots, h_l$. Under the piecewise continuous control strategy

$$u(t) = B_{i_k}^T e^{A_{i_k}^T (t_{k+1} - t)} a_{k+1} \quad t_k \leq t < t_{k+1} \quad k = 0, 1, \cdots, l \qquad (4.75)$$

where $t_0 = 0, t_{k+1} = t_k + h_k, \ k = 1, \cdots, l$, we have

$$x = -\left(e^{A_{i_l} h_l} \cdots e^{A_{i_0} h_0}\right)^{-1} \left[e^{A_{i_l} h_l} \cdots e^{A_{i_1} h_1} W_{h_0}^{i_0}, \cdots, e^{A_{i_l} h_l} W_{h_{l-1}}^{i_{l-1}}, W_{h_l}^{i_l}\right] a$$

where $a = [a_1^T, \cdots, a_{l+1}^T]^T$, and x is a controllable state. This equation sets up a connection between the controllable state x and the constant vector a which relates to the control input law.

Let $x_0 = 0$ and define matrix

$$L = [e^{A_{i_l} h_l} \cdots e^{A_{i_1} h_1} W_{h_0}^{i_0}, \cdots, e^{A_{i_l} h_l} W_{h_{l-1}}^{i_{l-1}}, W_{h_l}^{i_l}]. \qquad (4.76)$$

Condition (4.73) implies that the linear operator

$$\mathbf{L} \colon \mathbf{R}^{(l+1)n} \mapsto \mathbf{R}^n \quad \mathbf{L}(a) \stackrel{def}{=} La$$

is onto (surjective). This, together with the linearity of the operator, implies that, for any set W in $\mathbf{R}^{(l+1)n}$ containing the origin as an interior, the set

$$\mathbf{L}(W) = \{\mathbf{L}(a) \colon a \in W\}$$

also contains the origin as an interior in $\mathbf{R}^n$. On the other hand, from (4.75), it can be seen that the set of allowed constant vectors

$$\{a = [a_1^T, \cdots, a_{l+1}^T]^T \colon u(t) \in U \quad \forall\, t \in [0, \sum_{k=1}^{l} h_k]\}$$

contains the origin as an interior point in $\mathbf{R}^{(l+1)n}$.

The above analysis shows that, for the switched linear system with the input constraint, the reachable set $R(0, \sum_{k=0}^{l} h_k, \mathcal{U}, \mathcal{S})$ at the origin contains the origin as an interior point.

Finally, note that, for any $T_1 \leq T_2$, we have

$$R(0, T_1, \mathcal{U}, \mathcal{S}) \subset R(0, T_2, \mathcal{U}, \mathcal{S}). \qquad (4.77)$$

Indeed, suppose that $x \in R(0, T_1, \mathcal{U}, \mathcal{S})$, then, there exist a switching signal $\sigma(\cdot)$ and an input $u(\cdot)$ defined on $[0, T_1]$, with $u \in \mathcal{U}$, such that

$$x = \phi(0, T_1, u, \sigma).$$

Now, define another switching signal σ' and input u' on $[0, T_2]$ by

$$\sigma'(t) = \begin{cases} \sigma(0) & t \in [0, T_2 - T_1) \\ \sigma(t + T_1 - T_2) & \text{otherwise} \end{cases}$$

and

$$u'(t) = \begin{cases} 0 & t \in [0, T_2 - T_1) \\ u(t + T_1 - T_2) & \text{otherwise.} \end{cases}$$

It can be seen that

$$\phi(0, T_2, u', \sigma') = \phi(0, T_1, u, \sigma) = x.$$

As a result, (4.77) holds.

The above reasonings show that, for any $T > 0$, the set $R(0, T, \mathcal{U}, \mathcal{S})$ contains the origin as an interior point. $\square$

Note that, in the proof of the theorem, the design of input and the design of the switching signal are decoupled in the following sense. In the proof, what we need is an index sequence $i_0, \cdots, i_l$, and a duration sequence $h_0, \cdots, h_l$, such that

$$e^{A_{i_l} h_l} \cdots e^{A_{i_1} h_1} \mathcal{D}_{i_0} + \cdots + e^{A_{i_l} h_l} \mathcal{D}_{i_{l-1}} + \mathcal{D}_{i_l} = \mathcal{V}. \tag{4.78}$$

For the purpose of proving the theorem, any sequence $i_0, \cdots, i_l$ with this property suffices. This observation, together with the proof of Theorem 4.70, indicates that the following generalized conclusion can be made.

Theorem 4.73. *Suppose that directed graph G permits a spanning and closed walk, and U is a set in $\mathbf{R}^n$ containing the origin as an interior. Then, for the constrained switched linear system*

$$\dot{x}(t) = A_\sigma x(t) + B_\sigma u(t) \quad u(t) \in U \quad \sigma \in \mathcal{S}_G$$

the constrained T-reachable set $R(0, T, \mathcal{U}, \mathcal{S}_G)$ contains the origin as an interior for any $T > 0$ if and only if the unconstrained switched system is completely controllable.

This theorem applies to more practical situations where both the switching signal and the control input are subject to constraints. Of course, an important concern is whether the region of locally reachable set is large enough. In general, the locally reachable set is not necessarily convex or in a regular shape, even if set U is a closed convex set centered at the origin. Accordingly, it seems difficult to exactly determine the reachable set. Nevertheless, the set can be estimated in the following way.

Suppose that the unconstrained switched linear system is completely controllable. Then, for any fixed $T > 0$, there is an allowed switching signal σ with index sequence $i_0, \cdots, i_l$, and duration sequence $h_0, \cdots, h_l$, $\sum_{k=0}^{l} h_k = T$, such that (4.78) holds. Suppose also that

$$U \supseteq \{u \in \mathbf{R}^p \colon \|u\|_\infty \leq r\}.$$

Note that such an r always exists. Let

$$\mu = \max_{k=0}^{l}\{\|B_{i_k}e^{A_{i_k}h_k}\|_\infty\}.$$

Let $\kappa = n(l+1)$ and $\zeta_1, \cdots, \zeta_k$ be the unit vectors in $\mathbf{R}^k$. Next, compute $x_i = \frac{1}{\mu}L\zeta_i$ for each $i = 1, \cdots, \kappa$, where L is given in (4.76). Finally, let $\mathcal{X}_\sigma$ be the closed and convex closure of set $\{x_1, \cdots, x_\kappa\}$. Then, we have

$$\mathcal{X}_\sigma \subseteq R(0, T, \mathcal{U}, \mathcal{S}_G).$$

Let

$$\mathcal{X} = \cup_\sigma \mathcal{X}_\sigma. \qquad (4.79)$$

This set can be seen as a (conservative) estimation of the locally reachable set at the origin. Due to the finiteness of l, there is a finite number of allowed switching index sequences. However, the number of allowed switching time sequences is infinite. To approximate $\mathcal{X}$ in a tractable way, we can divide the interval $[0, T]$ into an appropriate number of sub-intervals and assume that switching only occurs at the discrete instants. In this way, the number of feasible switching signals is limited and the union in (4.79) is computable.

4.7.3 Local Controllability

Local controllability means the ability to control the system to nearby (local) states rather than global states in the state space.

Definition 4.74. *Nonlinear system (4.67) is:*

- locally controllable at x_0, *if x_0 is an interior point of $R(x_0)$; and*
- locally weakly controllable, *if x_0 is an interior point of $WR(x_0)$.*

Definition 4.75. *Nonlinear system (4.67) is:*

- locally (completely) controllable, *if it is locally controllable at any non-origin state; and*
- locally (completely) weakly controllable, *if it is locally weakly controllable at any non-origin state.*

First, we present a necessary and sufficient condition for locally complete (weak) controllability. Suppose that the system is locally completely controllable. This means that for any non-origin state x_0, $x_0 \in \text{int}R(x_0)$. As a special case, for each non-origin state x_0, there is a neighborhood of x_0, N_{x_0}, such that $N_{x_0} \subseteq R(x_0)$. In particular, we have

$$\mathbf{S}_1 \subset \cup_{x \in \mathbf{S}_1} N_x.$$

As the unit sphere is a compact set of $\mathbf{R}^n$, it follows from the Finite Covering Theorem that, there is a finite number of states $y_1, \cdots, y_s$ such that

$$\mathbf{S}_1 \subset \cup_{i=1}^s N_{y_i}.$$

On the other hand, for any positive real number r, the radially linear property (*c.f.* Proposition 1.7) guarantees that we can choose N_x with

$$N_{\lambda x} = \lambda N_x \quad \forall \, x \neq 0 \ \lambda \neq 0$$

such that

$$\mathbf{S}_r \subset \cup_{i=1}^s N_{ry_i}.$$

Therefore, the system is locally controllable at x_0 if and only if it is controllable at λx_0 for any nonzero λ.

For locally weak controllability, the above reasonings also make sense. Hence, we have the following simple proposition.

Proposition 4.76. *The switched linear system is locally completely (weakly) controllable if and only if, there exist a finite number s, and a set of states $\{y_i\}_{i=1}^s$ on $\mathbf{S}_1$, such that*

$$\mathbf{S}_1 \subset \cup_{i=1}^s \operatorname{int} R(y_i) \quad (\mathbf{S}_1 \subset \operatorname{int} WR(y_i) \ \ resp.)$$

where int *denotes the interior of a set w.r.t. the normal topology of* $\mathbf{R}^n$.

Next, we turn to the switched linear autonomous system $\Sigma(A_i)_M$. For this system, there is no control input, the only design variable is the switching signal. As the origin is always an equilibrium of the system under any switching signal, the origin itself forms an invariant set of the system and hence is not locally controllable. As an implication, the unforced system is not globally controllable in any sense. However, by means of the switching signal, it is still possible to make the system locally controllable in a certain area of the state space. In fact, in some cases the switched system is 'almost globally' controllable as exhibited in the following example.

Example 4.77. Consider the planar unforced switched linear system $\Sigma(A_i)_{\bar{2}}$ with

$$A_1 = I_2 \text{ and } A_2 = \begin{bmatrix} 0 & 1 \\ -1 & 0 \end{bmatrix}.$$

According to Theorem 4.18, the controllable subspace is null. Take any two non-origin states x_1 and x_2. If $\|x_1\| \leq \|x_2\|$, then it can be seen that the $x_2 \in R(x_1)$. Indeed, let $x_0 = \frac{\|x_1\|}{\|x_2\|} x_2$, then, $x_2 \in R(x_0)$ via A_1 as x_2 and x_0 are in the same direction, and $x_0 \in R(x_1)$ via A_2 as A_2 is a purely rotative matrix. This means that, for any two non-origin state x_1 and x_2 in $\mathbf{R}^n$, we have $x_2 \in WR(x_1)$, and hence the system is locally completely weakly controllable. In the same way, we can prove that, for the enlarged system $\Sigma(A_i)_{\bar{3}}$ with $A_3 = -I_2$, each non-origin state is locally completely controllable.

The example exhibits the independent interest of the local controllability for switched autonomous systems.

Let $\mathcal{L} = \{A_1 x, \cdots, A_m x\}_{LA}$ be the Lie algebra generated by $A_i x$. The algebra is spanned by all the vectors in the form

$$[A_{i_1} x, [A_{i_2} x, [\cdots, [A_{i_{l-1}} x, A_{i_l} x]]]] \quad i_j \in M$$

where $[\cdot, \cdot]$ denotes the Lie product. As

$$[A_1 x, A_2 x] = (A_2 A_1 - A_1 A_2) x$$

for any $x \in \mathbf{R}^n$, $\mathcal{L}(x)$ is in the form

$$\mathcal{L}(x) = \text{span} \ \{A_1 x, \cdots, A_m x, B_1 x, B_2 x, \cdots\}$$

where B_j is a commutator in the form

$$B_j = [A_{k_1}, [A_{k_2}, [\cdots [A_{k_{l-1}}, A_{k_l}]]]] \quad k_i \in M.$$

The following properties of the Lie algebra are readily obtained:

(i) $\mathcal{L}(0)$ is the null space.
(ii) $\dim \mathcal{L}(\lambda x) = \dim \mathcal{L}(x)$, for all state $x \in \mathbf{R}^n$ and real number $\lambda \neq 0$.
(iii) Let $r = \max\{\dim \mathcal{L}(x) \colon x \in R^n\}$, and $\Omega = \{x \in \mathbf{R}^n \colon \dim \mathcal{L}(x) = r\}$, then, Ω is an open and dense subset of $\mathbf{R}^n$ (c.f. The proof of Lemma 4.16).

Now, we establish a criterion for local controllability. For this, we need the following technical lemma.

Lemma 4.78. *Let U be a pathwise connected open set, and assume that $\dim \mathcal{L}(x) = n$ for all $x \in U$. Let $\mathcal{T}(x)$ denote the largest integral sub-manifold of $\mathcal{L}$ passing through x. Then, we have*

$$\mathcal{T}(x) \supseteq U \quad \forall \, x \in U.$$

Proof. If this is not true, then, there are states x and y such that $x \in U, y \in U$, and y is a boundary point of $\mathcal{T}(x)$. As $\mathcal{T}(x) \cup \mathcal{T}(y)$ is a connected integral sub-manifold of $\mathcal{L}$ passing through x, and y is an interior of $\mathcal{T}(x) \cup \mathcal{T}(y)$, $\mathcal{T}(x)$ is a strict subset of $\mathcal{T}(x) \cup \mathcal{T}(y)$, which contradicts the fact that $\mathcal{T}(x)$ is the largest integral sub-manifold of $\mathcal{L}$ passing through x. $\square$

Theorem 4.79. *For the unforced switched linear system $\Sigma(A_i)_M$, a state x is locally weakly controllable if and only if the Lie algebra $\mathcal{L}$ is of full rank at x, that is*

$$\text{rank} \, \mathcal{L}(x) = n. \tag{4.80}$$

If the system is symmetric, then state x is locally controllable if and only if Condition (4.80) holds.

Proof. If $\dim \mathcal{L}(x) = s < n$, then, by Theorem 2.16, $\mathcal{T}(x)$ is a sub-manifold of dimension s. As $WR(x) \subseteq \mathcal{T}(x)$, the system is not locally weakly controllable.

If $\dim \mathcal{L}(x) = n$, then, there is a neighborhood U of x such that $\dim \mathcal{L}(y) = n$ for all $y \in U$. By Lemma 4.78, we have $\mathcal{T}(x) \supseteq U$. It follows from the Generalized Chow's Theorem (Theorem 2.16) that state x is locally weakly controllable. $\square$

Corollary 4.80. *The unforced switched linear system $\Sigma(A_i)_M$ is locally completely weakly controllable if and only if $\operatorname{rank} \mathcal{L}(x) = n$ for all $x \in \mathbf{S}_1$.*

Example 4.81. Suppose that $n = 3$, $m = 2$, and

$$
A_1 = \begin{bmatrix} 1 & 0 & 0 \\ 0 & 1 & 0 \\ 0 & 0 & -1 \end{bmatrix} \text{ and } A_2 = \begin{bmatrix} 0 & 0 & 1 \\ 0 & -1 & 0 \\ -1 & 0 & 0 \end{bmatrix}.
$$

Simple computation gives

$$
\{A_1 x, A_2 x\}_{LA} = \operatorname{span}\left\{ \begin{bmatrix} x_1 \\ x_2 \\ -x_3 \end{bmatrix}, \begin{bmatrix} x_3 \\ -x_2 \\ -x_1 \end{bmatrix}, \begin{bmatrix} x_3 \\ 0 \\ x_1 \end{bmatrix}, \begin{bmatrix} x_1 \\ 0 \\ -x_3 \end{bmatrix} \right\}.
$$

It can be seen that

$$
\dim\{A_1 x, A_2 x\}_{LA} = \begin{cases} 0 & \text{if } x = 0 \\ 1 & \text{if } x_1 = x_3 = 0 \ \ x_2 \neq 0 \\ 2 & \text{if } x_2 = 0 \ \ x_1^2 + x_3^2 \neq 0 \\ 3 & \text{otherwise.} \end{cases}
$$

Accordingly, let

$$
\Omega_1 = \{x \colon x_2 > 0, x_1^2 + x_3^2 \neq 0\} \text{ and } \Omega_2 = \{x \colon x_2 < 0, x_1^2 + x_3^2 \neq 0\}
$$

the two sets are pathwise connected open sets in which the Lie algebra is the total space. By Theorem 4.79, both Ω_1 and Ω_2 are locally weakly controllable. In fact, it follows from Lemma 4.78 that

$$
WR(x) = \Omega_i \quad \forall \ x \in \Omega_i \ \ i = 1, 2.
$$

Similarly, let

$$
\Omega_3 = \{x \colon x_2 = 0, x_1^2 + x_3^2 \neq 0\}.
$$

Apply Theorem 4.79 to this two dimensional sub-manifold, Ω_3 is locally weakly controllable. In addition, on sub-manifold Ω_3, A_2 is purely rotational and hence it can reach any direction, and A_1 possesses a stable mode and an unstable mode. This means that Ω_3 is in fact locally controllable, and we have

$$R(x) = \Omega_3 \quad \forall \, x \in \Omega_3.$$

Finally, let

$$\Omega_4 = \{x \colon x_2 > 0, x_1^2 + x_3^2 = 0\} \text{ and } \Omega_5 = \{x \colon x_2 < 0, x_1^2 + x_3^2 = 0\}.$$

Then, both Ω_4 and Ω_5 are locally controllable.

It is interesting to note that, each region Ω_i is invariant under the locally weak controllability, and any two states within the same set are weakly controllable. In other words, each region is the largest integral sub-manifold of the Lie algebra.

4.7.4 Decidability of Reachability and Observability

Decidability involves the ability to verify a property in a finite time. For continuous-time switched linear systems, the reachable set is a subspace which can be explicitly determined in a finite time. Accordingly, the property of complete reachability is decidable, which means that we can assert a definite (yes or no) answer in a finite time to the question "Is this system completely reachable?". For discrete-time switched systems, the situation becomes quite challenging. As proved in Section 4.4.2, complete reachability is decidable for reversible discrete-time switched systems. For general (not necessarily reversible) discrete-time systems, we showed in Theorem 4.31 that system $\Sigma_d(A_i, B_i)_M$ is reachable if and only if there exist an integer $k < \infty$, and an index sequence $i_0, \cdots, i_k$, such that

$$A_{i_k} \cdots A_{i_1} \operatorname{Im} B_{i_0} + \cdots + A_{i_k} \operatorname{Im} B_{i_{k-1}} + \operatorname{Im} B_{i_k} = \mathbf{R}^n$$

or equivalently,

$$\operatorname{rank}[A_{i_k} \cdots A_{i_1} B_{i_0}, \cdots, A_{i_k} B_{i_{k-1}}, B_{i_k}] = n. \tag{4.81}$$

By this criterion, if we know an upper bound of k, say, $k \le l$ and l is known, then the complete reachability can be verified in a finite time by examining relationship (4.81) for all possible combinations of $i_0, \cdots, i_k$ with $i_j \in M$ and $k \le l$. Therefore, the decidability of the complete reachability depends on whether such an upper bound exists. Unfortunately, we do not currently know the answer and thus the decidability of the complete reachability is still open. Similarly, the decidability of complete controllability/observability/reconstructibility is also open for further investigation.

We now turn to another kind of reachability and observability, known as the pathwise reachability and the pathwise observability.

Definition 4.82. *Switched system* $\Sigma(C_i, A_i, B_i)_M$ *is said to be* pathwise reachable, *if for any switching path* σ*, the reachable set via* σ *is the total space*

$$\mathcal{R}_\sigma(C_i, A_i, B_i)_M = \mathbf{R}^n.$$

Definition 4.83. *Switched system* $\Sigma(C_i, A_i, B_i)_M$ *is said to be* pathwise observable, *if for any switching path σ, the corresponding unobservable set is the null space*

$$\mathcal{UO}_\sigma(C_i, A_i, B_i)_M = 0.$$

Note that pathwise reachability/observability require that the switched system is reachable/observable via each possible switching signal. As a result, each subsystem must be reachable/observable. For a continuous-time switched system, this is also sufficient for pathwise reachability/observability, as shown in the following proposition.

Proposition 4.84. *A continuous-time switched linear system is pathwise reachable/observable if and only if each of its subsystems is completely reachable/observable.*
Proof. We need only to prove the sufficiency.
Suppose that each of the subsystems is reachable. Let σ be any switching signal. Suppose that its switching sequence is

$$\{(t_0, i_0), (t_1, i_1), \cdots, (t_l, i_l)\}$$

where $0 \leq l \leq \infty$. The reachable set via σ is $(c.f.\ (4.5))$

$$\mathcal{R}_\sigma = \cup_{k=1}^{l}(e^{A_{i_k}(t_{k+1}-t_k)} \cdots e^{A_{i_1}(t_2-t_1)}\mathcal{D}_{i_0} + \cdots + \mathcal{D}_{i_k})$$

where $t_{l+1} \overset{def}{=} \infty$. Note that $\mathcal{D}_{i_k}$ itself is the total space, thus we have

$$\mathcal{R}_\sigma = \mathbf{R}^n.$$

The case of observability can be proven in the same manner. $\square$
For pathwise reachability of the discrete-time switched system, the decidability depends on the full rank property of any matrix in the form

$$[B_{i_k}, A_{i_k}B_{i_{k-1}}, \cdots, A_{i_k} \cdots A_{i_1}B_{i_0}]$$

where $i_0, \cdots, i_k \in M$. Suppose that there is a finite k, such that any rank defectiveness of the above matrix means that the system is not pathwise reachable, then the property of pathwise reachability is decidable. Fortunately, this idea does work and we can prove the following theorem.

Theorem 4.85. *The pathwise reachability/observability property is decidable.*
Proof. We prove that the pathwise reachability is decidable.
If the system is pathwise reachable, then, each subsystem (A_i, B_i) must be reachable. This means that

$$A_i^{n-1}\operatorname{Im}B_i + \cdots + \operatorname{Im}B_i = \mathbf{R}^n$$

which implies that

$$\operatorname{Im} A_i + \operatorname{Im} B_i = \mathbf{R}^n. \tag{4.82}$$

Accordingly, we assume that (4.82) holds.

Let Ξ be the set of subspaces of $\mathbf{R}^n$. Under the subset relationship, Ξ is partially ordered with the unique maximum element $\mathbf{1} \stackrel{def}{=} \mathbf{R}^n$ and the unique minimum element $\mathbf{0}$ which is the null subspace. In addition, the set is $(n+2)$-Nether as it does not contain any strictly decreasing sequence of length $n+2$. Define a finite set of maps

$$G = \{f_i \colon \Xi \mapsto \Xi\}_{i \in M}$$

by

$$f_i(\mathcal{Y}) = A_i \mathcal{Y} + \operatorname{Im} B_i \quad \mathcal{Y} \in \Xi \quad i \in M.$$

Then, we define a language in the alphabet $\{1, \cdots, m\}$ as

$$L_G = \{\omega = \omega_1 \cdots \omega_l \colon f_\omega(\mathbf{0}) < \mathbf{1}\}$$

where $f_\omega \stackrel{def}{=} f_{\omega_1} \circ f_{\omega_2} \circ \cdots \circ f_{\omega_l}$ (c.f. Section 2.11).

It can be seen that the pair (Ξ, G) is a monotone automaton. Indeed, it is clear that all maps in G are monotone with respect to the partial order in Ξ. Besides, it follows from (4.82) that $f_i(\mathbf{1}) = \mathbf{1}$ for all $f_i \in G$. By Lemma 2.17, there exists a number $F(n+2, m)$, such that, for any $w \in L_G$ with $|w| > F(n+2, m)$, we can find strings x, y and z satisfying the following properties:

(i) $w = xyz$;
(ii) $|xy| \le F(n+2, m)$;
(iii) $|y| \ge 1$; and
(iv) for all $i \in \mathbf{N}_+$, we have $xy^i z \in L_G$.

As an implication, if there exists a $w \in L_G$ with $|w| > F(n+2, m)$, then, $f_{xy^i z}(\mathbf{0}) < \mathbf{1}$ for all $i = 1, 2, \cdots$.

By the definition of G, for a word $\omega = \omega_1 \cdots \omega_k$, we have

$$f_\omega(0) = A_{\omega_1} A_{\omega_2} \cdots A_{\omega_{k-1}} \operatorname{Im} B_{\omega_k} + \cdots + A_{\omega_1} \operatorname{Im} B_{\omega_2} + \operatorname{Im} B_{\omega_1}$$

which is exactly the reachable set of the switched system along the switching path $\omega_1 \cdots \omega_k$.

Combining the above reasonings, we can conclude that, if the pathwise reachability cannot be achieved in time $F(n+2, m)$, then, the system must not be pathwise reachable. As a result, the pathwise reachability is decidable.
$\square$

4.8 Notes and References

The concepts of controllability and observability presented here are natural extensions of the standard ones from linear systems. The reader is referred to the standard textbooks, *e.g.*, [21] and [77]. We adopted Wonham's geometric approach [160] as a main tool for analysis. The controllability/observability criteria in Section 4.2 were first presented in [142]. See also [40, 150, 145, 162, 168, 161, 85] for related work. Studies for second-order switched linear systems can be found in [96, 163].

The counterpart of discrete-time systems has been studied for many years by Stanford and his co-workers. The material here was mainly selected from [127, 128, 26, 27]. Theorems 4.30, 4.36 and 4.37 were taken from [52].

The structure decomposition and canonical forms were adopted from [127, 26] for discrete time and [137, 139, 134] for continuous time. This part extends the related theory of linear systems to switched linear systems, and will play a key role in addressing the feedback stabilization problem in Chapter 5.

Sampling of dynamic systems is an old topic and the supporting lemma 4.53 was adopted from [153]. The main results presented here, as well as the discussion on digital control and regular switching, were taken from [138, 133].

In Section 4.7, the topics are natural extensions of the controllability/observability discussed in the previous sections. The first two subsections, 4.7.1, and 4.7.2, were adopted from the recent work [135]. Section 4.7.3 on local controllability was mainly adopted from [22]. Theorem 4.85 can be found in [56], see also a much detailed proof in [5]. Other parts of Section 4.7.4 were newly developed based on the results presented in the previous sections.

5
Feedback Stabilization

5.1 Introduction and Preliminaries

In this chapter, we address the problem of state/output feedback stabilization of switched linear control systems. As in the standard linear system theory, we present constructive design procedures based on the canonical forms presented in Chapter 4.

Feedback stabilization involves finding appropriate switching signals as well as state/output feedback controllers to make the closed-loop systems (asymptotically) stable. Once the feedback controllers are given, the closed-loop systems are force free, and the switching signal design can then be carried out using the approaches presented in Chapter 3. Thus, we mainly focus on the feedback control design issues in this chapter.

This chapter addresses several classes of switched linear systems. The first is a single process controlled/measured by multiple controllers/sensors. In this case, we present a concise and complete treatment of the problem of dynamic output stabilization. The second is a controllable system where the summation of the controllability subspaces of the individual subsystems is the total state space. For this class of systems, we propose a stabilizing state-feedback control scheme with dwell time. The third is in the general controllability canonical form, and for which we present criteria and design methods for state feedback stabilization.

Consider the switched linear control system given by

$$
\begin{aligned}
\dot{x}(t) &= A_\sigma x(t) + B_\sigma u_\sigma(t) \\
y(t) &= C_\sigma x(t)
\end{aligned}
\tag{5.1}
$$

where $x \in \mathbf{R}^n$ is the state, $u_k \in \mathbf{R}^{p_k}, k = 1, \cdots, m$ are piecewise continuous inputs, $y \in \mathbf{R}^q$ is the output, $\sigma \in M$ is the switching signal to be designed, and $A_i, B_i, C_i, i \in M$ are real matrices of compatible dimensions.

For the problem of stabilization of switched linear control systems, we seek both the switching signal and the control input to steer the switched system asymptotically stable. Accordingly, both the switching signal and control input are design variables. As we explained in Section 1.3, the switching signal may either be time-driven, event-driven, or mixed. Similarly, the control input may either be open-loop or closed-loop. As a result, there are several types of stabilizability, for example, the stabilizability via time-driven switching and open-loop input, and the stabilizability via state-feedback switching and output feedback control input, etc.

In this chapter, we assume that the switching signal is measurable on-line. Accordingly, we can incorporate the switching signal into the controller such that the control input is in the piecewise (gain-scheduling type) form

$$u_i(t) = \varphi_i(x(t)) \quad i \in M. \tag{5.2}$$

Definition 5.1. *System (5.1) is said to be (state) feedback stabilizable, if there exist a switching signal σ, and state feedback control inputs in the form (5.2), such that the closed-loop switched system*

$$\dot{x}(t) = A_\sigma x(t) + B_\sigma \varphi_\sigma(x(t))$$

is well-posed and uniformly asymptotically stable.

Note that the switched linear control system becomes a switched nonlinear autonomous system if the controller is nonlinear. Usually, switched nonlinear systems are much harder to handle than switched linear systems as the transition invariance properties (*c.f.* Section 1.3.5) do not hold for general switched nonlinear systems. In view of this, we restrict our attention to piecewise linear controllers as defined below.

Definition 5.2. *System (5.1) is said to be (piecewise) linear (state) feedback stabilizable, if there exist a switching signal σ, and piecewise linear state feedback control inputs*

$$u_i(t) = F_i x(t) \quad i \in M \tag{5.3}$$

such that the closed-loop switched system

$$\dot{x}(t) = (A_\sigma + B_\sigma F_\sigma)x(t)$$

is well-posed and uniformly asymptotically stable.

By Theorem 3.9, the piecewise linear state feedback stabilizability is equivalent to the exponential stabilizability by means of piecewise linear state feedback controllers.

The above definition of linear feedback stabilizability is still quite general. Whenever possible, we further restrict our attention to a special type of the stabilizability as follows.

Definition 5.3. *System (5.1) is said to be (piecewise linear state) quadratically stabilizable, if there exist a switching signal σ, and piecewise linear state* *feedback control inputs in form (5.3), such that the closed-loop switched system is well-posed and quadratically stable.*

We also need the concept of stabilization by dynamic output feedback control inputs.

Definition 5.4. *System (5.1) is said to be dynamic output feedback stabilizable, if there exist a dynamic output feedback control law of form*

$$u(t) = g(y(t), \hat{x}(t), \sigma)$$
$$\dot{\hat{x}}(t) = \bar{f}(\hat{x}(t), y(t), \sigma)$$

and a switching signal

$$\sigma(t+) = \psi(t, \sigma(t), y(t), \hat{x}(t))$$

such that the overall system

$$\dot{x}(t) = A_\sigma x(t) + B_\sigma g(y(t), \hat{x}(t), \sigma(t))$$
$$\dot{\hat{x}}(t) = \bar{f}(\hat{x}(t), y(t), \sigma(t))$$

is well-posed and uniformly asymptotically stable.

The following two lemmas are obtained directly from Chapter 3.

Lemma 5.5. *System (5.1) is quadratically stabilizable if there exist gain matrices F_i, $i \in M$ such that the matrix pencil*

$$\left\{ \sum_{i \in M} w_i (A_i + B_i F_i) \colon w_i \geq 0, \sum_{i \in M} w_i = 1 \right\}$$

contains a Hurwitz matrix.

Lemma 5.6. *System (5.1) is linear feedback stabilizable if there exist gain matrices F_i, $i \in M$, a time sequence $h_1, \cdots, h_l$, and an index sequence $j_1, \cdots, j_l$, such that matrix*

$$e^{(A_{j_l} + B_{j_l} F_{j_l}) h_l} \cdots e^{(A_{j_1} + B_{j_1} F_{j_1}) h_1}$$

is Schur.

5.2 Multiple Controller and Sensor Systems

In this section, we consider the switched linear control system described by

$$\begin{cases} \dot{x}(t) = Ax(t) + B_\sigma u_\sigma(t) \\ y(t) = C_\sigma x(t) \end{cases} \tag{5.4}$$

where $x(t) \in \mathbf{R}^n$ is the state, $u_i(t) \in \mathbf{R}^p$, $i \in M$ are the control inputs, $y(t) \in \mathbf{R}^q$ is the measurable output, σ is the switching signal to be designed, and A, B_i, C_i are fixed matrices of compatible dimensions.

System (5.4) represents a linear plant with multiple control/sensor devices. The description includes the multi-controller switching and multi-sensor scheduling as special cases [117].

5.2.1 State Feedback Stabilization

In this subsection, let us consider the problem of stabilization via state feedback controllers and time/state-driven switching signals.

Recall that any switched linear system permits the canonical decomposition. For system (5.4), we have the following special type of canonical decomposition.

Lemma 5.7. *Switched system (5.4) is equivalent, via some state transformation, $z = Tx$, to system*

$$\dot{z}(t) = \begin{bmatrix} \dot{z}_1 \\ \dot{z}_2 \\ \dot{z}_3 \end{bmatrix} = \bar{A}z(t) + \bar{B}_\sigma u_\sigma(t)$$

$$y(t) = \bar{C}_\sigma z(t) \tag{5.5}$$

where

$$\bar{A} = TAT^{-1} = \begin{bmatrix} \bar{A}_{11} & \bar{A}_{12} & \bar{A}_{13} \\ 0 & \bar{A}_{22} & \bar{A}_{23} \\ 0 & 0 & \bar{A}_{33} \end{bmatrix} \qquad \bar{B}_i = TB_i = \begin{bmatrix} \bar{B}_{i,1} \\ \bar{B}_{i,2} \\ 0 \end{bmatrix}$$

$$\bar{C}_i = C_i T^{-1} = [0 \ \bar{C}_{i,1} \ \bar{C}_{i,2}] \quad i \in M. \tag{5.6}$$

In (5.5), z_1 is controllable but unobservable, z_2 is controllable and observable, and z_3 is uncontrollable.

It can be seen from the decomposition (5.5) that z_3 is decoupled from z_1, z_2 and the control input. As a consequence, system (5.5) is stabilizable only if matrix $\bar{A}_{33}$ is Hurwitz, or equivalently, the unstable mode of A is controllable. Let

$$\beta(s) = \Pi_{j=1}^l (s + s_j)$$

be the monic minimal polynomial of matrix A. Define polynomial

$$\beta^+(s) = \Pi_{\Re s_j \geq 0}(s + s_j).$$

Note that the subspace

$$\mathrm{Ker}\,\beta^+(A) = \mathrm{Ker}(\Pi_{\Re s_j \geq 0}(A + s_j I_n))$$

is the unstable mode of A. Let $\mathcal{C}$ denote the controllable subspace of system (5.4), and $n_2 = \dim \mathcal{C}$.

Assumption 5.1. $\mathrm{Ker}\,\beta^+(A) \subseteq \mathcal{C}$.

This assumption ensures that the uncontrollable mode is stable, that is, matrix $\bar{A}_{33}$ in (5.6) is Hurwitz.

Now, let us consider system (5.5) in canonical form. Denote

$$\hat{A}_1 = \begin{bmatrix} \bar{A}_{11} & \bar{A}_{12} \\ 0 & \bar{A}_{22} \end{bmatrix} \quad \hat{B}_i = \begin{bmatrix} \bar{B}_{i,1} \\ \bar{B}_{i,2} \end{bmatrix} \quad i \in M$$

and

$$\hat{B} = [\hat{B}_1, \cdots, \hat{B}_m].$$

Note that the controllable subspace of switched system $\Sigma(\hat{A}_1, \hat{B}_i)_M$ is exactly the controllable subspace of the pair $(\hat{A}_1, \hat{B})$. As the former is controllable, matrix pair $(\hat{A}_1, \hat{B})$ is controllable. Therefore, for any arbitrarily given set of desired (symmetric) poles $\Lambda = \{\lambda_1, \cdots, \lambda_{n_2}\}$ with negative real parts, we can construct a feedback gain matrix $G \in \mathbf{R}^{mp \times n_2}$, such that matrix $\hat{A}_1 + \hat{B}G$ possesses eigenvalue set Λ. Let us partition G as

$$G = \begin{bmatrix} G_1 \\ \vdots \\ G_m \end{bmatrix} \quad G_i \in \mathbf{R}^{p \times n_2}.$$

Fix a set of weighted factors, $w_i > 0$, $i \in M$ with $\sum_{i \in M} w_i = 1$, and define

$$F_i = \frac{1}{w_i} G_i \quad \bar{F}_i = [F_i, 0] \in \mathbf{R}^{p \times n} \quad i \in M.$$

For system (5.5) with the piecewise linear feedback control inputs

$$u_i(t) = \bar{F}_i z(t) = F_i \begin{bmatrix} z_1(t) \\ z_2(t) \end{bmatrix} \quad i \in M \tag{5.7}$$

the overall system is given by

$$\dot{z}(t) = (\bar{A} + \bar{B}_\sigma \bar{F}_\sigma) z(t). \tag{5.8}$$

Denote $\tilde{A}_i = \bar{A} + \bar{B}_i \bar{F}_i$, $i \in M$, and furthermore, the average matrix

$$\tilde{A} = \sum_{i \in M} w_i \tilde{A}_i = \begin{bmatrix} \hat{A}_1 + \hat{B}G & \hat{A}_2 \\ 0 & \hat{A}_{33} \end{bmatrix}$$

where $\hat{A}_2 = \begin{bmatrix} \bar{A}_{13} \\ \bar{A}_{23} \end{bmatrix}$. Because both diagonal blocks of $\tilde{A}$ are Hurwitz, matrix $\tilde{A}$ itself is also Hurwitz. From Lemma 5.5, system (5.8) is quadratically stabilizable.

In what follows, we briefly introduce two switching strategies for switched system (5.8). One is based on the average method and leads to periodic switching paths, and the other is based on the Lyapunov approach and results in state-feedback switching laws.

First, according to Lemma 2.10, there is a positive real number η such that

$$\exp(\tilde{A}_m w_m \rho) \cdots \exp(\tilde{A}_1 w_1 \rho) = \exp\left(\rho \tilde{A} + \rho^2 \Upsilon_\rho\right) \qquad (5.9)$$

for any $\rho \le \eta$, where entries of matrix Υ_ρ are bounded and continuous.

For any matrix E, let $\psi(E)$ denote the maximum real part of all its eigenvalues. Fix a positive real number $\epsilon < -\psi(\tilde{A})$ and select a δ such that

$$\psi(\tilde{A} + \delta \Upsilon_\delta) \le \psi(\tilde{A}) + \epsilon.$$

The existence of such a δ is guaranteed by (5.9) and the continuity of eigenvalues.

Define the periodic switching path as

$$\sigma(t) = \begin{cases} 1 & \mod (t, \delta) \in [0, w_1 \delta) \\ 2 & \mod (t, \delta) \in [w_1 \delta, (w_1 + w_2)\delta) \\ \vdots & \\ m & \mod (t, \delta) \in [\sum_{i=1}^{m-1} w_i \delta, \delta) \end{cases} \qquad \forall\, t \ge t_0. \qquad (5.10)$$

Simple analysis shows that, under this switching path and the feedback control law (5.7), the switched system is exponentially stable with convergence rate $-\psi(\tilde{A}) - \epsilon$.

Second, fix a positive number τ such that $\tilde{A} + \tau I_n$ is still Hurwitz. Fix a positive-definite matrix Q, and denote by P the unique, positive-definite solution of the Lyapunov equation

$$(\tilde{A} + \tau I_n)^T P + P(\tilde{A} + \tau I_n) = -Q.$$

Denote

$$Q_k = (\tilde{A}_k + \tau I_n)^T P + P(\tilde{A}_k + \tau I_n) \quad k \in M.$$

For any initial state $x(t_0) = x_0$, set

$$\sigma(t_0) = \arg\min\{x_0^T Q_1 x_0, \cdots, x_0^T Q_m x_0\}.$$

The following switching time/index sequences are defined recursively by

$$t_{k+1} = \inf\left\{t > t_k : \; x^T(t) Q_{\sigma(t_k)} x(t) > 0\right\}$$
$$\sigma(t_{k+1}) = \arg\min_{i \in M} \left\{x^T(t_{k+1}) Q_i x(t_{k+1})\right\} \quad k = 0, 1, \cdots. \qquad (5.11)$$

Under this state-feedback switching law and feedback control law (5.7), the switched system is well-posed and exponentially convergent at rate τ, *i.e.*, there is a positive number N such that

$$\|x(t)\| \le N e^{-\tau(t - t_0)} \|x_0\|.$$

The above analysis is summarized in the following theorem.

Theorem 5.8. *System (5.4) satisfying Assumption 5.1 is linear feedback stabilizable.*

Remark 5.9. If system (5.4) is completely controllable, then, we can select the feedback control law and the switching signal, such that the rate of convergence can be arbitrarily assigned for the closed-loop system. Indeed, in this case, the convergence rate of the average matrix $\tilde{A}$ can be arbitrarily assigned, and the convergence rate of the closed-loop system can arbitrarily approach that of the average system by sufficiently high frequency switching.

Remark 5.10. The switching index sequence can be arbitrarily assigned. That is, we can first activate subsystem 1, then switch to subsystem 2, then subsystem 3, *etc.* Alternatively, we can first activate subsystem 2, then, subsystem 3, then subsystem 1, *etc.* This feature is crucial in practice when some subsystems must be activated before others, as is often the case encountered in workshops. However, different sequences of switching may require different switching frequencies to ensure stability of the closed-loop system.

Remark 5.11. The ratios among weighted factors can be arbitrarily assigned. The only requirement is that $w_i \neq 0$ for all $i \in M$. In contrast, the assumption of $\sum_{i \in M} w_i = 1$ is technical. This flexibility of choosing weighted factors would be beneficial in some circumstances. For example, some control devices may be more reliable than others. In this case, the ratios of these devices can be set higher than others. In particular, if we choose $w_1 = \cdots = w_m = \frac{1}{m}$, then, the switching signal (5.10) is periodic with the same duration interval for any subsystem at each period. In this case, the closed-loop system behaves in a multi-rate control manner which can be easily implemented in practice.

5.2.2 State Estimator

In the previous subsection, we introduced the state feedback control scheme under the assumption that the state variables are available. When the measurement of the state variables is not available, we need additional dynamics to estimate them.

In this subsection, we propose a state estimator to approximate the state variables. For stabilization, we only need to estimate the state variables which are used in the feedback loop, that is, the controllable part as illustrated in (5.7). For this, it is necessary for this part to be completely observable. Let $\mathcal{W}$ denote the unobservable subspace of system (5.4).

Assumption 5.2. $\mathcal{C} \cap \mathcal{W} = \{0\}$.

Under this assumption, the canonical form (5.5) can be rewritten as

$$\dot{z}(t) = \begin{bmatrix} \dot{z}_2 \\ \dot{z}_3 \end{bmatrix} = \begin{bmatrix} \bar{A}_{22} & \bar{A}_{23} \\ 0 & \bar{A}_{33} \end{bmatrix} z(t) + \begin{bmatrix} \bar{B}_{\sigma 2} \\ 0 \end{bmatrix} u_\sigma(t)$$

$$y(t) = [\bar{C}_{\sigma 2} \ \bar{C}_{\sigma 3}] z(t) \tag{5.12}$$

where z_2 is controllable and observable and z_3 is uncontrollable.

We propose the following estimator (observer) for z_2

$$\dot{\hat{z}}_2(t) = \bar{A}_{22}\hat{z}_2(t) + L_\sigma[y(t) - \bar{C}_{\sigma 2}\hat{z}_2] + \bar{B}_{\sigma 2}u_\sigma(t) \tag{5.13}$$

where $u_i(t)$, $y(t)$ and σ are the inputs, output and switching signal of system (5.4), respectively, and matrices $L_1, \cdots, L_m \in \mathbf{R}^{n_2 \times q}$ are to be determined later.

The dynamical equation of the estimator can be rewritten as

$$\dot{\hat{z}}_2(t) = (\bar{A}_{22} - L_\sigma\bar{C}_{\sigma 2})\hat{z}_2(t) + L_\sigma y(t) + \bar{B}_{\sigma 2}u_\sigma(t). \tag{5.14}$$

Note that the estimator itself is a switched system, but the switching signal is the same as in system (5.4). Hence, the switching signal here is not an independent design variable.

Define the difference between the real state and the estimated state

$$\tilde{z}_2 = z_2 - \hat{z}_2. \tag{5.15}$$

Subtracting (5.14) from (5.12), we obtain

$$\dot{\tilde{z}}_2 = (\bar{A}_{22} - L_\sigma\bar{C}_{\sigma 2})\tilde{z}_2 + (\bar{A}_{23} - L_\sigma\bar{C}_{\sigma 3})z_3$$
$$\dot{z}_3 = \bar{A}_{33}z_3. \tag{5.16}$$

Theorem 5.12. *Under Assumptions 5.1 and 5.2, system (5.16) is exponentially stabilizable via the periodic switching signal (5.10).*

Proof. Denote matrix $\bar{C}_2 = \begin{bmatrix} \bar{C}_{12} \\ \vdots \\ \bar{C}_{m2} \end{bmatrix}$. It follows from Assumption 5.2 that matrix pair $(\bar{C}_2, \bar{A}_{22})$ is completely observable. Therefore, for any arbitrarily given set of desired (symmetric) poles $\Psi = \{\psi_1, \cdots, \psi_{n_2}\}$ with negative real parts, we can construct a feedback gain matrix $\bar{L} \in \mathbf{R}^{n_2 \times mq}$ such that matrix $\bar{A}_{22} - \bar{L}\bar{C}_2$ possesses eigenvalue set Ψ. Partition $\bar{L}$ as

$$\bar{L} = \begin{bmatrix} \bar{L}_1, \cdots, \bar{L}_m \end{bmatrix} \quad \bar{L}_i \in \mathbf{R}^{n_2 \times q}. \tag{5.17}$$

Furthermore, define

$$L_i = \frac{1}{w_i}\bar{L}_i \quad i \in M.$$

Compute

$$\sum_{i \in M} w_i \begin{bmatrix} \bar{A}_{22} - L_i\bar{C}_{i2} & \bar{A}_{23} - L_i\bar{C}_{i3} \\ 0 & \bar{A}_{33} \end{bmatrix} = \begin{bmatrix} \bar{A}_{22} - \bar{L}\bar{C}_2 & \bar{A}_{23} - \bar{L}\bar{C}_3 \\ 0 & \bar{A}_{33} \end{bmatrix} \tag{5.18}$$

where $\bar{C}_3 \overset{def}{=} \begin{bmatrix} \bar{C}_{13} \\ \vdots \\ \bar{C}_{m3} \end{bmatrix}$. Note that $\bar{A}_{22} - \bar{L}\bar{C}_2$ is Hurwitz, and it follows from

Assumption 5.1 that $\bar{A}_{33}$ is also Hurwitz. Consequently, the average matrix in (5.18) is Hurwitz. By Lemma 3.26, system (5.16) is exponentially stabilizable via a periodic switching signal in form (5.10) with a sufficiently high switching frequency. This completes the proof of the theorem. □

Remark 5.13. Due to Theorem 5.12, we call estimator (5.13) the *asymptotic state estimator*. Note that the error dynamics do not rely on the input, and the estimated state can track the real state asymptotically whether the real state converges or not. However, stability (and convergence rate) of the error system does depend on the switching frequency of the switching signal. The estimated state may not approach the real state if the switching is not fast enough.

Remark 5.14. From the proof, we see that, if the system is completely controllable, then, we can assign any pole set for the average system by appropriately selecting gain matrix L. Furthermore, if the switching frequency is sufficiently high, then the error dynamical system converges exponentially at a rate near that of the average system. Hence, the estimator can approximate the real state at any given rate of convergence.

5.2.3 Separation Principle

In this subsection, we show that the design of the state feedback and the design of the state estimator can be carried out independently for the problem of dynamic output feedback stabilization. This separation property enables us to solve the problem in a clear and constructive way.

Using estimated state (5.13) to substitute for the real state in feedback controller (5.7), we obtain

$$\dot{z}(t) = \begin{bmatrix} \dot{z}_2 \\ \dot{z}_3 \end{bmatrix} = \begin{bmatrix} \bar{A}_{22} & \bar{A}_{23} \\ 0 & \bar{A}_{33} \end{bmatrix} z(t) + \begin{bmatrix} \bar{B}_{\sigma 2} \\ 0 \end{bmatrix} u_\sigma(t)$$

$$y(t) = [\bar{C}_{\sigma 2} \ \bar{C}_{\sigma 3}] z(t)$$

$$u_i(t) = F_i \hat{z}_2(t) \quad i \in M$$

$$\dot{\hat{z}}_2(t) = \bar{A}_{22}\hat{z}_2(t) + L_\sigma[y(t) - \bar{C}_{\sigma 2}\hat{z}_2] + \bar{B}_{\sigma 2}u_\sigma(t). \tag{5.19}$$

Substituting the inputs and output expressions into the differential equations, we obtain the overall system given by

$$\dot{z}_2 = \bar{A}_{22}z_2 + \bar{B}_{\sigma 2}F_\sigma\hat{z}_2 + \bar{A}_{23}z_3$$

$$\dot{\hat{z}}_2 = (\bar{A}_{22} + \bar{B}_{\sigma 2}F_\sigma - L_\sigma\bar{C}_{\sigma 2})\hat{z}_2 + L_\sigma\bar{C}_{\sigma 2}z_2 + L_\sigma\bar{C}_{\sigma 3}z_3$$

$$\dot{z}_3 = \bar{A}_{33}z_3. \tag{5.20}$$

Denote $\omega = \left[z_2^T, \hat{z}_2^T, z_3^T\right]^T$, and

$$\Omega_i = \begin{bmatrix} \bar{A}_{22} & \bar{B}_{i2}F_i & \bar{A}_{23} \\ L_i\bar{C}_{i2} & \bar{A}_{22} + \bar{B}_{i2}F_i - L_i\bar{C}_{i2} & L_i\bar{C}_{i3} \\ 0 & 0 & \bar{A}_{33} \end{bmatrix} \quad i \in M.$$

We can rewrite the overall switched system (5.20) as

$$\dot{\omega}(t) = \Omega_\sigma \omega(t). \tag{5.21}$$

The average matrix of Ω_i, $i \in M$ under weighted factors w_i, $i \in M$ can be computed as

$$\Omega = \sum_{i \in M} w_i \Omega_i = \begin{bmatrix} \bar{A}_{22} & \bar{B}_2G & \bar{A}_{23} \\ \bar{L}\bar{C}_2 & \bar{A}_{22} + \bar{B}_2G - \bar{L}\bar{C}_2 & \bar{L}\bar{C}_3 \\ 0 & 0 & \bar{A}_{33} \end{bmatrix}.$$

It can be seen that this matrix is similar to a block triangular matrix with stable diagonal blocks. Indeed, let

$$T = \begin{bmatrix} I_{n_2} & 0 & 0 \\ I_{n_2} & -I_{n_2} & 0 \\ 0 & 0 & I_{n-n_2} \end{bmatrix}.$$

Simple calculation gives

$$T\Omega T^{-1} = \begin{bmatrix} \bar{A}_{22} + \bar{B}_2G & -\bar{B}_2G & \bar{A}_{23} \\ 0 & \bar{A}_{22} - \bar{L}\bar{C}_2 & \bar{A}_{23} - \bar{L}\bar{C}_3 \\ 0 & 0 & \bar{A}_{33} \end{bmatrix}.$$

Therefore, the characteristic polynomial of the average matrix is the product of those of the state feedback, the state estimator and the uncontrollable mode. As a result, the average system

$$\dot{\omega}(t) = \Omega \omega(t) \tag{5.22}$$

is stable. By Lemma 3.26, we obtain the main result of this section.

Theorem 5.15. *Under Assumptions 5.1 and 5.2, system (5.4) is dynamical output feedback stabilizable.*

Remark 5.16. From the above analysis, the separation property holds for the average system in terms of eigenvalue assignment. That is, for the switched system, the design of the state feedback controller (5.7) and the design of the observer gain matrices (5.17) can be carried out independently. However, there is no such separation property for the design of the switching signal. In fact, a switching signal in the form of (5.10) stabilizes the overall system (5.20) if and only if it stabilizes both the state feedback system (5.8) and the state estimator system (5.14).

5.2.4 Design Procedure and Illustrative Examples

In this subsection, we summarize the results presented in the previous subsections and carry out a systematic design procedure for the problem of dynamic output feedback stabilization.

Step 0. (Initialization) Suppose that system (5.4) is given with known parameters A, B_i and C_i, $i \in M$.

Step 1. Choose a group of linear independent column vectors from matrix $[B_1, \cdots, B_m, \cdots, A^{n-1}B_1, \cdots, A^{n-1}B_m]$ with maximum number, then extend it to a basis of $\mathbf{R}^n$. Let T be the matrix of transition from the standard basis of $\mathbf{R}^n$ to this basis. Compute

$$\bar{A} = TAT^{-1} \quad \bar{B}_i = TB_i \quad \bar{C}_i = C_iT^{-1} \quad i \in M.$$

These matrices must be of the form

$$\bar{A} = TAT^{-1} = \begin{bmatrix} \hat{A}_1 & \hat{A}_2 \\ 0 & \bar{A}_{33} \end{bmatrix} \quad \bar{B}_i = TB_i = \begin{bmatrix} \hat{B}_i \\ 0 \end{bmatrix}$$

$$\bar{C}_i = C_iT^{-1} = [\bar{C}_{i,1} \ \bar{C}_{i,2}] \quad i \in M. \tag{5.23}$$

Step 2. Verify if $\bar{A}_{33}$ is Hurwitz. If not, then exit. Otherwise, continue with the following steps.

Step 3. Fix weighted factors w_i, $i \in M$. Denote

$$\hat{B} = [w_1\hat{B}_1, \cdots, w_m\hat{B}_m].$$

Design matrix F such that matrix $\hat{A}_1 + \hat{B}F$ is Hurwitz or possesses a pre-assigned eigen-structure. Partition F as

$$F = [F_1, \ \cdots, \ F_m].$$

Step 4. Let $\hat{C}_1 = \begin{bmatrix} w_1\bar{C}_{11} \\ \vdots \\ w_m\bar{C}_{m1} \end{bmatrix}$. Verify if the pair $(\hat{C}_1, \hat{A}_1)$ is observable. If not, then exit. Otherwise, continue with the following steps.

Step 5. Design matrix L such that matrix $\hat{A}_1 - L\hat{C}_1$ is Hurwitz or possesses a pre-assigned eigen-structure. Partition L as $L = \begin{bmatrix} L_1 \\ \vdots \\ L_m \end{bmatrix}$.

Step 6. Let $\Omega_i = \begin{bmatrix} \hat{A}_1 & \hat{B}_iF_i & \hat{A}_2 \\ L_i\bar{C}_{i1} & \hat{A}_1 + \hat{B}_iF_i - L_i\bar{C}_{i1} & L_i\bar{C}_{i2} \\ 0 & 0 & \bar{A}_{33} \end{bmatrix}$, $i = 1, \cdots, m$. Find a positive real number δ such that matrix $\exp(\Omega_mw_m\delta) \cdots \exp(\Omega_1w_1\delta)$ is Schur.

Based on this procedure, the following conclusions can be drawn:

(i) If the procedure stops at Step 2, then the system is not stabilizable by any control/switching laws.

(ii) If the procedure stops at Step 4, then the switched system is stabilizable via switching signal (5.10) with the piecewise linear state feedback

$$u_i(t) = [F_i \; 0]Tx(t) \quad i \in M.$$

However, the output feedback stabilization problem does not admit a solution in general.

(iii) Otherwise, the system is dynamic output feedback stabilizable. Dynamic output feedback

$$u_i(t) = F_i \hat{z}_2(t) \quad i \in M$$
$$\dot{\hat{z}}_2(t) = (\hat{A}_1 + \hat{B}_\sigma F_\sigma - L_\sigma \bar{C}_{\sigma 1})\hat{z}_2(t) - L_\sigma y(t) \tag{5.24}$$

and switching signal (5.10) provide a solution for the stabilization problem of system (5.4).

In what follows, we present two numerical examples for illustration. The former focuses on the state feedback control scheme and the main issue is to compare the state-feedback switching scheme with the periodic switching scheme. The latter focuses on the observer design and the dynamic output feedback stabilization.

Example 5.17. Consider system (5.4) with $n = 5$, $m = 3$, and

$$A = \begin{bmatrix} 0 & 1 & 0 & 0 & 2 \\ 2 & 1 & -3 & 2 & -1 \\ -1 & 2 & 3 & 1 & -4 \\ -2 & 3 & 1 & 5 & 0 \\ 0 & 0 & 0 & 0 & -2 \end{bmatrix} \quad B_1 = \begin{bmatrix} 0 \\ 1 \\ 0 \\ 0 \\ 0 \end{bmatrix} \quad B_2 = \begin{bmatrix} 0 \\ 0 \\ 1 \\ 0 \\ 0 \end{bmatrix} \quad B_3 = \begin{bmatrix} 0 \\ 0 \\ 0 \\ 1 \\ 0 \end{bmatrix}. \tag{5.25}$$

This system is already in controllability canonical form (5.5). It has a one-dimensional, uncontrollable, but stable mode with a pole at $\phi_1 = -2$, and a four-dimensional, controllable mode which possesses three unstable poles and one stable pole.

Fix the equi-weighted factors $w_1 = w_2 = w_3 = \frac{1}{3}$. Let

$$B = [w_1 B_1, w_2 B_2, w_3 B_3].$$

For the average system pair (A, B), by assigning its controllable mode poles to $\phi_2 = \{-2.5, -3, -3.5, -4\}$, we obtain a feedback gain matrix as

$$F = \begin{bmatrix} F_1 \\ F_2 \\ F_3 \end{bmatrix} = \begin{bmatrix} 28.5 & 19.5 & -9 & 6 & 0 \\ -3 & 6 & 19.5 & 3 & 0 \\ -6 & 9 & 3 & 27 & 0 \end{bmatrix}. \tag{5.26}$$

The average matrix can be computed to be

$$A_0 = A + BF = \begin{bmatrix} 0 & 1 & 0 & 0 & 2 \\ 2 & 1 & -3 & 2 & -1 \\ -1 & 2 & 3 & 1 & -4 \\ -2 & 3 & 1 & 5 & 0 \\ 0 & 0 & 0 & 0 & -2 \end{bmatrix} \qquad (5.27)$$

which possesses pole set $\phi_1 \cup \phi_2$.

For the switched system, define the piecewise linear feedback control law as

$$u_i(t) = F_i x(t) \quad i \in M.$$

The switching signal $\sigma(t)$ will be designed later. The autonomous system with this control law becomes

$$\dot{x} = (A + B_\sigma F_\sigma)x(t). \qquad (5.28)$$

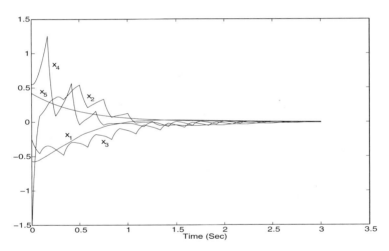

Fig. 5.1. State trajectory under periodic switching

Let us design switching signals for system (5.28).

First, applying the average method of Section 5.2.1 with $\epsilon = 0.5$, we can compute that any frequency higher than 3.98 will work, that is,

$$\psi(e^{A_3 w_1 \tau} e^{A_2 w_2 \tau} e^{A_1 w_3 \tau}) \leq e^{\varrho \tau} \qquad (5.29)$$

for any $\tau < \frac{1}{3.98}$, where

$$A_i = A + B_i F_i \quad i \in M \text{ and } \varrho = \max\left(\max(\phi_2) + \epsilon, \phi_1\right) = -2.$$

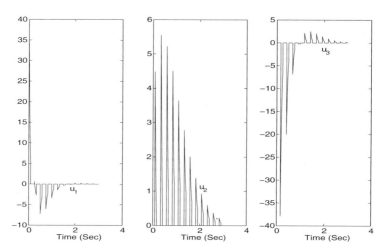

Fig. 5.2. Input trajectories under periodic switching

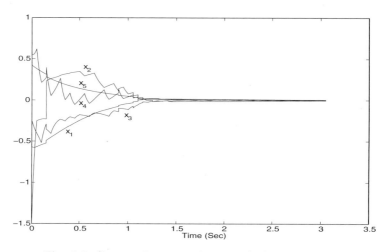

Fig. 5.3. State trajectory under state-feedback switching

Fix $\delta = 0.25$, and define a periodic switching path as in (5.10). The overall system is exponentially stable with the convergence rate of ϱ. Figure 5.1 depicts the state trajectory initialized at

$$x_0 = [-0.5703, \ -1.4986, \ -0.2517, \ 0.5530, \ 0.4175]^T.$$

It can be seen that the trajectory is not smooth at the switching instants. However, the uncontrollable mode, x_5, is always smooth because it does not rely on the switching signal. The bounded control effort of each subsystem is shown in Figure 5.2.

Second, applying the state-feedback strategy, we obtain a switching law as in (5.11). The state and input evolutions are given in Figures 5.3 and 5.4,

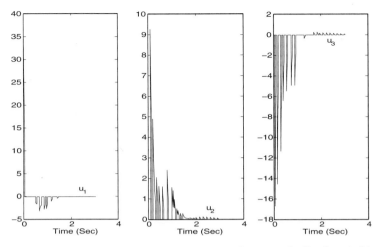

Fig. 5.4. Input trajectories of subsystems under state-feedback switching

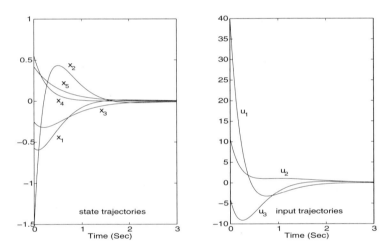

Fig. 5.5. State and input trajectories of the average system

respectively. For comparison, the state and input trajectories of the average system are depicted in Figure 5.5. It seems that there is no significant difference among the convergence rates. However, the control inputs are quite different.

Figure 5.6 shows the switching signals of both switching schemes. It clearly illustrates that the former is regular, while the latter is quite irregular. Surprisingly, the (average) frequency of the state-feedback switching is higher than that of the periodic switching. This indicates that the average method does not always lead to higher frequency of switching in comparison with other well-established switching strategies.

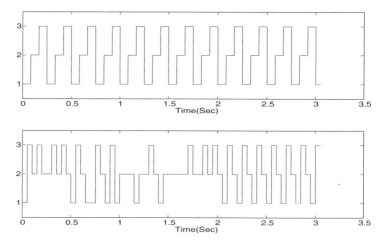

Fig. 5.6. Switching signals of the two schemes

Example 5.18. Consider system (5.4) with $n = 4$, $m = 2$, and

$$A = \begin{bmatrix} 1 & -1 & 1 & 2 \\ 0 & 0 & 1 & -1 \\ 0 & 3 & 2 & -1 \\ 0 & 0 & 0 & -2 \end{bmatrix} \quad B_1 = \begin{bmatrix} 1 \\ 0 \\ 0 \\ 0 \end{bmatrix} \quad B_2 = \begin{bmatrix} 0 \\ 0 \\ 1 \\ 0 \end{bmatrix}$$

$$C_1 = [1 \ -1 \ 0 \ 1] \quad C_2 = [0 \ 1 \ 0 \ -2]. \tag{5.30}$$

This system is already in canonical form (5.5). It has a one-dimensional, un-controllable, but stable mode with a pole at $\phi_1 = -2$, and a three-dimensional, controllable mode which possesses three unstable poles.

Partition the system matrices as in (5.23) and fix the equally weighted factors $w_1 = w_2 = \frac{1}{2}$. Let $\hat{B} = [w_1 \hat{B}_1, w_2 \hat{B}_2]$. For the matrix pair $(\hat{A}_1, \hat{B})$, by assigning its poles to $\phi_2 = \{-2.5, -3, -3.5\}$, we obtain a feedback gain matrix as

$$F = \begin{bmatrix} F_1 \\ F_2 \end{bmatrix} = \begin{bmatrix} -8.00 & 0.58 & -2.47 \\ -0.017 & -23.5 & -16 \end{bmatrix}.$$

Similarly, set $\hat{C}_1 = \begin{bmatrix} w_1 \bar{C}_{11} \\ w_2 \bar{C}_{21} \end{bmatrix}$ and $\phi_3 = \{-4, -4.5, -5\}$. Find a gain matrix L such that $\hat{A}_1 - L\hat{C}_1$ has the pole set ϕ_3. This yields

$$L = [L_1, L_2] = \begin{bmatrix} 17.38 & 22.0 \\ 6.39 & 22.0 \\ 41.25 & 90.0 \end{bmatrix}.$$

The average matrix can be computed to be

$$\Omega = \begin{bmatrix} 1.0 & -1.0 & 1.0 & -4.0 & 0.29 & -1.24 & 0 \\ 0 & 0 & 1.0 & 0 & 0 & 0 & 0 \\ 0 & 3.0 & 2.0 & -0.0 & -11.75 & -8.0 & 0 \\ 8.69 & 2.31 & 0 & -11.69 & -3.02 & -0.24 & -13.3 \\ 3.19 & 7.81 & 0 & -3.19 & -7.81 & 1.0 & -18.81 \\ 20.62 & 24.38 & 0 & -20.63 & -33.13 & -6.0 & -69.38 \\ 0 & 0 & 0 & 0 & 0 & 0 & -2.0 \end{bmatrix}$$

which possesses the pole set $\phi_1 \cup \phi_2 \cup \phi_3$.

For the switched system, define the state estimator as

$$\dot{\hat{z}}_2(t) = (\hat{A}_1 + \hat{B}_\sigma F_\sigma - L_\sigma \bar{C}_{\sigma 1})\hat{z}_2(t) - L_\sigma y(t)$$

and the feedback control law as

$$u_i(t) = F_i \hat{z}_2(t) \quad i \in M.$$

The switching signal $\sigma(t)$ will be designed later. The overall system with this control law becomes

$$\dot{\omega} = \Omega_\sigma \omega(t). \tag{5.31}$$

Now, we design the switching signal for system (5.31).

First, let us require that the closed-loop system has an average convergence rate of $r = -2.0$. Numerical computation shows that any periodic switching with base period less than 0.77 will work, that is, matrix $\exp(\Omega_2 w_2 \delta)\exp(\Omega_1 w_1 \delta)$ has spectral radius less than $e^{-r\delta}$ for any $\delta < 0.77$.

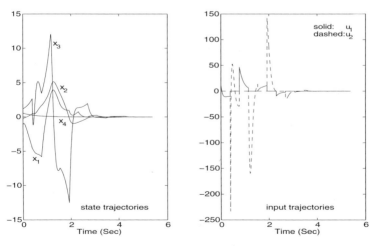

Fig. 5.7. State and input trajectories with $\delta = 0.77$

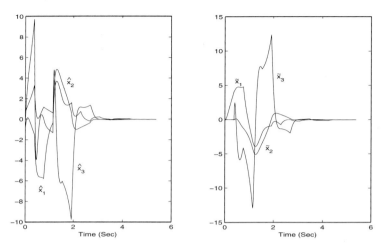

Fig. 5.8. Trajectories of the estimator and the error dynamics

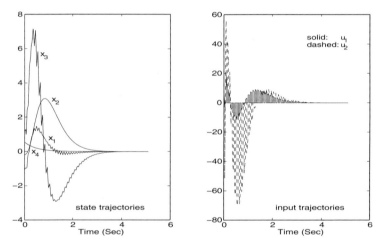

Fig. 5.9. State and input trajectories of the overall system with $\delta = 0.1$

Let us fix the period at $\delta = 0.77$, and define a periodic switching law as in (5.10). Figure 5.7 depicts the state and input trajectories initialized at

$$\omega_0 = [-1.1398, -0.2111, 1.1902, -1.1162, 0.6353, -0.6014, 0.5512]^T.$$

It can be seen that the state trajectory is not smooth at switching instants. However, the uncontrollable mode x_4 is always smooth because it does not rely on the switching law. The trajectories of the estimator and the error dynamics are shown in Figure 5.8.

For the switching signal, a smaller period would result in a system behavior that better resembles that of the average system. Figure 5.9 shows the state and input trajectories with period $\delta = 0.1$. Compared with the previous

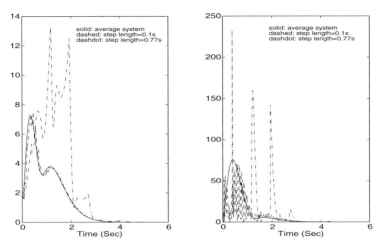

Fig. 5.10. Norms of the state trajectories (left) and input trajectories (right)

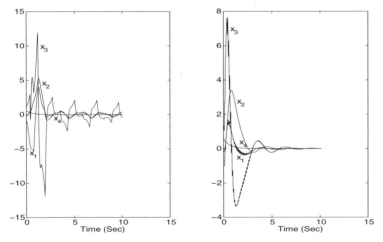

Fig. 5.11. State trajectories of the disturbed system with $\delta = 0.77$ (left) and $\delta = 0.10$ (right)

figures, this one has more switches but appears much more like a linear time-invariant system. For comparison, norms of the state and output trajectories of the average system, and of the switched system with $\delta = 0.10$ and $\delta = 0.77$, respectively, are depicted in Figure 5.10. It can be seen that the system with the larger period deviates (from the average system) more than that of the smaller one in the transient process.

Finally, it is interesting to simulate the robust performance of close-loop system. Consider the disturbed switched system given by

$$\dot{x}(t) = A_\sigma x(t) + f(x(t)) \tag{5.32}$$

where $f(x) = [0.5\sin(x_1), 0.5\sin(x_2), 0.5\sin(x_3), 0.5\sin(x_4)]^T$. The gain matrices F and L are the same as in the above. The state trajectories under periodic switching paths with $\delta = 0.77$ and $\delta = 0.10$ are shown in Figure 5.11. For this example, it is clear that the system with the smaller period performs better in attenuating disturbance.

5.3 A Stabilizing Strategy with Dwell Time

In this section, we investigate the problem of stabilization for a switched linear system where the summation of controllable subspaces of the individual subsystems is the total state space.

Consider a switched linear system given by

$$\dot{x}(t) = A_\sigma x(t) + B_\sigma u_\sigma(t). \tag{5.33}$$

Let $\mathcal{D}_i$ denote the controllable subspace of the ith subsystem, $i.e.$,

$$\mathcal{D}_i = \sum_{j=0}^{n-1} A_i^j \operatorname{Im} B_i.$$

Assumption 5.3. $\sum_{i \in M} \mathcal{D}_i = \mathbf{R}^n$.

By this assumption, the summation of the controllable subspaces of the individual subsystems is the total state space. As a result, the switched system is controllable.

Define a sequence of subspaces of $\mathbf{R}^n$ described by

$$\mathcal{W}_1 = \cap_{i \in M} \mathcal{D}_i$$
$$\mathcal{W}_k = \sum_{i_1, \cdots, i_{k-1} \in M} (\cap_{j \notin \{i_1, \cdots, i_{k-1}\}} \mathcal{D}_j) \quad k = 2, \cdots, m.$$

It is obvious that $\mathcal{W}_1 \subset \mathcal{W}_2 \subset \cdots \subset \mathcal{W}_m = \sum_{i \in M} \mathcal{D}_i = \mathbf{R}^n$.

A basis of $\mathbf{R}^n$ can be constructed according to the following procedure. First, choose a group of base vectors $\gamma_1, \cdots, \gamma_{s_1}$ in $\mathcal{W}_1$. Then, expand them to $\gamma_1, \cdots, \gamma_{s_1}, \gamma_{s_1+1}, \cdots, \gamma_{s_2}$, which form a basis of $\mathcal{W}_2$. Continuing with this process, we finally find a basis $\gamma_1, \cdots, \gamma_n$ of $\mathcal{W}_m = \mathbf{R}^n$. Let T be the matrix that transforms the standard basis of $\mathbf{R}^n$ to this basis.

Let $z(t) = T^{-1}x(t)$. System (5.33) can be re-written as

$$\dot{z} = T^{-1}A_\sigma T z + T^{-1}B_\sigma u_\sigma. \tag{5.34}$$

Suppose that $\dim \mathcal{D}_i = k_i$ and let indices $j_1^i < j_2^i < \cdots < j_{k_i}^i$ be such that

$$\operatorname{span}\{\gamma_{j_1^i}, \cdots, \gamma_{j_{k_i}^i}\} = \mathcal{D}_i.$$

Let

$$y_i = [z_{j_1^i}, \cdots, z_{j_{k_i}^i}]^T \quad i \in M$$

$$y^i = [z_1, \cdots, z_{j_1^i-1}, z_{j_1^i+1}, \cdots, z_{j_2^i-1}, \cdots, z_{j_{k_i}^i+1}, \cdots, z_n]^T \quad i \in M.$$

Let $\lambda_{i,1}, \cdots, \lambda_{i,n-k_i}$ be the real parts of the eigenvalues of A_i corresponding to the uncontrollable mode of system (A_i, B_i). Define

$$\lambda_i = \max\{\lambda_{i,1}, \cdots, \lambda_{i,n-k_i}\} \quad i \in M.$$

If $\lambda_i < 0$ for some i, then the i-th subsystem is stabilizable, and the constant switching signal $\sigma \equiv i$ will make the switched system stable. In this case, the switched system degenerates into a stabilizable linear time-invariant system. To avoid this trivial case, we assume that $\lambda_i \geq 0$ for each $i \in M$. Define

$$\eta = \max\{\|A_i\|_\infty : i \in M\}.$$

Fix an $r > m\eta + 1$. It is routine from linear system theory that we can construct feedback gain matrices F_i, $i \in M$, such that each of the real parts of the eigenvalues of $A_i + B_i F_i$ corresponding to the controllable mode of (A_i, B_i) is less than $-r$. Applying linear feedback input

$$u_i(t) = F_i x(t) \quad i \in M \tag{5.35}$$

Equation (5.34) becomes

$$\dot{z} = T^{-1}(A_\sigma + B_\sigma F_\sigma)Tz. \tag{5.36}$$

For notational convenience, let $\| \cdot \|$ denote the l_∞ norm of a vector. As y_i is in the controllable subspace of subsystem (A_i, B_i), the real parts of the corresponding poles are less than $-r$. Therefore, if $\sigma(t) = i$ for $t \in [t_1, t_2]$, then

$$\|y_i(t_2)\| \leq c_i e^{-r(t_2-t_1)}\|y_i(t_1)\| \tag{5.37}$$

for some positive constant c_i. On the other hand, for any $t_2 > t_1$, we have

$$\|z(t_2)\| \leq e^{\eta(t_2-t_1)}\|z(t_1)\|. \tag{5.38}$$

Let $c = \max\{c_1, \cdots, c_m, 1\}$. Fix a positive real number $\tau > \dfrac{(m+1)\ln c}{r - m\eta}$. We are now ready to formulate a stabilizing switching strategy for system (5.36).

For any given initial state $z_0 = z(t_0)$, there exists an integer i_0, $i_0 \in M$, such that $\|y_{i_0}(t_0)\| = \|z_0\|$. Define recursively the following switching time/index sequences by

$$t_k = \inf\{t : t \geq t_{k-1} + \tau, \|y_{i_{k-1}}(t)\| \leq \|y^{i_{k-1}}(t)\|\}$$
$$i_k = \arg_{k \in M}\{\|y_k(t_k)\| = \|z(t_k)\|\} \quad k = 1, 2, \cdots. \tag{5.39}$$

Theorem 5.19. *Suppose that switched system (5.33) satisfies Assumption 5.3. Then, under the piecewise linear state feedback control law (5.35) and the switching law (5.39), the closed-loop system is asymptotically stable.*

Proof. For any fixed $j \geq 0$, consider the interval $[t_j, t_{j+m+1}]$. We are to prove that the rate of convergence of $\|z(t)\|$ in this interval is less than

$$\delta \stackrel{def}{=} \frac{(m+1)\ln c - (r - m\eta)\tau}{(m+1)\tau} < 0.$$

Accordingly, the closed-loop system is asymptotically stable.

To this end, denote

$$\Lambda = \{l : j \leq l \leq j+m, t_{l+1} - t_l > \tau\}.$$

From (5.37) and (5.38), it follows that

$$\|z(t_{l+1})\| \begin{cases} = \|y_{k_l}(t_{l+1})\| \leq ce^{-rh_l}\|z(t_l)\| & \text{if } l \in \Lambda \\ \leq e^{\eta h_l}\|z(t_l)\| = e^{\eta\tau}\|z(t_l)\| & \text{otherwise} \end{cases} \quad j \leq l \leq j+m \quad (5.40)$$

where $h_l = t_{l+1} - t_l$, $l = j, \cdots, j+m$.

On one hand, if $\Lambda \neq \emptyset$, then from (5.40), and the fact that

$$\delta > \frac{(m+1)\ln c + \rho\eta\tau - r\sum_{j\in\Lambda} h_j}{t_{j+m+1} - t_j}$$

we have

$$\|z(t_{j+m+1})\| \leq c^{m+1}e^{(\rho\eta\tau - r\sum_{j\in\Lambda} h_j)}\|z(t_j)\| \leq e^{\delta(t_{j+m+1}-t_j)}\|z(t_j)\|$$

where ρ is the number of elements of set $\{l : j \leq l \leq j+m, h_l = \tau\}$.

On the other hand, if $\Lambda = \emptyset$, then $h_l = \tau$, $l = j, \cdots, j+m$. Because $i_l \in \{1, \cdots, m\}$ for $l = j, \cdots, j+m$, there exist integers l_1 and l_2, $j \leq l_1 < l_2 \leq j+m$, such that $k_{l_1} = k_{l_2}$. From (5.40) and (5.38), it follows that

$$\begin{aligned}
\|z(t_{j+m+1})\| &\leq c^{j+m+1-l_2}e^{(j+m+1-l_2)\eta\tau}\|z(t_{l_2})\| \\
&= c^{j+m+1-l_2}e^{(j+m+1-l_2)\eta\tau}\|y_{k_{l_2}}(t_{l_2})\| \\
&\leq c^{j+m-l_1}e^{(j+m-l_1)\eta\tau}\|y_{k_{l_2}}(t_{l_1+1})\| \\
&\leq c^{j+m+1-l_1}e^{(j+m-l_1)\eta\tau-r\tau}\|y_{k_{l_1}}(t_{l_1})\| \\
&= c^{j+m+1-l_1}e^{(j+m-l_1)\eta\tau-r\tau}\|z(t_{l_1})\| \\
&\leq c^{m+1}e^{m\eta\tau-r\tau}\|z(t_j)\| \\
&= e^{\delta(m+1)\tau}\|z(t_j)\| \\
&= e^{\delta(t_{j+m+1}-t_j)}\|z(t_j)\|. \quad \square
\end{aligned}$$

Remark 5.20. Note that for each subsystem, the convergence rate of the controllable sub-dynamics dominates the divergence rate of uncontrollable sub-dynamics. Consequently, any periodic switching with a sufficiently large dwell time can also lead to asymptotic stability. That is, for a sufficiently large dwell time τ, define the periodic switching sequence as

$$\{(0,1), \cdots, ((m-1)\tau, m), (m\tau, 1), \cdots, ((2m-1)\tau, m), \cdots\} \quad (5.41)$$

then, the closed-loop system of (5.33), (5.35) and (5.41) is asymptotically stable.

Example 5.21. Let $n = 5, m = 3$, and

$$A_1 = \begin{bmatrix} 0 & 0 & 0 & 1 & 0 \\ 0 & 1 & 0 & 0 & 0 \\ 0 & 0 & 0 & 0 & 0 \\ 0 & 0 & 0 & 0 & 0 \\ 0 & 0 & 0 & 0 & 1 \end{bmatrix} \quad B_1 = \begin{bmatrix} 0 \\ 0 \\ 0 \\ 1 \\ 0 \end{bmatrix} \quad A_2 = \begin{bmatrix} 0 & 1 & 0 & 0 & 0 \\ 0 & 0 & 0 & 0 & 0 \\ 0 & 0 & 0 & 0 & 0 \\ 0 & 0 & 0 & 1 & 0 \\ 0 & 0 & 0 & 0 & 2 \end{bmatrix} \quad B_2 = \begin{bmatrix} 0 & 0 \\ 1 & 0 \\ 0 & 1 \\ 0 & 0 \\ 0 & 0 \end{bmatrix}$$

$$A_3 = \begin{bmatrix} 0 & 0 & 0 & 0 & 1 \\ 1 & 0 & 0 & 0 & 0 \\ 0 & 0 & -1 & 0 & 0 \\ 0 & 0 & 0 & 1 & 0 \\ 0 & 0 & 0 & 0 & 0 \end{bmatrix} \quad B_3 = \begin{bmatrix} 0 \\ 0 \\ 0 \\ 0 \\ 1 \end{bmatrix}.$$

It can be verified that Assumption 5.3 holds. According to Theorem 5.19, this switched system is asymptotically stabilizable.

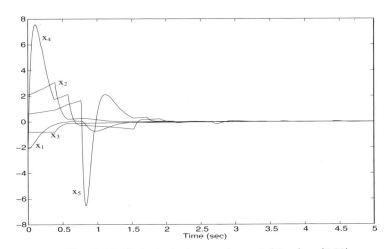

Fig. 5.12. State trajectory under switching law (5.39)

By fixing $r = 10$ and $\tau = 0.2$, a stabilizing state feedback and switching strategy can be obtained accordingly. Let the initial state be given by

$$x(0) = [-2.0452, 2.0757, -0.7796, -2.7625, 0.6311]^T.$$

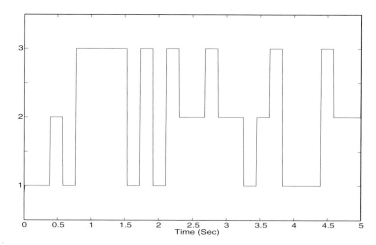

Fig. 5.13. Switching signal under switching law (5.39)

Figure 5.12 shows the convergence of the state, while Figure 5.13 gives the corresponding switching path. As shown in Figure 5.13, neither the switching index sequence nor the switching time sequence is periodic.

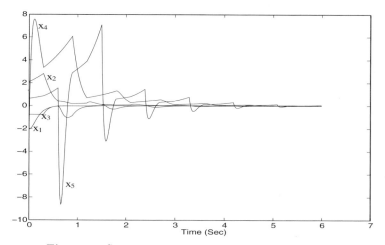

Fig. 5.14. State trajectory under a periodic switching path

As pointed in Remark 5.20, for a sufficiently large dwell time, a periodic switching path also results in a stable closed-loop system. Figure 5.14 shows the convergence of the state with dwell time $\tau = 0.3$. Simulation study reveals that, for any dwell time smaller than 0.25, the state trajectory of the closed-loop system diverges to infinity exponentially.

5.4 General Controllable Systems

In this section, we discuss the relationship between controllability and stabilizability. By definition, complete controllability means that the system is
stabilizable (in a finite time) via appropriate open-loop control input and time-
driven switching signal, see Section 4.3.3 for constructive design procedures.
It is natural to ask: Does controllability imply feedback stabilizability or even
quadratic stabilizability? As we will see, the question is quite complicated and
challenging.

5.4.1 General Results

Consider a controllable switched linear system given by

$$\dot{x}(t) = A_\sigma x(t) + B_\sigma u_\sigma(t). \tag{5.42}$$

It is well known that a controllable linear system is both linear state feedback stabilizable and quadratically stabilizable. For switched linear systems,
the relationship between the controllability and the stabilizability is much
more complicated. In the following, we present an example which is controllable but not linear state feedback stabilizable. For this, we need a supporting
lemma which sets a necessary condition for linear feedback stabilizability.

Lemma 5.22. *Switched linear system (5.42) is not piecewise linear feedback
stabilizable, if for each sequence of gain matrices F_i, $i \in M$, there exists a
one-dimensional subspace $\mathcal{W}$ of $\mathbf{R}^n$, such that $\mathcal{W}$ is $(A_i + B_i F_i)$-invariant
and $(A_i + B_i F_i)|_{\mathcal{W}}$ is unstable for all $i \in M$.*
Proof. As $\mathcal{W}$ is of dimension one, $(A_i + B_i F_i)|_{\mathcal{W}}$ is in fact a scalar matrix with
a positive real entry. Since $\mathcal{W}$ is invariant under $(A_i + B_i F_i)$ for all $i \in M$, any
state trajectory initialized from $\mathcal{W}$ will remain in the subspace and hence diverge under arbitrary switching signal. This implies that the switched system
is not stabilizable by means of gain matrices F_i, $i \in M$. The assumption of arbitrariness of the gain matrices in the theorem clearly excludes the possibility
of the linear state feedback stabilizability of the switched system. $\square$

Example 5.23. Consider system (5.42) with $n = 3$, $m = 2$, and

$$A_1 = \begin{bmatrix} 0\,0\,0 \\ 1\,1\,0 \\ 0\,0\,1 \end{bmatrix} \quad B_1 = \begin{bmatrix} 1 \\ 0 \\ 0 \end{bmatrix} \quad A_2 = \begin{bmatrix} 0\,0\,0 \\ 0\,1\,0 \\ 1\,0\,1 \end{bmatrix} \quad B_2 = \begin{bmatrix} 0 \\ 0 \\ 0 \end{bmatrix}. \tag{5.43}$$

It can be verified that the system is completely controllable.
    For any gain matrices F_1 and F_2 with

$$F_i = [f_{i1}, f_{i2}, f_{i3}] \quad i = 1, 2$$

it is clear that

$$A_1 + B_1 F_1 = \begin{bmatrix} f_{11} & f_{12} & f_{13} \\ 1 & 1 & 0 \\ 0 & 0 & 1 \end{bmatrix} \text{ and } A_2 + B_2 F_2 = \begin{bmatrix} 0 & 0 & 0 \\ 0 & 1 & 0 \\ 1 & 0 & 1 \end{bmatrix}.$$

If $f_{12}^2 + f_{13}^2 \neq 0$, it can be verified that subspace

$$\mathcal{W} = \text{span}\{ \begin{bmatrix} 0 \\ -f_{13} \\ f_{12} \end{bmatrix} \}$$

is invariant under $A_i + B_i F_i$ for $i = 1, 2$. Otherwise, subspace

$$\mathcal{W} = \text{span}\{ \begin{bmatrix} 0 \\ 1 \\ 0 \end{bmatrix} \}$$

is invariant under $A_i + B_i F_i$ for $i = 1, 2$. In either case, we have

$$(A_i + B_i F_i)|_{\mathcal{W}} = 1 \quad i = 1, 2$$

which are unstable. By Lemma 5.22, the switched linear control system is not linear feedback stabilizable. □

It is interesting to notice that the unstable sub-dynamics (subspace) rely on the (parameters of) gain matrices. In other words, different sequences of gain matrices may correspond to different unstable sub-dynamics. This means that piecewise linear gain matrices are not always able to eliminate unstable common sub-dynamics. To overcome this intrinsic problem, we can either search for nonlinear feedback controllers or extend the scheme of piecewise linear feedback stabilization (for example, a subsystem is associated with more than one linear controller). This is an important subject for further investigation.

Next, we assume that the system is single-input and in the controllable normal form. That is, we restrict systems $\sum(A_i, B_i)_M$ with

$$B_1 = [1, 0, \cdots, 0]^T \quad B_j = 0 \quad j \neq 1.$$

The following theorem establishes a simple sufficient condition for feedback stabilizability.

Theorem 5.24. *For a single-input switched system $\sum(A_i, B_i)_M$, suppose that there exists a sequence of real numbers w_i, $i \in M$, such that matrix pair $(\sum_{i \in M} w_i A_i, B_1)$ is controllable. Then, the switched system is quadratically stabilizable.*

Proof. Without loss of generality, we assume that each w_i is nonnegative and $w_1 > 0$ (see Remark 5.26 below). Let $A_0 = \sum_{i \in M} w_i A_i$ and $B_0 = w_1 B_1$. As (A_0, B_1) is controllable, (A_0, B_0) is also controllable. Therefore, we can find a feedback gain matrix F_1 such that $A_0 + B_0 F_1$ is Hurwitz. Introducing the linear state feedback control input

$$u_1(t) = F_1 x(t)$$

the closed-loop switched system is $\sum (A_i + B_i F_i)_M$ with $F_j = 0$, $j \neq 1$. Note that

$$\sum_{i \in M} w_i (A_i + B_i F_i) = A_0 + B_0 F_1$$

is Hurwitz. By Lemma 5.5, the switched system is quadratically stabilizable. □

Remark 5.25. From the proof, the eigenvalues of the average matrix $A_0 + B_0 F_1$ can be arbitrarily (symmetrically) assigned by appropriately choosing F_1. On the other hand, the convergence rate of the switched system can arbitrarily approach that of the average system by a suitable switching signal. As a result, the convergence rate of the switched system can be arbitrarily pre-assigned.

Remark 5.26. Note that the controllability of $(\sum_{i \in M} w_i A_i, B_1)$ for a sequence $w_1, \cdots, w_m$ implies the controllability of $(\sum_{i \in M} w_i A_i, B_1)$ for almost any sequence $w_1, \cdots, w_m$ in $\mathbf{R}^m$. Here "for almost all parameter values" is to be understood as "for all parameter values except for those in some proper algebraic variety in the parameter space". In other words, the set

$$\{(w_1, \cdots, w_m) \colon (\sum_{i \in M} w_i A_i, B_1) \text{ is not controllable}\}$$

is a variety in $\mathbf{R}^m$. This comes from the fact that the controllability is a generic property (see, *e.g.*, [160, 35]). As a result, controllability is preserved in an open and dense set of $\mathbf{R}^m$. This fact plays a crucial role in the design of the stabilizing strategy for general switched systems in the next section.

5.4.2 Low Dimensional Systems

In this subsection, we apply Theorem 5.24 to low-dimensional controllable switched systems.

For $n = 2$, a controllable system satisfies either of the two cases:

(i) rank$[B_1, A_1 B_1] = 2$;
(ii) there is a $j \in M$, $j \neq 1$ such that rank$[B_1, A_j B_1] = 2$.

In the former case, the condition of Theorem 5.24 is satisfied with $w_1 = 1$ and $w_i = 0$, $i = 2, \cdots, m$. In the latter case, the condition of Theorem 5.24 is satisfied with $w_j = 1$ and $w_i = 0$, $i \neq j$. As a result, any controllable planar switched system is quadratically stabilizable. This is summarized in the following theorem.

Theorem 5.27. *A second-order controllable switched linear system is quadratically stabilizable with any pre-assigned rate of convergence.*

For the single-input third-order system with two subsystems, as classified in Section 4.5.2, there are five normal forms as follows:

(a) $\bar{A}_1 = \begin{bmatrix} 0\,0\,0 \\ 1\,0\,0 \\ 0\,1\,0 \end{bmatrix}$ and $\bar{A}_2 = \begin{bmatrix} *\,*\,* \\ *\,*\,* \\ *\,*\,* \end{bmatrix}$;

(b) $\bar{A}_1 = \begin{bmatrix} 0\,0\,0 \\ 1\,0\,* \\ 0\,0\,* \end{bmatrix}$ and $\bar{A}_2 = \begin{bmatrix} 0\,*\,* \\ 0\,*\,* \\ 1\,*\,* \end{bmatrix}$;

(c) $\bar{A}_1 = \begin{bmatrix} 0\,0\,0 \\ 1\,0\,* \\ 0\,0\,* \end{bmatrix}$ and $\bar{A}_2 = \begin{bmatrix} *\,0\,* \\ *\,0\,* \\ 0\,1\,* \end{bmatrix}$;

(d) $\bar{A}_1 = \begin{bmatrix} 0\,0\,0 \\ 0\,0\,* \\ 0\,1\,* \end{bmatrix}$ and $\bar{A}_2 = \begin{bmatrix} 0\,*\,* \\ 1\,*\,* \\ 0\,*\,* \end{bmatrix}$;

(e) $\bar{A}_1 = \begin{bmatrix} 0\,0\,0 \\ 0\,*\,* \\ 0\,0\,* \end{bmatrix}$ and $\bar{A}_2 = \begin{bmatrix} 0\,0\,* \\ 1\,0\,* \\ 0\,1\,* \end{bmatrix}$.

It can be verified that, for forms (a), (c), (d) and (e), the condition of Theorem 5.24 is always satisfied; for forms (b), the condition of Theorem 5.24 is violated if and only if the normal form is

$$\bar{A}_1 = \begin{bmatrix} 0 & 0 & 0 \\ 1 & 0 & 0 \\ 0 & 0 & v_1 \end{bmatrix} \text{ and } \bar{A}_2 = \begin{bmatrix} 0 & v_2 & v_3 \\ 0 & v_4 & 0 \\ 1 & -v_1 & v_4 \end{bmatrix} \tag{5.44}$$

where $v_1, \cdots, v_4$ are arbitrary real numbers.

For switched system in the form (5.44), a detailed analysis based on Lemma 5.5 shows that, the system is quadratically stabilizable if and only if either $v_1 < 0$ or $v_4 < 0$. In other words, the system is not quadratically stabilizable for the case when $v_1 \geq 0$ and $v_4 \geq 0$.

For the case where the system is not quadratically stabilizable, it may be linear feedback stabilizable as illustrated in the following.

To stabilize systems in the form (5.44), we seek a linear feedback control input such that the closed-loop system is stabilizable by means of periodic switching signals. By Lemma 5.6, it suffices to find a feedback gain vector $f_1 = [f_{11}, f_{12}, f_{13}]$, and two positive real numbers h_1 and h_2, such that matrix

$$\exp\left(A_2 h_2\right) \exp\left((A_1 + b_1 f_1) h_1\right) \tag{5.45}$$

is Schur.

Fix a positive real number h_1. Let $f_{11} = -2\rho$, $f_{12} = -\rho^2$, and $f_{13} = \eta$, where ρ and η are real numbers to be determined. It can be computed that

$$\exp\left(\left(A_1 + b_1 f_1\right)h_1\right) = \exp\left(\begin{bmatrix} -2\rho & -\rho^2 & \eta \\ 1 & 0 & 0 \\ 0 & 0 & v_1 \end{bmatrix} h_1\right) =$$

$$\begin{bmatrix} \left(1 - h_1\rho\right)e^{-h_1\rho} & -h_1\rho^2 e^{-h_1\rho} & \dfrac{\eta\left(h_1\rho^2 e^{-h_1\rho}+h_1 v_1\rho e^{-h_1\rho}-v_1 e^{-h_1\rho}+v_1 e^{h_1 v_1}\right)}{(\rho+v_1)^2} \\[2ex] h_1 e^{-h_1\rho} & \left(1 + h_1\rho\right)e^{-h_1\rho} & -\dfrac{\eta\left(h_1\rho e^{-h_1\rho}+h_1 v_1 e^{-h_1\rho}+e^{-h_1\rho}-e^{h_1 v_1}\right)}{(\rho+v_1)^2} \\[2ex] 0 & 0 & e^{h_1 v_1} \end{bmatrix}.$$

The analytic expression of $\exp\left(A_2 h_2\right)$ can also be computed as

$$e^{A_2 h_2} = \begin{bmatrix} v_5 & v_6 & v_7 \\ 0 & e^{v_4 h_2} & 0 \\ v_8 & v_9 & v_{10} \end{bmatrix}$$

where $v_5, \cdots, v_{10}$ are analytic functions of h_2. Note that $v_1 v_8 + v_9$ is a nonzero function if and only if $v_2 \neq v_1 v_4$. As the function is analytic, it is nonzero for almost any h_2 (except for possibly isolated points) when $v_2 \neq v_1 v_4$.
Suppose that $v_2 \neq v_1 v_4$. Fix a positive h_2 such that $v_1 v_8 + v_9 \neq 0$. Let

$$\eta = \frac{-v_{10}}{v_1 v_8 + v_9}\left(\rho + v_1\right)^2.$$

With some manipulation, we can express $\left(A_2 h_2\right)\exp\left(\left(A_1 + b_1 f_1\right)h_1\right)$ in the form

$$\Lambda(\rho) = \begin{bmatrix} q_1(\rho)e^{-\rho h_1} & q_2(\rho)e^{-\rho h_1} & \dfrac{r_1+q_3(\rho)e^{-\rho h_1}}{(\rho+v_1)^2} \\[2ex] q_4(\rho)e^{-\rho h_1} & q_5(\rho)e^{-\rho h_1} & \dfrac{r_2+q_6(\rho)e^{-\rho h_1}}{(\rho+v_1)^2} \\[2ex] q_7(\rho)e^{-\rho h_1} & q_8(\rho)e^{-\rho h_1} & \dfrac{q_9(\rho)e^{-\rho h_1}}{(\rho+v_1)^2} \end{bmatrix}$$

where r_1, r_2 are fixed real numbers and $q_i(\cdot)$, $i = 1, \cdots, 9$ are polynomials of ρ whose degrees are less than 3.
It is clear that

$$\Lambda_\infty \overset{def}{=} \lim_{\rho\to\infty} \Lambda(\rho) = \begin{bmatrix} 0 & 0 & r_1 \\ 0 & 0 & r_2 \\ 0 & 0 & 0 \end{bmatrix}.$$

The spectral radius of this matrix is zero. As a result, for any given positive number $c < 1$, there is a ρ_ϵ such that

$$\mathrm{sr}\left(\Lambda(\rho)\right) \leq \epsilon \quad \forall\, \rho \geq \rho_\epsilon.$$

Pick such a ρ, the closed-loop switched system is exponentially stable with the convergence rate not less than $\dfrac{\ln \epsilon}{h_1 + h_2}$.

Example 5.28. For the controllable single-input normal system $\sum(A_i, b_i)_{\bar{2}}$ with

$$A_1 = \begin{bmatrix} 0 & 0 & 0 \\ 1 & 0 & 0 \\ 0 & 0 & 1 \end{bmatrix} \text{ and } A_2 = \begin{bmatrix} 0 & 0 & 0 \\ 0 & 1 & 0 \\ 1 & -1 & 1 \end{bmatrix}. \tag{5.46}$$

this corresponds to the form (5.44) with $v_2 \neq v_1 v_4$.

Let $h_1 = h_2 = 0.1$ and $\rho = 10$. By the above-mentioned design procedure, we have

$$f_1 = [-20.0000, -100.0000, 328.9121].$$

It can be verified that matrix $e^{A_1 h_1} e^{A_2 h_2}$ has spectral radius 0.4359 and hence is Schur stable. Therefore, system $\sum(A_1+b_1 f_1, A_2)$ is stable under the periodic switching signal

$$\sigma(t) = \begin{cases} 1 & \text{if mod } (t, 0.2) \leq 0.1 \\ 2 & \text{otherwise.} \end{cases}$$

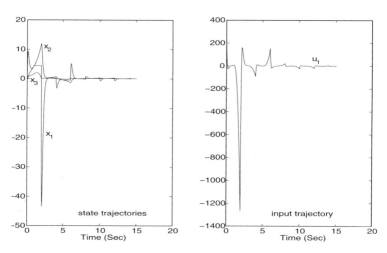

Fig. 5.15. State and input trajectories of system (5.46)

The above analysis shows that the system is linear feedback stabilizable. A sample state trajectory and the corresponding input trajectory are shown in Figure 5.15 where the initial state is

$$x(0) = [1.4435, -0.3510, 0.6232]^T. \quad \square$$

Next, we turn to the case that $v_2 = v_1 v_4$ in form (5.44). In this case, we can transform system $\sum(\bar{A}_i, \bar{B}_i)_{\bar{2}}$ into an equivalent form $\sum(\hat{A}_i, \hat{B}_i)_{\bar{2}}$ with

$$\hat{A}_1 = T\bar{A}_1 T^{-1} + T\bar{B}_1 f_1 = \begin{bmatrix} 0 & 0 & 0 \\ 1 & v_1 & 0 \\ 0 & 0 & v_1 \end{bmatrix}$$

$$\hat{A}_2 = T\bar{A}_2 T^{-1} = \begin{bmatrix} 0 & 0 & v_3 \\ 0 & v_4 & 0 \\ 1 & 0 & v_4 \end{bmatrix} \tag{5.47}$$

where

$$T = \begin{bmatrix} 1 & -v_1 & 0 \\ 0 & 1 & 0 \\ 0 & 0 & 1 \end{bmatrix} \text{ and } f_1 = [v_1 \ v_1^2 \ 0].$$

For system $\sum(\hat{A}_i, \bar{B}_i)_{\bar{2}}$, let

$$\bar{f}_1 = [-\bar{\rho}_1 - \bar{\rho}_2 + v_1, -\bar{\rho}_1\bar{\rho}_2, \bar{f}_{13}]$$

where $\bar{\rho}_1$, $\bar{\rho}_2$ and $\bar{f}_{13}$ will be determined later. Simple calculation gives

$$\exp\left((\hat{A}_1 + \bar{B}_1\bar{f}_1)\bar{h}_1\right) = \begin{bmatrix} \omega_1 & \omega_2 & \omega_3 \\ \omega_4 & \omega_5 & \omega_6 \\ 0 & 0 & 1 \end{bmatrix} e^{\bar{h}_1 \, v_1}$$

where

$$\omega_1 = \frac{-\bar{\rho}_2 \, e^{-\bar{h}_1 \bar{\rho}_2} + e^{-\bar{h}_1 \bar{\rho}_1}\bar{\rho}_1}{-\bar{\rho}_2 + \bar{\rho}_1} \qquad \omega_2 = \frac{\left(e^{-\bar{h}_1 \bar{\rho}_1} - e^{-\bar{h}_1 \bar{\rho}_2}\right)\bar{\rho}_2\bar{\rho}_1}{-\bar{\rho}_2 + \bar{\rho}_1}$$

$$\omega_3 = -\frac{\bar{f}_{13}\left(e^{-\bar{h}_1 \bar{\rho}_1} - e^{-\bar{h}_1 \bar{\rho}_2}\right)}{-\bar{\rho}_2 + \bar{\rho}_1} \qquad \omega_4 = -\frac{e^{-\bar{h}_1 \bar{\rho}_1} - e^{-\bar{h}_1 \bar{\rho}_2}}{-\bar{\rho}_2 + \bar{\rho}_1}$$

$$\omega_5 = \frac{-\bar{\rho}_2 \, e^{-\bar{h}_1 \bar{\rho}_1} + e^{-\bar{h}_1 \bar{\rho}_2}\bar{\rho}_1}{-\bar{\rho}_2 + \bar{\rho}_1}$$

$$\omega_6 = \frac{\bar{f}_{13}\left(-e^{-\bar{h}_1 \bar{\rho}_2}\bar{\rho}_1 + \bar{\rho}_2 \, e^{-\bar{h}_1 \bar{\rho}_1} - \bar{\rho}_2 + \bar{\rho}_1\right)}{\bar{\rho}_1\bar{\rho}_2 \left(-\bar{\rho}_2 + \bar{\rho}_1\right)}.$$

Denote

$$e^{\hat{A}_2 \bar{h}_2} = \begin{bmatrix} \bar{v}_5 & 0 & \bar{v}_6 \\ 0 & e^{v_4 \bar{h}_2} & 0 \\ \bar{v}_7 & 0 & \bar{v}_8 \end{bmatrix}.$$

Suppose that $v_3 > 0$, then we have $\bar{v}_8 > e^{v_4 \bar{h}_2}$. Let

$$\bar{\rho}_1 = \frac{1}{\bar{h}_1} \ln \frac{\bar{v}_8 - e^{v_4 \bar{h}_2}}{\bar{v}_8 e^{\bar{v}_1 \bar{h}_1}}$$

and

$$\bar{f}_{13} = \frac{1}{\bar{v}_7}(\bar{v}_8 e^{v_1 \bar{h}_1} + e^{v_4 \bar{h}_2}).$$

When $\bar{\rho}_2 \to \infty$, the spectral radius of matrix

$$\exp\left(\hat{A}_2 \bar{h}_2\right) \exp\left((\hat{A}_1 + \bar{B}_1 \bar{f}_1)\bar{h}_1\right)$$

approaches zero. As a result, for any given positive number $\epsilon < 1$, there is a ρ_ϵ such that

$$\mathrm{sr}\left(\exp\left(\hat{A}_2 \bar{h}_2\right) \exp\left((\hat{A}_1 + \bar{B}_1 \bar{f}_1)\bar{h}_1\right)\right) \leq \epsilon \quad \forall \, \bar{\rho}_2 > \rho_\epsilon.$$

Pick such a $\bar{\rho}_2$, the closed-loop switched system is exponentially stable with the convergence rate not less than $\dfrac{\ln \epsilon}{\bar{h}_1 + \bar{h}_2}$.

Now, suppose that $v_3 < -\frac{v_4^2}{4}$. In this case, let

$$\bar{f}_1 = [-2\rho + v_1, \ -\rho^2, \ 0]$$

where ρ is a positive real number to be determined. At the same time, let $\bar{h}_2$ be such that

$$v_4 \sin\left(\bar{h}_2 \sqrt{-v_3 + \frac{v_4^2}{4}}\right) + \sqrt{-4v_3 + v_4^2} \cos\left(\bar{h}_2 \sqrt{-v_3 + \frac{v_4^2}{4}}\right) = 0.$$

Simple manipulation yields

$$\bar{h}_2 = \frac{2}{\sqrt{-4v_3 + v_4^2}}\left(-\arctan(\frac{\sqrt{-4v_3 + v_4^2}}{v_4})\right) + k\pi \quad k = 1, 2, \cdots.$$

It can be calculated that matrix $\exp(\hat{A}_2 \bar{h}_2)$ is of form $\begin{bmatrix} * & 0 & * \\ 0 & * & 0 \\ * & 0 & 0 \end{bmatrix}$. From this, we can establish that

$$\lim_{\rho \to \infty} \exp(\hat{A}_2 \bar{h}_2) \exp(\hat{A}_1 \bar{h}_1) = \begin{bmatrix} 0 & 0 & * \\ 0 & 0 & 0 \\ 0 & 0 & 0 \end{bmatrix}.$$

As a result, for sufficiently large ρ, the closed-loop system is exponentially stable with any pre-assigned rate of convergence.

Example 5.29. Consider controllable single-input normal system $\sum(\hat{A}_i, \bar{B}_i)_{\bar{2}}$ with

$$\hat{A}_1 = \begin{bmatrix} 0 & 0 & 0 \\ 1 & 1 & 0 \\ 0 & 0 & 1 \end{bmatrix} \text{ and } \hat{A}_2 = \begin{bmatrix} 0 & 0 & 2 \\ 0 & 1 & 0 \\ 1 & 0 & 1 \end{bmatrix}. \tag{5.48}$$

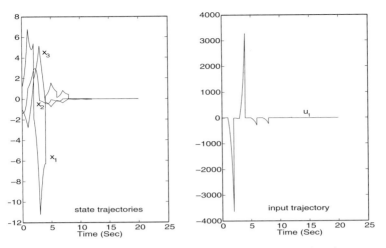

Fig. 5.16. State and input trajectories of system (5.48)

This corresponds to the form (5.47) with $v_3 > 0$.

Let $\bar{h}_1 = \bar{h}_2 = 1$ and $\bar{\rho}_2 = 500$. From the above-mentioned design procedure, we have

$$\bar{f}_1 = [-498.2269, 386.5457, -1078.5935].$$

It can be verified that matrix $e^{\hat{A}_1 \bar{h}_1} e^{\hat{A}_2 \bar{h}_2}$ has spectral radius 0.2592 and hence is Schur stable. As a result, system $\sum(\hat{A}_1 + \bar{B}_1 \bar{f}_1, \hat{A}_2)$ is stable under the periodic switching signal

$$\sigma(t) = \begin{cases} 1 & \text{if mod } (t, 2) \leq 1 \\ 2 & \text{otherwise.} \end{cases}$$

The above analysis shows that the system is feedback stabilizable. A sample state trajectory and the corresponding input trajectory are shown in Figure 5.16, where the initial state is

$$x(0) = [1.4435, -0.3510, 0.6232]^T. \quad \square$$

When $v_3 = 0$, it can be verified that Lemma 5.22 applies and hence the system is not linear feedback stabilizable. In the case that $-\frac{v_4^2}{4} \leq v_3 < 0$, the stabilizability of system (5.47) is still an open problem.

Finally, we consider multi-input switched systems. For a switched system $\Sigma(A_i, B_i)_{\bar{2}}$ with

$$\text{rank } B_1 + \text{rank } B_2 \geq 2$$

without loss of generality, we assume that

$$\text{rank } B_2 \leq \text{rank } B_1 \leq 2.$$

Let b_1 be the first column of B_1. According to the proof of Theorem 4.51, there exist a nonsingular matrix T, and feedback gain matrices F_1 and F_2, such that the single-input switched system $\Sigma(\bar{A}_i, \bar{b}_i)_{\bar{2}}$ is in the controllability normal form, where

$$\bar{A}_i = T(A_i + B_i F_i)T^{-1} \quad i = 1,2 \quad \bar{b}_1 = Tb_1 = e_1 \quad \bar{b}_2 = 0.$$

Let $\bar{B}_i = TB_i$ for $i = 1,2$. It is clear that the stabilizability of system $\Sigma(A_i, B_i)_{\bar{2}}$ is equivalent to the stabilizability of $\Sigma(\bar{A}_i, \bar{B}_i)_{\bar{2}}$. For the latter, it has been proven that the system is quadratically stabilizable except for the case when $\bar{A}_1$ and $\bar{A}_2$ are in the form (5.44). As a result, we only need to address this special case.

If rank $B_1 = 2$, then, there is a $\bar{b}_3 \notin \text{span}\{e_1\}$ such that

$$\bar{B}_1 = [\bar{b}_1, \bar{b}_3].$$

It can be verified that, there exist a gain matrix $\bar{F}_1$, and nonnegative real numbers w_1 and w_2, such that matrix

$$w_1(\bar{A}_1 + \bar{B}_1 \bar{F}_1) + w_2 \bar{A}_2$$

is Hurwitz, which means that system $\Sigma(\bar{A}_i, \bar{B}_i)_{\bar{2}}$ is quadratically stabilizable. Similarly, if rank $B_1 = $ rank $B_2 = 1$, it can be verified that, there always exist a gain matrix $\bar{F}_2$, and nonnegative real numbers w_1 and w_2, such that matrix

$$w_1 \bar{A}_1 + w_2(\bar{A}_2 + \bar{B}_2 \bar{F}_2)$$

is Hurwitz, which also means that system $\Sigma(\bar{A}_i, \bar{B}_i)_{\bar{2}}$ is quadratically stabilizable.

Summarizing the above analysis, we have the following theorem.

Theorem 5.30. *For a three dimensional controllable switched linear control system with two subsystems, $\Sigma(A_i, B_i)_{\bar{2}}$, we have the following conclusions:*

(i) if rank $B_1 + $ rank $B_2 \geq 2$, then it is quadratically stabilizable;
(ii) if the system is single-input, then it is quadratically stabilizable if it is not equivalent to normal form (5.44);
(iii) for normal system (5.44), it is quadratically stabilizable if and only if either $v_1 < 0$ or $v_4 < 0$; and
(iv) for normal system (5.44), it is linear feedback stabilizable if either $v_2 \neq v_1 v_4$ or $v_3 \notin [-\frac{v_4^2}{4}, 0]$.

5.5 General Systems in Controllability Canonical Form

In this section, we address the stabilizability of switched linear control systems which are not necessarily controllable.

Consider the general switched system in controllability canonical form

$$\dot{x}_1 = A_{1\sigma}x_1 + A_{2\sigma}x_2 + B_{1\sigma}u_\sigma$$
$$\dot{x}_2 = A_{3\sigma}x_2 \tag{5.49}$$

where $\sum(A_{1i}, B_{1i})_M$ is completely controllable.

To stabilize the system, we need to find a switching signal σ, and feedback control inputs u_i, $i \in M$, such that the closed-loop system is asymptotically stable. For this, a necessary condition for stabilizability is that the uncontrollable mode system $\sum(A_{3i})_M$ is stabilizable via suitable switching signal. In addition, a common switching signal has be be sought to stabilize both the controllable part and the uncontrollable part. This, of course, brings new challenges to the problem.

To utilize Lemma 5.5, we make a more restrictive assumption as follows.

Assumption 5.4. *For system (5.49), matrices $A_{31}, \cdots, A_{3m}$ have a stable convex combination.*

For the controllable mode $\sum(A_{1i}, B_{1i})_M$, it can be transformed into a controllable single-input system $\sum(\bar{A}_i, b_i)_M$ via a coordinate change $z_1 = T^{-1}x_1$ and a feedback reduction $u_i = F_i x_1 + G_i v_i$, $i \in M$.

Combining Theorem 5.24 and and Lemma 5.5, we obtain the following result.

Theorem 5.31. *For system (5.49), suppose that Assumption 5.4 holds. If there exists a sequence of real numbers w_i, $i \in M$, such that matrix pair $(\sum_{i \in M} w_i \bar{A}_i, b_1)$ is controllable, then the system is quadratically stabilizable.*
Proof. By Assumption 5.4, there is a stable convex combination of A_{3i}, namely, $A_{30} = \sum_{i \in M} \mu_i A_{3i}$, where μ_i, $i \in M$ are nonnegative real numbers. It is clear that there is a sufficiently small positive real number η, such that $\sum_{i \in M} \nu_i A_{3i}$ is still stable for any $\nu_i \in (\mu_i, \mu_i + \eta)$. On the other hand, the assumption of the theorem ensures that matrix pair $(\sum_{i \in M} w_i \bar{A}_i, b_1)$ is controllable for almost any sequence $w_1, \cdots, w_m$ (c.f. Remark 5.26). As a result, there is a sequence of positive real numbers w_i, $w_i \in (\mu_i, \mu_i + \eta)$, such that matrix pair $(\sum_{i \in M} w_i \bar{A}_i, b_1)$ is controllable. Let gain matrix f_1 be such that matrix $\sum_{i \in M} w_i \bar{A}_i + w_1 b_1 f_1$ is stable. Back to the original system (5.49), we introduce the linear feedback control inputs

$$u_i = (F_i + G_i f_i T^{-1})x_1 \stackrel{def}{=} \tilde{F}_i x_1 \quad i \in M$$

where $f_i = 0$ for $i \neq 1$. The closed-loop system reads as

$$\dot{x}_1 = (A_{1\sigma} + B_{1\sigma}\tilde{F}_i)x_1 + A_{2\sigma}x_2$$
$$\dot{x}_2 = A_{3\sigma}x_2.$$

Computing the convex combination with weighted factors ω_i gives

$$\sum \omega_i \begin{bmatrix} (A_{1\sigma} + B_{1\sigma}\tilde{F}_i\, A_{2\sigma} & A_{2\sigma} \\ 0 & A_{3\sigma} \end{bmatrix} = \begin{bmatrix} T(\sum \omega_i \bar{A}_i + b_1 f_1)T^{-1} & A_{2i} \\ 0 & \sum \omega_i A_{3i} \end{bmatrix}.$$

As both the diagonal sub-matrices are Hurwitz, the above matrix is also Hurwitz. By Lemma 5.5, the original system (5.49) is quadratically stabilizable. □

Corollary 5.32. *For system (5.49), suppose that Assumption 5.4 holds. If the controllable subspace is of dimension one or two, then the system is quadratically stabilizable.*

Example 5.33. Consider system $\sum (A_i, B_i)_{\bar{2}}$ with

$$A_1 = \begin{bmatrix} 0 & 0 & 0 & -1 & 2 \\ 1 & 0 & 0 & 3 & 0 \\ 0 & 0 & 1 & 1 & 0 \\ 0 & 0 & 0 & 1 & 2 \\ 0 & 0 & 0 & -2 & -3 \end{bmatrix} \quad B_1 = \begin{bmatrix} 1 \\ 0 \\ 0 \\ 0 \\ 0 \end{bmatrix}$$

$$A_2 = \begin{bmatrix} 0 & -3 & 1 & -2 & 0 \\ 0 & 1 & 0 & -1 & 3 \\ 1 & 0 & 1 & 2 & -2 \\ 0 & 0 & 0 & -2 & 3 \\ 0 & 0 & 0 & 0 & 1 \end{bmatrix} \quad B_2 = 0. \tag{5.50}$$

It is clear that the system is in the controllability canonical form and the controllable mode is in the single-input controllable normal form. Partition the system matrices as in (5.49). It can be seen that the third-order controllable mode is not in form (5.44). Simple verification shows that $\mu A_{31} + (1-\mu)A_{32}$ is stable for $\mu \in (\frac{1}{4}, \frac{2}{3})$. According to Theorem 5.31, the system is quadratically stabilizable.

Next, fix $\omega = 0.40$. By assigning the poles of pair $(\omega A_{31} + (1-\omega)A_{32}, \omega b_1)$ at $\{-1.0, -1.1, -1.2\}$, we obtain a gain matrix

$$f_1 = [13.2500, 99.0000, -116.08333].$$

By introducing the state feedback control law $u_1 = f_1 x_1$, the system is transformed into an autonomous system, which possesses a stable convex combination. Applying the state-feedback switching strategy (3.21), we obtain a stabilizing switching signal. Take initial condition

$$x(0) = [-1.0106, 0.6145, 0.5077, 1.6924, 0.5913]^T.$$

Figure 5.17 shows the state and input trajectories. It can be seen that the state trajectory looks quite 'smooth' in appearance, which indicates that the switching frequency is quite high.

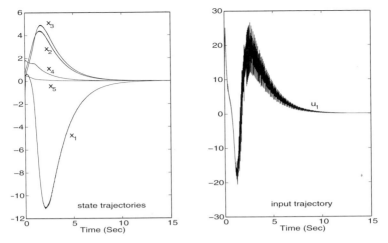

Fig. 5.17. State and input trajectories of system (5.50)

5.6 Stabilization of Discrete-time Systems

For the discrete-time switched linear system

$$x_{k+1} = A_\sigma x_k + B_\sigma u_k \tag{5.51}$$

where $x_k \in \mathbf{R}^n$, $u_k \in \mathbf{R}^p$, the problem of state feedback stabilization is to seek a switching signal and a state feedback input to steer the switched system asymptotically stable.

In this section, we present two schemes for the feedback stabilization problem. One is the quadratic stabilization by means of piecewise linear state feedback input. The other is the deadbeat control for controllable switched systems.

5.6.1 Piecewise Linear Quadratic Stabilization

The switched linear system is *piecewise linear quadratically stabilizable*, if there exist a switching signal σ, and a mode-dependent linear feedback input law of the form

$$u_k = F_\sigma x_k \tag{5.52}$$

such that the switched system

$$x_{k+1} = (A_\sigma + B_\sigma F_\sigma)x_k \tag{5.53}$$

is quadratically stable.

Recall that the quadratic stability of a system means that there is a quadratic Lyapunov candidate $V(x) = x^T P x$, such that the function strictly decreases along each non-trivial system trajectory, *i.e.*,

$$V(x_{k+1}) - V(x_k) < 0 \quad k = 0, 1, \cdots . \tag{5.54}$$

Combining (5.53) and (5.54) leads to

$$\min_{i \in M} \{ x^T \left((A_i + B_i F_i)^T P(A_i + B_i F_i) - P \right) x \} < 0 \quad \forall\, x \neq 0. \tag{5.55}$$

On the other hand, suppose that for switched system (5.51), there exist a set $\{F_i\}_{i \in M}$ of gain matrices and a positive definite matrix P satisfying (5.55), then the state-feedback switching law

$$\sigma(k) = \arg \min_{i \in M} \{ x_k^T \left((A_i + B_i F_i)^T P(A_i + B_i F_i) - P \right) x_k \}$$

$$k = 0, 1, \cdots \tag{5.56}$$

quadratically stabilizes the system (5.53), which means that switched system (5.51) is piecewise linear quadratically stabilizable. The above observation leads naturally to the following definition and lemma.

Definition 5.34. *The set* $\{H_1, \cdots, H_s\}$ *of symmetric* $n \times n$ *matrices is said to be* positive definite, *denoted by* $\{H_i\}_{i=1}^s > 0$, *if* $\forall\, x \in \mathbf{R}^n$, $x \neq 0$, *there is an index* i, $1 \leq i \leq s$, *such that* $x^T H_i x > 0$.

In the literature, the positive definiteness of a set of matrices (or functions) is also called *completeness* [124, 174].

Lemma 5.35. *Switched system (5.51) is piecewise linear quadratically stabilizable, if and only if, there exist a set* $\{F_i\}_{i \in M}$ *of gain matrices, and a positive definite matrix* P, *such that the set* $\{P - (A_i + B_i F)^T P(A_i + B_i F_i)\}_{i \in M}$ *is positive definite. Furthermore, in this case, (5.56) provides a stabilizing switching strategy.*
Proof. It is obvious from the above analysis. $\square$

By this lemma, the piecewise linear quadratic stabilizability is reduced to the existence of the matrices P and $\{F_i\}_{i \in M}$ satisfying (5.55). However, the simultaneous searching of the $m + 1$ matrices is usually very difficult. In what follows, we prove that we can reduce the problem to the searching of only one matrix. For this, we need a technical lemma.

Fix a positive definite matrix P in $\mathbf{R}^{n \times n}$. For a subspace $\mathcal{W}$ of $\mathbf{R}^n$, let

$$\mathcal{W}_P^\perp = \{ y \in \mathbf{R}^n : y^T P x = 0 \quad \forall\, x \in \mathcal{W} \}.$$

Lemma 5.36. *Suppose that* $A \in \mathbf{R}^{n \times l}$ *and* $\mathcal{W} = \mathrm{Im}\, A$. *Then, each* $z \in \mathbf{R}^n$ *can be written as* $z = z_1 + z_2$, *where*

$$z_1 = A(A^T P A)^+ A^T P z \in \mathcal{W}$$

and

$$z_2 = (I_n - A(A^T P A)^+ A^T P) z \in \mathcal{W}_P^\perp$$

where A^+ is the Moore-Penrose pseudo-inverse of A.

Proof. Since $\mathbf{R}^n = \mathcal{W} \oplus \mathcal{W}_P^\perp$, there exist a unique $w_1 \in \mathcal{W}$ and a unique $w_2 \in \mathcal{W}_P^\perp$, such that $z = w_1 + w_2$. Let $v \in \mathbf{R}^n$ be such that $w_1 = Av$. As $w_2 \in \mathcal{W}_P^\perp$, we have

$$A^T P(z - Av) = A^T P w_2 = 0.$$

Thus, equation $A^T P A x = A^T P z$ has a unique solution $x = v$. Note that

$$\mu = (A^T P A)^+ A^T P z$$

is also a solution of the same equation. This implies that

$$z - A\mu \in \mathrm{Ker}(A^T P). \tag{5.57}$$

We claim that $\mathrm{Ker}(A^T P) = \mathcal{W}_P^\perp$. Indeed, this can be seen from the following relationships

$$(y \in \mathcal{W}_P^\perp) \Longleftrightarrow (y^T P A \xi = 0 \quad \forall \xi \in \mathbf{R}^l) \Longleftrightarrow (A^T P y = 0).$$

It is clear that $A\mu \in \mathcal{W}$. Therefore, we have

$$z = A\mu + (z - A\mu)$$

with

$$A\mu = A(A^T P A)^+ A^T P z = z_1 \in \mathcal{W}$$

and

$$z - A\mu = (I_n - A(A^T P A)^+ A^T P)z = z_2 \in \mathcal{W}_P^\perp. \quad \square$$

The following theorem reduces the problem of stabilizability to the search of only one matrix, hence greatly improve the practical utilization of the approach.

Theorem 5.37. *Suppose that switched system (5.51) is reversible. Then, the system is piecewise linear quadratically stabilizable, if and only if, there exists a positive definite matrix P such that the set*

$$\{P - (A_i + B_i K_i)^T P(A_i + B_i K_i)\}_{i \in M}$$

is positive definite, where

$$K_i = -(B_i^T P B_i)^+ B_i^T P A_i \quad i \in M.$$

In this case, a stabilizing control law is

$$u_k = K_\sigma x_k \quad k = 0, 1, \cdots$$

and a stabilizing switching law is defined in (5.56).

Proof. The sufficiency is directly from Lemma 5.35. Hence, we only need to prove the necessity.

Suppose that the system is piecewise linear quadratically stabilizable. Then, by Lemma 5.35, there exist a set of gain matrices $\{F_i\}_{i \in M}$, and a positive definite matrix P, such that

$$\max_{i \in M}\{x^T(P - (A_i + B_iF_i)^T P(A_i + B_iF_i))x\} > 0 \quad \forall\, x \neq 0.$$

This is equivalent to

$$\min_{i \in M}\{x^T((A_i + B_iF_i)^T P(A_i + B_iF_i))x\} < x^T Px\} \quad \forall\, x \neq 0.$$

To prove that the set

$$\{P - (A_i + B_iK_i)^T P(A_i + B_iK_i)\}_{i \in M}$$

is positive definite, it suffices to prove, for any $i \in M$, the inequality

$$(A_i + B_iK_i)^T P(A_i + B_iK_i) \leq (A_i + B_iF_i)^T P(A_i + B_iF_i). \qquad (5.58)$$

For this, fix an $i \in M$ and let $A = A_i$, $B = B_i$, and $\mathcal{W} = \operatorname{Im} B$. By Lemma 5.36, any $x \in \mathbf{R}^n$ can be written as $x = x_{\mathcal{W}} + x_{\mathcal{W}}^{\perp}$, where

$$x_{\mathcal{W}} = A(A^T PA)^+ A^T Px \in \mathcal{W}$$

and

$$x_{\mathcal{W}}^{\perp} = (I_n - A(A^T PA)^+ A^T P)x \in \mathcal{W}_P^{\perp}.$$

Fix an $x \in \mathbf{R}^n$ and let $y = Ax$. For any $F \in \mathbf{R}^{p \times n}$, we have

$$x^T(A + BF)^T P(A + BF)x =$$
$$y^T Py + (y_{\mathcal{W}} + BFA^{-1}y)^T P(y_{\mathcal{W}} + BFA^{-1}y).$$

Note that, the first term in the right side is independent of F, and the second term is 0 if and only if $y_{\mathcal{W}} + BFA^{-1}y = 0$. By Lemma 5.36, this is exactly the case when $F = -(B^T PB)^+ B^T PA$. This means that (5.58) holds and hence the proof is complete. $\square$

By this theorem, the piecewise linear quadratic stabilizability of the reversible switched system is equivalent to the feasibility of the matrix inequality

$$\{P - (A_i - B_i(B_i^T PB_i)^+ B_i^T PA_i)^T P(A_i - B_i(B_i^T PB_i)^+ B_i^T PA_i)\}_{i \in M} < 0$$

with respect to P. Unfortunately, the inequality is generally non-convex and a constructive method to check its feasibility is yet to be known.

5.6.2 Deadbeat Control

For continuous-time switched linear systems with piecewise linear feedback control inputs, the origin itself forms an isolated invariant set, that is, a non-origin state cannot reach the set in a finite time. Accordingly, to steer any non-origin state to the origin in a finite time, nonlinear feedback control laws must be exploited. For discrete-time switched systems, however, it is possible to achieve finite time stabilizability by means of piecewise linear feedback control laws. This motivates the following concept.

Definition 5.38. *The discrete-time switched linear system $\sum(A_i, B_i)_M$ is said to be (piecewise linear state) feedback deadbeat stabilizable, if there are gain matrices F_i, $i \in M$, a natural number k, and an index sequence $i_1, \cdots, i_k$, such that*

$$H(i_1, \cdots, i_k) \stackrel{def}{=} (A_{i_k} + B_{i_k} F_{i_k}) \cdots (A_{i_2} + B_{i_2} F_{i_2})(A_{i_1} + B_{i_1} F_{i_1}) = 0.$$

In this case, the gain matrices F_i, $i \in M$ are said to be deadbeat gain matrices.

Recall that any controllable linear time-invariant system (A, B) is feedback equivalent to the Brunovski normal form. That is, there exist a nonsingular matrix T, and a gain matrix F, such that

$$T^{-1}(A + BF)T = \begin{bmatrix} 0 & 1 & 0 & \cdots & 0 \\ 0 & 0 & 1 & \cdots & \\ & & & \ddots & \\ 0 & 0 & 0 & \cdots & 1 \\ 0 & 0 & 0 & \cdots & 0 \end{bmatrix} \stackrel{def}{=} \tilde{A} \text{ and } T^{-1}B = \begin{bmatrix} 0 \\ 0 \\ \vdots \\ 0 \\ 1 \end{bmatrix} \stackrel{def}{=} \tilde{B}.$$

Note that $\tilde{A}^n = 0$, and hence $(A + BF)^n = 0$ and F is a deadbeat gain matrix.

In the following, we assume that the system is reversible and the input matrices $B_i \in \mathbf{R}^{n \times p}$, $i \in M$ are of full column rank, that is, rank$B_i = p$ for all $i \in M$.

Theorem 5.39. *Suppose that $p = n - 1$ and switched system $\sum(A_i, B_i)_M$ is completely controllable. Then, the system is feedback deadbeat stabilizable.*

Proof. If one of the subsystems is completely controllable, then, it is clear that the controllable subsystem (and hence the switched system) is feedback deadbeat stabilizable. Otherwise, none of the subsystems are controllable, and it can be seen that, there are at least two subsystems, say, the first and the second, such that rank$[B_1, B_2] = n$. This means that, there is a column b_2 of B_2, such that $T = [B_1, b_2]$ is square and nonsingular. By means of appropriate feedback gain matrices F_1 and F_2, we have

$$T^{-1}(A_1 + B_1 F_1)T = \begin{bmatrix} 0 & 0 & \cdots & 0 \\ & & \ddots & \\ 0 & 0 & \cdots & 0 \\ 0 & 0 & \cdots & 1 \end{bmatrix} \text{ and } T^{-1}(A_2 + B_2 F_2)T = \begin{bmatrix} * & * & \cdots & * \\ & & \ddots & \\ * & * & \cdots & * \\ 0 & 0 & \cdots & 0 \end{bmatrix}$$

where '*' denotes possible nonzero entries. Note that the multiplication of the above two matrices is the zero matrix. As a result, the switched system is feedback deadbeat stabilizable. □

Theorem 5.40. *Suppose that $p = 1$, $m = n$, and switched system $\sum(A_i, B_i)_M$ is completely controllable and irreducible. Then, the system is feedback deadbeat stabilizable.*

Proof. The assumptions of the theorem imply that $T \stackrel{def}{=} [B_1, \cdots, B_n]$ is square and nonsingular. Furthermore, the controllable subspace of each pair (A_i, B_j) is a subspace of $\text{Im}[B_i, B_j]$ for all $i, j \in M$. The latter implies that, for each $T^{-1}A_iT$, its off-diagonal entries are 0 except possibly those in the ith row. The ith row, however, can be made to be zero by means of a suitable feedback transformation. In other words, we can find a feedback gain matrix, such that matrix $T^{-1}(A_i + B_iF_i)T$ has zero off-diagonal entries and zero ith row. Accordingly, the multiplication of all these matrices is the zero matrix, hence the switched system is feedback deadbeat stabilizable. □

We now turn to three-dimensional switched systems. In view of Theorems 5.39 and 5.40, we only need to consider the case when $p = 1$ and $m = 2$.

Theorem 5.41. *Suppose that $p = 1$, $m = 2$ and $n = 3$, and switched system $\sum(A_i, B_i)_M$ is completely controllable. Then, the system is feedback deadbeat stabilizable.*

Proof. Without loss of generality, we assume that neither subsystem is controllable.

First, suppose that B_1 and B_2 are linearly dependent, that is, $B_2 = \lambda B_1$. In this case, the controllability of the switched system implies that $T = [B_1, A_1B_1, A_2B_1]$ is nonsingular. Denote

$$T^{-1}A_1T = \begin{bmatrix} 0 & a_1 & a_2 \\ 1 & a_3 & a_4 \\ 0 & 0 & a_5 \end{bmatrix} \text{ and } T^{-1}A_2T = \begin{bmatrix} 0 & b_1 & b_2 \\ 0 & b_3 & 0 \\ 1 & b_4 & b_5 \end{bmatrix}.$$

Let

$$F_1 = [0, -a_1, -a_2 - a_5b_5]T$$

and

$$F_2 = \frac{1}{\lambda}[-a_4, -b_1 - a_3b_3 - a_4b_4, -b_2 - a_4b_5]T.$$

It can be verified that

$$T^{-1}(A_2 + B_2F_2)(A_1 + B_1F_1)(A_2 + B_2F_2)T = 0$$

which implies that the switched system is feedback deadbeat stabilizable.

Second, suppose that B_1 and B_2 are linearly independent, that is, $\text{rank}[B_1, B_2] = 2$. By interchanging the subsystems' indices, we have either $T =$

$[B_1, B_2, A_1B_1]$ or $T = [B_1, B_2, A_1B_2]$ is nonsingular. In the former case, denote

$$T^{-1}A_1T = \begin{bmatrix} 0 & a_1 & a_2 \\ 0 & a_3 & 0 \\ 1 & a_4 & a_5 \end{bmatrix} \text{ and } T^{-1}A_2T = \begin{bmatrix} b_1 & b_2 & b_3 \\ b_4 & b_5 & b_6 \\ b_7 & b_8 & b_9 \end{bmatrix}.$$

If $b_1 + a_5b_7 \neq 0$, then, let

$$F_1 = [-\frac{b_7 + a_5b_9}{b_1 + a_5b_7}, -a_1 - \frac{a_3b_2 + a_4b_3 + a_3a_5b_8 + a_4a_5b_9}{b_1 + a_5b_7},$$
$$-a_2 - \frac{a_5b_3 + a_5^2b_9}{b_1 + a_5b_7}]T$$

and

$$F_2 = -[b_4, b_5, b_6]T.$$

It can be verified that

$$T^{-1}(A_1 + B_1F_1)(A_2 + B_2F_2)(A_1 + B_1F_1)T = 0$$

which implies that the switched system is feedback deadbeat stabilizable.
    If $b_1 + a_5b_7 = 0$, then, let

$$F_1 = [-a_5, -a_1, -a_2 - a_5^2]T$$

and

$$F_2 = -[b_4, b_5, b_6]T.$$

It can be verified that

$$T^{-1}(A_1 + B_1F_1)^2(A_2 + B_2F_2)T = 0$$

which implies that the switched system is feedback deadbeat stabilizable.
    In the later case, i.e., $T = [B_1, B_2, A_1B_1]$, we can prove in a similar way that the switched system is feedback deadbeat stabilizable. $\square$

5.7 Notes and References

Stability and stabilization are primary issues of dynamic systems. The problem of feedback stabilization of switched linear systems has been addressed for a long time. Early work includes [147, 148] for switched systems governed by random processes. Feedback stabilizing design for deterministic switched systems was addressed in [38, 39, 40], where necessary conditions and sufficient conditions were obtained for switched systems under periodic switching strategies. For switched linear discrete-time systems, feedback stabilization and state deadbeat control were addressed in [126, 127, 26, 8]. In particular, in [126, 8] a feedback design framework was established based on the Lyapunov approach. The same scheme has been applied to the continuous-time systems in [124, 118]. This approach provides a sufficient condition for feedback stabilizability, but it is usually not easy to find the required Lyapunov functions.

The canonical decomposition and normal forms presented in the previous chapter provide a rigorous approach for addressing the problem of feedback stabilization. A major advantage of this approach is that it is constructive and permits efficient and flexible design procedures. The material presented in this chapter mainly follows this approach. Section 5.2 was adopted from [139], Section 5.3 from [145], and Sections 5.4 and 5.5 from [137]. From Sections 5.4 and 5.5 we can see that, though the theory is far from complete, the scheme parallels standard linear system theory (see, *e.g.*, [21, 77]) in many aspects.

The stabilization problem for discrete-time switched systems is quite different from its continuous-time counterpart. In fact, most results for the latter depend more or less on the average method, which does not apply to discrete-time systems. As a result, the discrete-time stabilization theory is not as rich as the continuous-time theory. Nevertheless, by means of the Lyapunov approach, we can obtain some sufficient conditions such as those presented in Section 5.6.1. Most results presented in Section 5.6.1 were taken from [126]. The results for feedback deadbeat stabilization, presented in Section 5.6.2, were adopted from [26].

6

Optimization

6.1 Introduction

If the individual subsystems are given, then the behavior of a switched system depends on the switching signal. Usually, different switching strategies produce different system behaviors and hence lead to different system performances. A well-known example is the switched server system which is able, not only to produce regular stable behavior, but also to produce highly unstable behavior such as chaos and multiple limit circles. In this situation, the choice of a suitable switching law to optimize certain performance index becomes an important and well-motivated problem.

Optimization over switching signals is indeed a challenging problem. As a switching signal is a discontinuous function of time and possibly highly nonlinear, the optimization is extremely intricate and non-convex in nature.

In this chapter, we focus on two types of optimization problems for switched linear systems with/without control inputs.

First, we investigate how one particular choice of the switching signal affects the system performance. The aim is to find an optimal switching strategy that causes an unforced switched system to behave optimally according to a certain performance index. The performance index here is either the convergence rate or the infinite horizon integrand of the cost function. For the former, an optimal switching strategy is developed to minimize the convergence rate. For the latter, the finiteness and other basic properties are established for the optimal cost.

Second, we address several mixed optimization problems for switched linear control systems where both the switching signal and the control input are design variables. A two-stage optimization method is developed to solve the optimal switching/control problem by means of a set of differential and algebraic equations, and to solve the linear feedback suboptimal control problem by means of a locally convergent algorithm.

6.2 Optimal Convergence Rate

In this section, we consider the unforced switched linear system given by

$$\dot{x}(t) = A_\sigma x(t) \quad x(t_0) = x_0 \tag{6.1}$$

where the real constant matrices $A_i \in \mathbf{R}^{n \times n}$, $i \in M$ are given, and the switching signal $\sigma \in M$ is the design variable.

Without loss of generality, we assume that $t_0 = 0$.

6.2.1 Definitions and Preliminaries

In this subsection, we introduce the notion of optimal convergence rate for unforced switched linear systems. Generally speaking, the optimal convergence rate captures the largest possible convergence rate by means of appropriate switching.

Given a linear time-invariant system

$$\dot{z}(t) = Bz(t) \quad z(t_0) = z_0$$

it is well known that $z(t) = e^{B(t-t_0)}z_0$ and the state norm satisfies

$$\|z(t)\| \le e^{\|B\|(t-t_0)}\|z_0\| \quad \forall\, t \ge t_0 \tag{6.2}$$

and

$$\|z(t)\| \le \alpha(t)e^{\lambda_{mr}(t-t_0)}\|z_0\| \quad \forall\, t \ge t_0$$

for some $\alpha(\cdot) \in \mathcal{P}$, where $\mathcal{P}$ is the set of polynomial functions of time, and λ_{mr} is the convergence rate (the maximum real part of the eigenvalues) of B.

For switched system (6.1), the solution of the state equation can be computed to be

$$\phi(t; t_0, x_0, \sigma) = e^{A_{i_k}(t-t_k)} \cdots e^{A_{i_0}(t_1-t_0)}x_0 \quad t_k < t \le t_{k+1}$$

where $\{t_0, t_1, \cdots\}$ is the switching time sequence and $\{i_0, i_1, \cdots\}$ is the switching index sequence. It is clear that the state transition matrix is given by

$$\Psi(t_1, t_2, \sigma) = \Psi_0(t_1, t_0, \sigma)(\Psi_0(t_2, t_0, \sigma))^{-1} \quad t_1, t_2 \ge t_0$$

where

$$\Psi_0(t, t_0, \sigma) = e^{A_{i_k}(t-t_k)}e^{A_{i_{k-1}}(t_k-t_{k-1})} \cdots e^{A_{i_0}(t_1-t_0)} \quad t_k < t \le t_{k+1}.$$

From (6.2), it can be seen that

$$\|\phi(t; t_0, x_0, \sigma)\| \le e^{\rho(t-t_0)}\|x_0\| \quad t \ge t_0 \quad \sigma \in \mathcal{S}$$

where $\rho = \max\{\|A_1\|, \cdots, \|A_m\|\}$.

Definition 6.1. *The* convergence rate *of switched system (6.1) at x_0 under switching path σ is defined as*

$$\rho_\sigma(\Sigma(A_k)_M, x_0) = \inf\{\omega \colon \exists\, \alpha(\cdot) \in \mathcal{P} \ s.t. \ \|\phi(t, t_0, x_0, \sigma)\| \leq \alpha(t)e^{\omega(t-t_0)}\|x_0\| \quad \forall\, t \geq t_0\}.$$

In the literature, the convergence rate defined here was termed as the *stability index* for continuous-time systems [172] and the *Lyapunov indicator* for discrete-time systems [6].

Listed below are some simple facts on the convergence rate:

1) For a linear time-invariant system, the convergence rate is exactly given by the largest real part of its poles.
2) Given any real number $\beta > \rho_\sigma(\Sigma(A_k)_M, x_0)$, there exists a positive real constant δ such that

$$\|\phi(t; t_0, x_0, \sigma)\| \leq \delta e^{\beta(t-t_0)}\|x_0\| \quad \forall\, t \geq t_0$$

3) $\limsup_{t \to \infty} \dfrac{\ln(\|\phi(t; t_0, x_0, \sigma)\|/\|x_0\|)}{t - t_0} = \rho_\sigma(\Sigma(A_k)_M, x_0) \quad \forall\, x_0 \neq 0.$
4) If $\rho_\sigma(\Sigma(A_k)_M, x_0) < 0$, then switching path σ makes the switched system exponentially convergent starting from x_0.

Definition 6.2. *For switched system (6.1), the* optimal convergence rate *is defined as*

$$\rho^*(\Sigma(A_k)_M, \mathcal{S}) = \sup_{x_0 \neq 0} \inf_{\sigma \in \mathcal{S}} \{\rho_\sigma(\Sigma(A_k)_M, x_0)\}.$$

The following result collects known bounds for the convergence rates.

Proposition 6.3. *Denote by λ_k^{min} and λ_k^{max} the minimum and the maximum real parts of the eigenvalues of matrix $A_k, k \in M$, respectively. Then, we have*

$$\min_{k \in M} \lambda_k^{min} \leq \rho^*(\Sigma(A_k)_M, \mathcal{S}) \leq \min_{k \in M} \lambda_k^{max}.$$

Note that the optimal convergence rate is defined over the set of well-defined switching paths, $\mathcal{S}$. Similarly, we can define the convergence rate over any specific class of switching paths. For example, let $\mathcal{S}_\tau$ denote all the switching paths with dwell time τ, then we can define

$$\rho^*(\Sigma(A_k)_M, \mathcal{S}_\tau) = \sup_{x_0 \in \mathbf{R}^n} \inf_{\sigma \in \mathcal{S}_\tau} \{\rho_\sigma(\Sigma(A_k)_M, x_0)\}.$$

For simplicity, we denote ρ^* for $\rho^*(\Sigma(A_k)_M, \mathcal{S})$.

The notion of the optimal convergence rate is closely related to the stabilization problem studied in Chapter 3. In fact, if $\rho^*(\Sigma(A_k)_M, \mathcal{S}) < 0$, then there is a switching signal that makes the switched system stable, as indicated in the following proposition.

Proposition 6.4. *For switched system (6.1), the following statements are equivalent:*

(i) the system is asymptotically stabilizable;
(ii) the system is exponentially stabilizable; and
(iii) $\rho^* < 0$.

6.2.2 Triangularizable Systems

Although the notion of the convergence rate is very natural and fundamental from both theoretical and practical viewpoints, the computation of the optimal convergence rate is very difficult in general. That is understandable because, by Proposition 6.4, the computation of the optimal convergence rate is more difficult than verifying whether a switched system is stabilizable or not, which is already very intricate. Here, we focus on triangular systems whose component systems possess triangular structures.

Triangular systems are interesting because they have simple structures, and many non-triangular systems can be made to be triangular by means of equivalent transformations (simultaneous triangularization) [113]. For these systems, the optimal convergence rate can be explicitly formulated in terms of the system eigenvalues. The main point is that, as the optimal convergence rate is invariant under equivalent transformations, the computation of the convergence rates for a switched system can be reduced to that of an equivalent normal system with a simpler structure.

Switched system (6.1) is said to be simultaneously (upper) triangularizable, if the matrix set $\mathcal{A} = \{A_1, \cdots, A_m\}$ is simultaneously triangularizable, that is, there exists a complex nonsingular matrix $T \in \mathbf{C}^{n \times n}$ such that $B_k \overset{def}{=} T^{-1}A_kT$ are of upper triangular form:

$$\begin{bmatrix} b_k(1,1) & \cdots & b_k(1,n) \\ & \ddots & \\ 0 & \cdots & b_k(n,n) \end{bmatrix} \in \mathbf{C}^{n \times n} \quad k \in M. \tag{6.3}$$

For a simultaneously triangularizable matrix set, we can transform it into the following real normal form.

Theorem 6.5. *Suppose that* $\mathcal{A} = \{A_1, \cdots, A_m\}$ *is simultaneously triangularizable. Then, there exists a real nonsingular matrix* G, *such that* $G^{-1}A_kG$ *is of the form*

$$\bar{A}_k \overset{def}{=} G^{-1}A_kG = \begin{bmatrix} A_{1k} & * & \cdots & * \\ 0 & A_{2k} & \cdots & * \\ \vdots & \vdots & \ddots & \vdots \\ 0 & 0 & \cdots & A_{lk} \end{bmatrix} \tag{6.4}$$

where $l \leq n$, A_{jk} is either a 1×1 or 2×2 block, and the size of the jth block, A_{jk}, is the same for all $k \in M$. In addition, if A_{jk} is of 2×2, then it is in the form

$$A_{jk} = \begin{bmatrix} \mu_{jk} & \omega_{jk} \\ -\omega_{jk} & \mu_{jk} \end{bmatrix}.$$

Proof. As matrix set $\{A_1, \cdots, A_m\}$ is simultaneously triangularizable, for any polynomial $p(y_1, \cdots, y_m)$ over $\mathbf{R}$, the eigenvalues of $p(A_1, \cdots, A_m)$ are $p(b_1(i,i), \cdots, b_m(i,i))$, $i = 1, \cdots, n$. This shows that the matrix set possesses Property III in [49, pp.442]. By Theorems 1 and 9 in [49], there is an orthogonal matrix $H \in \mathbf{R}^{n \times n}$ such that

$$H^{-1} A_k H = \begin{bmatrix} B_{1k} & * & \cdots & * \\ 0 & B_{2k} & \cdots & * \\ \vdots & \vdots & \ddots & \vdots \\ 0 & 0 & \cdots & B_{lk} \end{bmatrix} \tag{6.5}$$

where $l \leq n$, and for fixed $j \leq l$, we have

(i) $B_{j1}, \cdots, B_{jm}$ are 1×1 or

(ii) $B_{j1}, \cdots, B_{jm}$ are 2×2, with one of these matrices, say, B_{jq}, of the form

$$B_{jq} = \begin{bmatrix} r_{jq} & u_{jq} \\ -v_{jq} & r_{jq} \end{bmatrix} \quad u_{jq} > 0 \;\; v_{jq} > 0$$

and each of B_{jk}, $k \in M$ is a real linear polynomial g_{jk} in B_{jq}, that is, $B_{jk} = g_{jk}(B_{jq})$.

As B_{jq} in (ii) possesses a pair of (conjugated) complex eigenvalues, it follows from standard matrix theory that there exists a real nonsingular matrix $T_j \in \mathbf{R}^{2 \times 2}$ such that

$$T_j^{-1} B_{jq} T_j = \begin{bmatrix} \mu_{jq} & \omega_{jq} \\ -\omega_{jq} & \mu_{jq} \end{bmatrix} \quad \mu_{jq}, \omega_{jq} \in \mathbf{R}.$$

Furthermore, since any polynomial of the matrix $\begin{bmatrix} \mu_{jq} & \omega_{jq} \\ -\omega_{jq} & \mu_{jq} \end{bmatrix}$ is still in the same form, we have

$$T_j^{-1} B_{jk} T_j = g_{jk}(T_j^{-1} B_{jq} T_j) = \begin{bmatrix} \mu_{jk} & \omega_{jk} \\ -\omega_{jk} & \mu_{jk} \end{bmatrix} \quad \mu_{jk}, \omega_{jk} \in \mathbf{R} \;\; k \in M.$$

Define $K = \mathrm{diag}[K_1, \cdots, K_l]$, where $K_j = 1$ if the corresponding block in (6.5) is 1×1, and $K_j = T_j$ if the block is 2×2. Let $G = HK$ and the theorem follows. $\square$

Remark 6.6. Simultaneous triangularization of matrix sets has been investigated extensively, see, for example, [104, 121, 113] and the references therein. In particular, the following classes of matrix sets (and the corresponding switched systems) have been proven to be simultaneously triangularizable:

a) the system matrices are commutative pairwise, that is $A_i A_j = A_j A_i$, $i, j \in M$ [108];
b) the Lie-algebra generated by the system matrices is solvable [91]; and
c) $\mathcal{A} = \{A_1, A_2\}$ and $\mathrm{rank}(A_1 A_2 - A_2 A_1) = 1$ [86].

6.2.3 Main Result

Suppose that switched system (6.1) is simultaneously triangularizable. That is, there exists a real nonsingular matrix T such that

$$\bar{A}_k = T^{-1} A_k T = \begin{bmatrix} A_{1k} & * & \cdots & * \\ 0 & A_{2k} & \cdots & * \\ \vdots & \vdots & \ddots & \vdots \\ 0 & 0 & \cdots & A_{lk} \end{bmatrix}$$

where A_{ik} is either 1×1 or 2×2.

Note that $A_{1k}(1,1), \cdots, A_{lk}(1,1)$ are (ordered) real parts of the eigenvalues of A_k (neglecting the multiplicity for the conjugated pairs). Let

$$s_i = [A_{i1}(1,1), \cdots, A_{im}(1,1)] \quad i = 1, \cdots, l$$

and

$$\Lambda = \{r = [r_1, \cdots, r_m]^T : r_k \geq 0, \sum_{k \in M} r_k = 1\}.$$

Theorem 6.7. *Suppose that switched system (6.1) is simultaneously triangularizable. Then, its optimal convergence rate is*

$$\rho^* = \min_{r \in \Lambda} \left(\max_{i=1}^{l} (s_i \cdot r) \right). \tag{6.6}$$

Proof. First, fix an $r = [r_1, \cdots, r_m] \in \Lambda$, and consider the periodic switching path given by

$$\sigma(t) = \begin{cases} 1 & \mathrm{mod}\ (t, 1) \in [0, r_1) \\ 2 & \mathrm{mod}\ (t, 1) \in [r_1, r_1 + r_2) \\ \vdots & \\ m & \mathrm{mod}\ (t, 1) \in [\sum_{i=1}^{m-1} r_i, 1) \end{cases} \quad t \geq t_0$$

where $\mathrm{mod}\ (a, b)$ denotes the remainder of a divided by b.

Denote

$$\Phi \stackrel{def}{=} \Psi(1,0,\sigma) = e^{A_m r_m} e^{A_{m-1} r_{m-1}} \cdots e^{A_1 r_1} = T \begin{bmatrix} B_1 & \cdots & * \\ \vdots & \ddots & \vdots \\ 0 & \cdots & B_l \end{bmatrix} T^{-1}$$

where $B_i = e^{A_{im} r_m} \cdots e^{A_{i1} r_1}$. If $A_{ik} \in \mathbf{R}^{1 \times 1}$, then, it can be seen that $B_i = \exp(s_i \cdot r)$. If $A_{ik} \in \mathbf{R}^{2 \times 2}$ with $A_{ik} = \begin{bmatrix} \mu_{ik} & \omega_{ik} \\ -\omega_{ik} & \mu_{ik} \end{bmatrix}$, then it can be verified that the spectral radius of B_i is

$$e^{\mu_{im} r_m} \cdots e^{\mu_{i1} r_1} = \exp(s_i \cdot r).$$

Therefore, the spectral radius of Φ is

$$\max\{\exp(s_1 \cdot r), \cdots, \exp(s_l \cdot r)\} = \max_{i=1}^{l}(s_i \cdot r).$$

By the standard matrix theory, there exists an $\alpha_0 \in \mathcal{P}$ such that

$$\|\Phi^j\| \le \alpha_0(j) \left(\exp\left(\max_{i=1}^{l}(s_i \cdot r) \right) \right)^j \quad j = 1, 2, \cdots$$

which implies that the state transition matrix satisfies

$$\|\Psi(j,0,\sigma)\| \le \alpha_0(j) \exp\left(j \max_{i=1}^{l}(s_i \cdot r) \right) \quad j = 1, 2, \cdots.$$

Furthermore, for any $t \ge t_0$, there is a nonnegative integer j_0 such that $t \in [j_0, j_0 + 1)$. As the switching path is periodic with period 1, we have

$$\Psi(t,0,\sigma) = \Psi(t - j_0, 0, \sigma) \cdot \Psi(j_0, 0, \sigma).$$

Define

$$\nu_1 = \max_{q \in [0,1]} \|\Psi(q,0,\sigma)\| \text{ and } \nu_2 = \max_{q \in [0,1]} \exp\left(-q \max_{i=1}^{l}(s_i \cdot r) \right).$$

Then, we have

$$\|\Psi(t,0,\sigma)\| \le \alpha(t) \exp\left(t \max_{i=1}^{l}(s_i \cdot r) \right)$$

where $\alpha = \nu_1 \nu_2 \alpha_0 \in \mathcal{P}$. By Definition 6.1, we have

$$\rho_\sigma(\Sigma(A_k)_M, x_0) \le \max_{i=1}^{l}(s_i \cdot r) \quad \forall \, x_0 \in \mathbf{R}^n.$$

As r is arbitrary from Λ, by Definition 6.2, we have

$$\rho^* \le \min_{r \in \Lambda} \left(\max_{i=1}^{l}(s_i \cdot r) \right).$$

Assume that the above inequality is strict, that is

$$\rho^* < \min_{r \in \Lambda} \left(\max_{i=1}^{l} (s_i \cdot r) \right). \tag{6.7}$$

Then, there exists a switching path $\sigma_0 \in \mathcal{S}$ such that

$$\rho^* \le \sup_{x_0 \ne 0} \rho_{\sigma_0}(\Sigma(A_k)_M, x_0) < \min_{r \in \Lambda} \left(\max_{i=1}^{l} (s_i \cdot r) \right).$$

Note that

$$\limsup_{t \to \infty} \frac{\ln(\|\phi(t, 0, \sigma_0)\| / \|x_0\|)}{t} = \rho_{\sigma_0}(x_0) \quad x_0 \ne 0$$

which implies that

$$\limsup_{t \to \infty} \frac{\ln(\|\Psi(t, 0, \sigma_0)\|)}{t} = \sup_{x_0 \ne 0} \rho_{\sigma_0}(x_0). \tag{6.8}$$

Let

$$\varepsilon = \frac{1}{2} \left(\min_{r \in \Lambda} \left(\max_{i=1}^{l} (s_i \cdot r) \right) - \sup_{x_0 \ne 0} \rho_{\sigma_0}(x_0) \right). \tag{6.9}$$

By (6.8), there exists a time $t_* \ge t_0$ such that

$$\|\Psi(t_*, 0, \sigma_0)\| \le \exp \left(\left(\sup_{x_0 \ne 0} \rho_{\sigma_0}(x_0) + \varepsilon \right) t_* \right). \tag{6.10}$$

Let $t_0 < t_1 < \cdots < t_j$ be the switching time sequence of σ_0 in $[t_0, t_*]$. Simple computation gives

$$\Psi(t_*, 0, \sigma_0) = \begin{bmatrix} \varphi_1 & \cdots & * \\ \vdots & \ddots & \vdots \\ 0 & \cdots & \varphi_l \end{bmatrix}$$

where

$$\varphi_1 = \exp \left(\sum_{\mu=0}^{j} A_{1\sigma_0(t_\mu)} (t_{\mu+1} - t_\mu) \right)$$

$$\vdots$$

$$\varphi_l = \exp \left(\sum_{\mu=0}^{j} A_{l\sigma_0(t_\mu)} (t_{\mu+1} - t_\mu) \right)$$

with $t_{j+1} = t_*$. Let r_k denote the active duration ratio of the kth subsystem, that is,

$$r_k = \frac{\sum_{\sigma_0(t_j)=k}(t_{j+1}-t_j)}{t_*} \quad k \in M.$$

It can be seen that $r \overset{def}{=} [r_1, \cdots, r_m]^T \in \Lambda$. As in the first part of the proof, the spectral radius of $\Psi(t_*, 0, \sigma_0)$ is computed to be $\exp\left(t_*(\max_{i=1}^l(s_i \cdot r))\right)$. As the spectral radius of a matrix is smaller than or equal to its 2-norm, by (6.10), we have

$$\exp\left(t_*(\max_{i=1}^l(s_i \cdot r))\right) \leq \|\Psi(t_*, 0, \sigma_0)\| \leq \exp\left(t_*(\sup_{x_0 \neq 0} \rho_{\sigma_0}(x_0) + \varepsilon)\right).$$

This, together with the definition of ε in (6.9), implies that

$$\max_{i=1}^l(s_i \cdot r) \leq \sup_{x_0 \neq 0} \rho_{\sigma_0}(x_0) + \varepsilon < \min_{r \in \Lambda}\left(\max_{i=1}^l(s_i \cdot r)\right)$$

which is a contradiction. This shows that the assumption (6.7) is not true. As a result, we have

$$\rho^* = \min_{r \in \Lambda}\left(\max_{i=1}^l(s_i \cdot r)\right). \quad \square$$

Note that Theorem 6.7 implies Theorem 3.13 as a special case.

It can be seen that the optimal convergence rate is related to not only the set of the eigenvalues, but also the order of the eigenvalues in the triangular normal form.

6.2.4 Computational Procedure and Example

First, we briefly discuss the computational procedure to obtain the optimal convergence rate. It is clear that the set Λ is a $(m-1)$-dimensional polygon of $\mathbf{R}^m$. Define

$$\Lambda_i = \{r \in \Lambda \colon s_i \cdot r = \max_{j=1}^l(s_j \cdot r)\} \quad i = 1, \cdots, l. \tag{6.11}$$

Note that some Λ_i's may be empty. It can be seen that each nonempty Λ_i is connected and convex, and $\cup_{i=1}^l \Lambda_i = \Lambda$. Thus, each nonempty Λ_i is also a polygon. Let Υ_i denote the set of the extremal points of Λ_i. Then, we have

$$\rho^* = \min_{r \in \Lambda}(\max_{i=1}^l(s_i \cdot r)) = \min_{j=1}^l(\min_{r \in \Lambda_j}(s_j \cdot r)) = \min_{j=1}^l(\min_{r \in \Upsilon_j}(s_j \cdot r)). \tag{6.12}$$

Second, as seen in the proof of Theorem 6.7, we indicate a way to design switching signals with optimal convergence rate. For a fixed period h, define a periodic switching path by

$$\sigma(t) = \begin{cases} 1 & \mathrm{mod}\,(t,h) \in [0, r_1 h) \\ 2 & \mathrm{mod}\,(t,h) \in [r_1 h, (r_1 + r_2)h) \\ \vdots & \\ m & \mathrm{mod}\,(t,h) \in [(\sum_{i=1}^{m-1} r_i)h, h). \end{cases} \tag{6.13}$$

Suppose that $r = [r_1, \cdots, r_m]^T$ is chosen corresponding to ρ^*, $i.e.$,

$$\rho^* = \max_{i=1}^{l}(s_i \cdot r).$$

Then, the switching path steers the switched system convergent (or divergent) at the optimal convergence rate. Note that, although the optimal convergence rate is the same for all period h, different periods may result in different transient performances for the switched system.

Example 6.8. For the block triangular switched system $\Sigma(A_i)_{\bar{2}}$ with

$$A_1 = \begin{bmatrix} 1 & -2 & 0 & -1 \\ 0 & -3 & 1 & 2 \\ 0 & 0 & -2 & 4 \\ 0 & 0 & -4 & -2 \end{bmatrix} \text{ and } A_2 = \begin{bmatrix} 2 & 3 & -2 & 1 \\ 0 & -4 & -2 & 1 \\ 0 & 0 & 1 & -2 \\ 0 & 0 & 2 & 1 \end{bmatrix}$$

we have

$$s_1 = [1, 2] \quad s_2 = [-3, -4] \quad s_3 = [-2, 1].$$

To compute the optimal convergence rate, it follows from (6.11) that

$$\Lambda_1 = \Lambda \text{ and } \Lambda_2 = \Lambda_3 = \emptyset.$$

It can be verified that the extreme points of set Λ_1 are

$$r^1 = [1, 0]^T \text{ and } r^2 = [0, 1]^T.$$

The optimal convergence rate can be calculated from (6.12) to be

$$\rho^* = \min_j \{s_1 \cdot r^j\} = 1.$$

This shows that the switched system is not asymptotically stabilizable. On the other hand, if we interchange the first and second diagonal entries of A_2, that is, let

$$\bar{A}_1 = A_1 \text{ and } \bar{A}_2 = \begin{bmatrix} -4 & 3 & -2 & 1 \\ 0 & 2 & -2 & 1 \\ 0 & 0 & 1 & -2 \\ 0 & 0 & 2 & 1 \end{bmatrix}$$

then, we have

$$s_1 = [1, -4] \quad s_2 = [-3, 2] \quad s_3 = [-2, 1].$$

Similar computation gives

$$\rho^* = s_1 \cdot \left[\frac{5}{8}, \frac{3}{8} \right]^T = -\frac{7}{8}.$$

This shows that switched system $\Sigma(\bar{A}_i)_{\bar{2}}$ is asymptotically stabilizable.

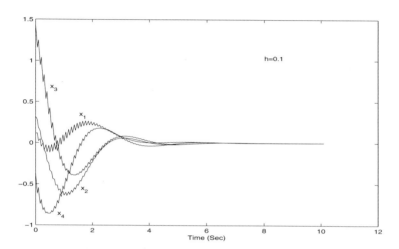

Fig. 6.1. The state trajectory with $h = 0.1$

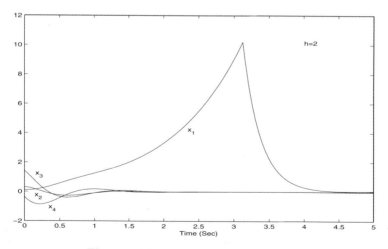

Fig. 6.2. The state trajectory with $h = 2$

Figures 6.1 and 6.2 show the state trajectories of system $\Sigma(\bar{A}_i)_{\bar{2}}$ under switching path (6.13) with $h = 0.1$ and $h = 2$, respectively. In both figures, the initial state is chosen to be

$$x_0 = [0.1184, 0.3148, 1.4435, -0.3509]^T.$$

It can be seen that, though the state converges to zero at the same (optimal) rate in both cases, their transient performances are quite different.

6.3 Infinite-time Optimal Switching

6.3.1 Finiteness of Optimal Cost

Suppose that $\{p_i\}_{i \in M}$ is a set of functions defined on $\mathbf{R}^n$. Assume that each of the functions is continuous, positive definite and radially unbounded. Accordingly, there are two sets of functions of class $\mathcal{K}$, $\{\gamma_i\}_{i \in M}$ and $\{\gamma^i\}_{i \in M}$, such that

$$\gamma_i(\|x\|) \le p_i(x) \le \gamma^i(\|x\|) \quad \forall \, i \in M \quad x \in \mathbf{R}^n. \tag{6.14}$$

For technical reasons, in certain situations, we further impose one or more of the following assumptions:

Assumption 6.1. γ^i, $i \in M$ are locally Lipschitz at zero, that is, there are positive real numbers ϵ_i and κ_i such that

$$|\gamma^i(t)| \le \kappa_i t \quad \forall \, t \in [0, \epsilon_i] \quad i \in M.$$

Assumption 6.2. There exist positive real numbers ε_i and ν_i such that

$$|\gamma_i(t)| \ge \nu_i t \quad \forall \, t \in [0, \varepsilon_i] \quad i \in M.$$

Assumption 6.3. For any $i \in M$, there exist a polynomial φ_i of time with $\varphi_i(0) = 0$, and positive real numbers μ_i, such that

$$|p_i(x) - p_i(y)| \le \varphi_i(\|x - y\|) \quad \forall \, x, y \in \mathbf{R}^n \quad x - y \in \mathbf{B}_{\mu_i}.$$

Note that Assumption 6.3 implies Assumption 6.1. Generally speaking, the assumptions are quite mild and are satisfied in many situations, for example, when $p_i(x)$ are homogeneous and of even degrees.

Consider the infinite horizon cost function given by

$$J(x_0, \theta) = \int_0^\infty p_\theta(x(t))dt \tag{6.15}$$

where $x(t) = \phi(t; 0, x_0, \theta)$ is the state trajectory starting from $x(0) = x_0$ under switching path θ.

The problem of optimal switching is to find, for a given x_0, a switch path θ_{x_0} that minimizes the cost function

$$J(x_0, \theta_{x_0}) = \min_{\theta \in \mathcal{S}_{[0,\infty)}} J(x_0, \theta).$$

The switching path θ_{x_0} is said to be an *optimal switching path at* x_0. If an optimal switching path exists at each $x_0 \in \mathbf{R}^n$, then, the switching law that generates θ_{x_0} at x_0 for any $x_0 \in \mathbf{R}^n$ is said to be an *optimal switching law*. The optimal state trajectory can be defined in the same manner.

As the problem may not admit any optimal switching path/law, we propose separately the notion of optimal cost.

Definition 6.9. *The optimal cost associated with the cost function at* x_0 *is*

$$J(x_0) = \inf_{\theta \in \mathcal{S}_{[0,\infty)}} J(x_0, \theta).$$

The worst case optimal cost (in the unit ball) associated with the cost function is

$$J_* = \sup_{\|x_0\| \le 1} J(x_0).$$

The worst case optimal cost reflects the worst possible cost in the unit ball. In many practical situations, the initial state is not exactly measurable in advance, and the worst case optimal cost measures the worst case over the initial states in the unit ball.

The following simple facts are easily verified:

- $J(x_0) \ge 0$ for all $x_0 \in \mathbf{R}^n$, and $J(x_0) = 0 \iff x_0 = 0$;
- if $x : [0, \infty) \mapsto \mathbf{R}^n$ is an optimal state trajectory, then, for any $s \in [0, \infty)$, the state trajectory $y : [0, \infty) \mapsto \mathbf{R}^n$ with $y(\cdot) = x(\cdot + s)$, is also optimal; and
- if each $p_i(\cdot)$ is homogeneous of degree k, *i.e.*,

$$p_i(\lambda x) = \lambda^k p_i(x) \quad \forall \lambda \in \mathbf{R} \ \ x \in \mathbf{R}^n \ \ i \in M$$

then $J(\cdot)$ is also homogeneous of degree k. In addition, if θ_{x_0} is an optimal switching path at x_0, then, it is also an optimal switching path at λx_0 with $\lambda \in \mathbf{R}$.

An important issue is the connection between the optimal problem and the stabilization problem. The following theorem sets up a connection in a clear way.

Theorem 6.10. *Suppose that Assumption 6.1 holds. Then, the following statements are equivalent:*

(i) the optimal cost $J(x_0)$ is finite for any given $x_0 \in \mathbf{R}^n$;
(ii) the worst case optimal cost is finite; and
(iii)the switched system is asymptotically stabilizable.

Proof. We establish the equivalence between (i) and (iii). The equivalence between (ii) and (iii) can be proven in the same manner.

To prove that $(i) \implies (iii)$, suppose that the optimal cost $J(x_0)$ is finite for any given x_0. This means that for each $x_0 \in \mathbf{R}^n$, there is a switching path $\theta_{[0,\infty)}$ such that

$$\int_0^\infty p_{\theta(t)}(\phi(t; 0, x_0, \theta))dt < \infty.$$

From (6.14), it can be seen that

$$p_{\theta(t)}(\phi(t; 0, x_0, \theta)) \geq \underline{\gamma}(\|\phi(t; 0, x_0, \theta)\|)$$

where the monotone function $\underline{\gamma}$ is defined by

$$\underline{\gamma}(t) = \min_{i \in M}\{\gamma_i(t)\} \quad t \in [0, \infty).$$

Accordingly, we have

$$\int_0^\infty \underline{\gamma}(\|\phi(t; 0, x_0, \theta)\|)dt < \infty. \tag{6.16}$$

This, together with the fact that

$$\|\frac{d}{dt}\phi(t; 0, x_0, \theta)\| \leq \eta\|\phi(t; 0, x_0, \theta)\| \quad \forall\, t \in (0, \infty) \tag{6.17}$$

where $\eta = \max\{\|A_i\|\}_{i \in M}$, implies that

$$\lim_{t \to \infty} \phi(t; 0, x_0, \theta) = 0. \tag{6.18}$$

Indeed, suppose that (6.18) does not hold. Accordingly, there exist a real number ε, and a time sequence $\{t_i\}_{i=1}^\infty$ with $\lim_{i \to \infty} t_i = \infty$ such that

$$\|\phi(t_i; 0, x_0, \theta)\| \geq \varepsilon \quad i = 1, 2, \cdots.$$

Without loss of generality, we assume that $t_{i+1} - t_i \geq 1$ for all $i \in \mathbf{N}_+$. By (6.17), it is clear that

$$\|\phi(t; 0, x_0, \theta)\| \geq \varepsilon e^{-\eta} \quad \forall\, t \in [t_i, t_i + 1].$$

Therefore, we have

$$\int_0^\infty \underline{\gamma}(\|\phi(t;0,x_0,\theta)\|)dt \geq \sum_{i=1}^\infty \int_{t_i}^{t_i+1} \underline{\gamma}(\varepsilon e^{-\eta})dt = \infty$$

which contradicts (6.16). This means that the state trajectory $\phi(t;0,x_0,\theta)$ is convergent. By Theorem 3.9, the switched system is asymptotically stabilizable.

Next, we show that $(iii) \implies (i)$. Suppose that the switched system is asymptotically stabilizable. Then, by Theorem 3.9, the system is exponentially stabilizable. That is, there are positive real numbers α and β, such that for any $x_0 \in \mathbf{R}^n$, there exists a switching signal σ satisfying

$$\|\phi(t;0,x_0,\sigma)\| \leq \beta e^{-\alpha t}\|x_0\| \quad \forall\, t \geq 0. \tag{6.19}$$

As a result, we have

$$J(x_0) \leq J(x_0,\sigma) = \int_0^\infty p_\sigma(\phi(t;0,x_0,\sigma))dt$$
$$\leq \int_0^\infty \bar{\gamma}(\beta e^{-\alpha t}\|x_0\|)dt$$

where function $\bar{\gamma}$ is defined by

$$\bar{\gamma}(t) = \max_{i \in M}\{\gamma^i(t)\} \quad t \in [0,\infty).$$

By Assumption 6.1, the γ^is are locally Lipschitz. As a result, for any $s \geq 0$, there is a real number ρ_s, such that the linear function $\hat{\gamma}(t) = \rho_s t$ overwhelms $\bar{\gamma}$ on $[0,s]$, that is,

$$\bar{\gamma}(t) \leq \hat{\gamma}(t) = \rho_s t \quad t \in [0,s].$$

Therefore, we have

$$J(x_0) \leq \int_0^\infty \bar{\gamma}(\beta e^{-\alpha t}\|x_0\|)dt \leq \int_0^\infty \rho_{\beta\|x_0\|}\beta e^{-\alpha t}\|x_0\|dt$$
$$= \frac{\beta}{\alpha}\rho_{\beta\|x_0\|}\|x_0\| \tag{6.20}$$

which sets an upper bound for the optimal cost. □

Note that if we know an exponentially stabilizing switching signal and the associated constants α and β as in (6.19), then (6.20) provides an upper bound for the optimal cost. The following example illustrates this idea.

Example 6.11. Consider the optimal switching problem for switched linear system $\Sigma(A_i)_{\{1,2\}}$ with

$$A_1 = \begin{bmatrix} 1 & -4 & 3 \\ 4 & 1 & -2 \\ 0 & 0 & -2 \end{bmatrix} \text{ and } A_2 = \begin{bmatrix} -3 & 1 & 5 \\ -1 & -3 & 2 \\ 0 & 0 & 2 \end{bmatrix}. \tag{6.21}$$

Let

$$p_1(x) = x_1^2 + x_2^2 + x_3 \sin x_2 + 2x_3^2$$
$$p_2(x) = 2x_1^4 + x_2^4 + x_1^3 \operatorname{sat}(x_3) + x_3^4$$

where $\operatorname{sat}(\cdot)$ is the standard saturation functions with unit limits. Set

$$\gamma^1(t) = 3t^2 \text{ and } \gamma^2(t) = 3t^4 \quad \forall\, t \in \mathbf{R}.$$

Accordingly, we can choose $\rho_s = 3 \max\{s, s^3\}$ for all $s \in \mathbf{R}$.

On the other hand, it can be seen that both A_1 and A_2 are in the upper block triangular form. From Theorem 6.7, it can be calculated that the optimal convergence rate is $\rho^* = -\frac{1}{2}$, and this convergence rate can be achieved by the periodic switching path

$$\theta(t) = \begin{cases} 1 & \operatorname{mod}(t, h) \in [0, w_1 h) \\ 2 & \operatorname{mod}(t, h) \in [w_1 h, h) \end{cases}$$

where h is any positive real number, and $w_1 = \frac{5}{8}$.

Let $h = 1$, $\alpha = -\rho^* = 0.5$, and

$$\beta = \max\left\{ \max_{t \in [0, w_1 h]} \{\|e^{A_1 t}\| \|e^{\alpha t}\|\}, \max_{t \in [w_1 h, h]} \{\|e^{A_2(t - w_1 h)} e^{A_1 w_1 h}\| \|e^{\alpha(t - w_1 h)}\|\} \right\}$$
$$< 3.300.$$

By the theory presented in Section 6.2, we have

$$\|\phi(t; 0, x_0, \theta)\| \le \beta e^{-\alpha t} \quad \forall\, t \ge 0.$$

By (6.20), an upper bound of the optimal cost is

$$J(x_0) \le \frac{\beta}{\alpha} \rho_{\beta\|x_0\|} < 65.34 \max\{\|x_0\|, 10.89\|x_0\|^3\}.$$

The worst case optimal cost is bounded by

$$J_* \le 65.34 * 10.89 \approx 711.55.$$

6.3.2 Basic Properties

Given an $s \ge 0$, define

$$J_s = \sup\{J(x_0) \colon \|x_0\| \le s\}.$$

It follows from (6.20) that

$$J_s \le \frac{\beta s}{\alpha} \rho_{\beta s} \tag{6.22}$$

which provides a useful upper bound for J_s.

Theorem 6.12. *Suppose that the switched system is asymptotically stabilizable, and the cost function satisfies Assumptions 6.1 and 6.2. Then, any optimal state trajectory is exponentially convergent.*

Proof. Suppose that $x(\cdot)$ is an optimal state strategy starting from $x(0) = x_0 \neq 0$. Let $\theta(\cdot)$ be the associated switching path, that is,

$$\dot{x}(t) = A_{\theta(t)}x(t) \quad \text{for almost all } t \in [0, \infty).$$

It is clear that

$$J(x_0) = J(x_0, \theta) = \int_0^\infty p_{\theta(t)}(x(t))dt < \infty. \tag{6.23}$$

From the proof of Theorem 6.10, it can be seen that

$$\int_0^\infty p_{\theta(t)}(x(t))dt \geq \int_0^\infty \underline{\gamma}(x(t))dt. \tag{6.24}$$

By Assumption 6.2, for any $t > 0$, there exist positive real numbers ν_i^t, such that

$$\gamma_i(s) \geq \nu_i^t s \quad \forall \, s \in [0, t].$$

Define $\nu = \min\{\nu_i^{\|x_0\|}\}_{i \in M}$ for all $t \geq 0$. Let $r_1 = \frac{\|x_0\|}{2}$. As pointed out in the proof of Theorem 6.10, the state trajectory must be convergent. Hence, the time

$$t_1 = \min\{t \geq 0 \colon x(t) \in \mathbf{B}_{r_1}\}$$

exists and is finite. As $x(t) \notin \mathbf{B}_{r_1}$ for $t \in [0, t_1)$, we have

$$\int_0^{t_1} \underline{\gamma}(x(t))dt \geq \int_0^{t_1} \nu r_1 dt = \frac{1}{2}\nu\|x_0\|t_1.$$

This, together with inequality (6.24), implies that

$$t_1 \leq \frac{2J(x_0)}{\nu\|x_0\|}.$$

Furthermore, it follows from (6.22) that

$$t_1 \leq \frac{2\beta}{\nu\alpha}\rho_{\beta\|x_0\|} \stackrel{def}{=} \delta.$$

This means that for any optimal state trajectory $x(\cdot)$, there is a time $t_1 \leq \delta$, such that $x(t_1) \in \mathbf{B}_{\frac{\|x(0)\|}{2}}$. On the other hand, let

$$y(t) = x(t + t_1) \quad t \in [0, \infty).$$

By the optimality of $x(\cdot)$, the state trajectory $y(\cdot)$ is also an optimal state trajectory starting from $y(0) = x(t_1) \in \mathbf{B}_{\|x(0)\|}$. Accordingly, there is a time $s_2 \leq \delta$ such that $y(s_2) \in \mathbf{B}_{\frac{\|y(0)\|}{2}}$, which implies that

$$x(t_2) \in \mathbf{B}_{\frac{\|x(t_1)\|}{2}}$$

where $t_2 = t_1 + s_2$. Continuing with this process, we can find a monotone time sequence $0 = t_0 < t_1 < t_2 < \cdots$ such that

(i) $t_{i+1} - t_i \leq \delta \quad \forall\, i \in \mathbf{N}_+$;
(ii) $x(t_{i+1}) \in \mathbf{B}_{\frac{\|x(t_i)\|}{2}} \quad \forall\, i \in \mathbf{N}_+$; and
(iii) $\|x(t)\| \leq e^{\eta\delta}\|x(t_i)\| \quad \forall\, t \in [t_i, t_{i+1}] \quad i \in \mathbf{N}_+$, where $\eta = \max_{i \in M}\{\|A_i\|\}$.

These facts suffice to conclude that there exist positive numbers α_1 and β_1, which may be dependent on x_0, such that

$$\|x(t)\| \leq \beta_1 e^{-\alpha_1 t}\|x_0\| \quad \forall\, t \in [0, \infty). \qquad \square$$

Theorem 6.13. *Suppose that the cost function satisfies Assumption 6.3, and the switched system is asymptotically stabilizable. Then, the optimal cost $J(x_0)$ is a continuous function of x_0.*
Proof. First, by (6.22), it can be seen that, for any $\varepsilon > 0$, there is $\delta > 0$ such that $J_\delta < \varepsilon$. This implies that $J(x)$ is continuous at $x = 0$. Select δ such that $\delta < 1$ and $\delta \leq \mu \overset{def}{=} \min\{\mu_i : i \in M\}$, where μ_i is given in Assumption 6.3.

Then, for $x_0 \neq 0$, let $x(\cdot)$ be a state trajectory starting from x_0 and satisfying

$$\left| \int_0^\infty p_{\theta(t)}(x(t))dt - J(x_0) \right| \leq \varepsilon$$

where $\theta(\cdot)$ is the associated switching path. As the state trajectory is convergent, there is a time t_1 such that

$$x(t) \in \mathbf{B}_{\frac{\delta}{2}} \quad \forall\, t \geq t_1.$$

On the other hand, there exists a time t_2 such that

$$\left| \int_0^{t_2} p_{\theta(t)}(x(t))dt - \int_0^\infty p_{\theta(t)}(x(t))dt \right| \leq \varepsilon.$$

Let $t^* = \max\{t_1, t_2\}$.

Next, let $\Phi(s_1, s_2, \theta)$ be the transition matrix corresponding to θ. It is clear that

$$\|\Phi(s_1, s_2, \theta)\| \leq e^{\eta(s_1 - s_2)} \quad \forall\, s_1 \geq s_2 \geq 0 \qquad (6.25)$$

where $\eta = \max\{\|A_i\| : i \in M\}$. Let

$$\varrho = \frac{\delta}{2}e^{-\eta t^*}.$$

Then, we have

$$\phi(t; 0, y_0, \theta) - x(t) = \Phi(t, 0, \theta)(y_0 - x_0) \in \mathbf{B}_{r(t)} \quad \forall \, \|y_0 - x_0\| \leq \varrho \ \ t \in [0, t^*]$$

where

$$r(t) = \frac{\delta}{2}e^{-\eta(t^*-t)} \quad t \in [0, t^*].$$

Fix such a y_0 and let $y(\cdot)$ be the state trajectory starting from y_0 via switching path θ, *i.e.*,

$$y(t) = \phi(t; 0, y_0, \theta) \quad t \in [0, \infty).$$

By Assumption 6.3, we have

$$\left| \int_0^{t^*} p_{\theta(t)}(x(t))dt - \int_0^{t^*} p_{\theta(t)}(y(t))dt \right| \leq \int_0^{t^*} \varphi(\|x(t) - y(t)\|)dt \quad (6.26)$$

where $\varphi = a_1 t + a_2 t^2 + \cdots + a_j t^j$ is the polynomial defined by

$$\varphi(t) = \sum_{i \in M} \varphi_i(t) \quad \forall \, t \geq 0.$$

It can be seen that

$$\int_0^{t^*} \varphi(\|x(t) - y(t)\|)dt \leq \int_0^{t^*} \sum_{i=1}^{j} |a_i| \frac{\delta^i}{2^i} e^{-i\eta(t^*-t)}dt$$

$$\leq \left(\sum_{i=1}^{j} \frac{\mu^{i-1}|a_i|}{i 2^i \eta} \right) \delta \stackrel{def}{=} \omega \delta.$$

Finally, note that $y(t^*) \in \mathbf{B}_\delta$. This means that $J(y(t^*)) < \varepsilon$. Let ϑ be a switching path such that

$$J(y(t^*), \vartheta) \leq J(y(t^*)) + \varepsilon \leq 2\varepsilon.$$

Let $\bar{\theta}$ denote the concatenation of $\theta_{[0,t^*]}$ and $\vartheta_{[0,\infty)}$, that is,

$$\bar{\theta}(t) = \begin{cases} \theta(t) & \text{if } t \in [0, t^*] \\ \vartheta(t - t^*) & \text{if } t \in (t^*, \infty). \end{cases}$$

Combining the above reasonings, we have

$$|J(y_0, \bar{\theta}) - J(x_0)| \leq |J(y_0, \bar{\theta}) - J(x_0, \theta)| + \varepsilon$$

$$\leq \left| \int_0^{t^*} \left(p_{\theta(t)}(y(t)) - p_{\theta(t)}(x(t)) \right) dt \right| + \int_{t^*}^{\infty} p_{\bar{\theta}(t)}(y(t))dt + 2\varepsilon$$

$$\leq \int_0^{t^*} \varphi(\|x(t) - y(t)\|)dt + J(y(t^*), \vartheta) + 2\varepsilon$$

$$\leq \omega\delta + 4\varepsilon = \bar{\varepsilon}.$$

As both ε and δ can be chosen to be arbitrarily small, $\bar{\varepsilon}$ can be made to be arbitrarily small. This implies that, if y_0 is sufficiently close to x_0, then $J(y_0) \leq J(x_0) + \bar{\varepsilon}$. By interchanging the roles of x_0 and y_0, we have $J(x_0) \leq J(y_0) + \bar{\varepsilon}$. This means that $\lim_{y_0 \to x_0} J(y_0) = J(x_0)$ which leads to the conclusion. $\square$

6.3.3 Optimal State Trajectory and Optimal Process

In this subsection, we establish a relationship between optimal state trajectory of the switched linear system and optimal process of the relaxed differential inclusion.

Suppose that $p(\cdot)$ is a function defined on $\mathbf{R}^n$. Assume that the function is continuous, positive definite, and that there exist a polynomial φ of time with $\varphi(0) = 0$, and a positive real number μ, such that

$$|p(x) - p(y)| \leq \varphi(\|x - y\|) \quad \forall \, x, y \in \mathbf{R}^n \quad \|x - y\| \leq \mu. \tag{6.27}$$

For the relaxed differential inclusion

$$\dot{x}(t) \in \mathrm{co}\{A_1 x(t), \cdots, A_m x(t)\} \tag{6.28}$$

where 'co' denotes the convex hull, consider the infinite horizon cost function given by

$$K(x(\cdot)) = \int_0^\infty p(x(t)) dt \tag{6.29}$$

where $x(\cdot)$ is a feasible state trajectory (process) of the relaxed differential inclusion.

The *optimal cost* associated with the cost function at x_0 is

$$K_{x_0} = \inf_{x(0)=x_0} K(x(\cdot)).$$

The *(worst case) optimal cost* associated with the cost function is

$$K_* = \sup_{\|x_0\| \leq 1} K_{x_0}.$$

A process $x^*(\cdot)$ with $K(x^*(\cdot)) = K_{x^*(0)}$ is said to be an *optimal process*.

Let $J(x_0)$ and J_* denote the optimal cost and the worst case optimal cost of the optimal switching problem for switched system (6.1) under the same cost function, respectively. As each state trajectory of the switched linear system is also a feasible process of the relaxed differential inclusion, it is clear that $K_{x_0} \leq J(x_0)$ and $K_* \leq J_*$.

Theorem 6.14. *Suppose that the switched linear system is asymptotically stabilizable. Then, for any $x_0 \in \mathbf{R}^n$, we have*

$$K_{x_0} = J(x_0).$$

Proof. We follow a similar procedure as in the proof of Theorem 6.13.

First, for any $\varepsilon > 0$, there is a $\delta \leq \min\{1, \mu\}$ such that

$$J(x_0) \leq \varepsilon \quad \forall \ x_0 \in \mathbf{B}_\delta.$$

Second, fix an $x_0 \neq 0$. For ε, there is a process $x(\cdot)$ of the relaxed differential inclusion, such that $x(0) = x_0$, and

$$|K(x(\cdot)) - K_{x_0}| \leq \varepsilon.$$

Let t_1 be a time such that

$$\|x(t)\| \leq \frac{\delta}{2} \quad \forall \ t \geq t_1.$$

By Lemma 2.12, for any continuous function $r \colon [0, t_1] \mapsto \mathbf{R}$ satisfying $r(t) > 0$ for all $t \in [0, t_1]$, there exists a state trajectory $y(\cdot)$ of the switched linear system with $y(0) = x_0$, such that

$$\|y(t) - x(t)\| \leq r(t) \quad \forall \ t \in [0, t_1].$$

Let $r(t) \leq \frac{\delta}{2} e^{-t}$. From (6.27), it is clear that

$$|\int_0^{t_1} p(y(t)) - p(x(t))dt| \leq \int_0^{t_1} \varphi(\|y(t) - x(t)\|)$$

where $\varphi = a_1 t + a_2 t^2 + \cdots + a_j t^j$ is the polynomial defined in (6.27). It can be seen that

$$\int_0^{t_1} \varphi(\|x(t) - y(t)\|)dt \leq \int_0^{t_1} \sum_{i=1}^{j} |a_i| \delta^i e^{-i\eta t} dt$$

$$\leq \left(\sum_{i=1}^{j} \frac{\mu^{i-1}}{i\eta} |a_i| \right) \delta \overset{def}{=} \omega\delta.$$

Third, as $\|y(t_1)\| \leq \|x(t_1)\| + r(t_1) \leq \delta$, we have

$$J(y(t_1)) \leq \varepsilon.$$

As a result, there is a switching path $\vartheta_{[0,\infty)}$ such that

$$\int_0^\infty p(\phi(t; 0, y(t_1), \vartheta))dt \leq J(y(t_1)) + \varepsilon \leq 2\varepsilon.$$

Let $z(\cdot)$ be the state trajectory concatenated by $y(\cdot)$ over $[0, t_1]$ and $\phi(\cdot; 0, y(t_1), \vartheta)$ over $(0, \infty)$, that is

$$z(t) = \begin{cases} y(t) & t \in [0, t_1] \\ \phi(t - t_1; 0, y(t_1), \vartheta) & t \in (t_1, \infty). \end{cases}$$

It can be seen that

$$\int_0^\infty p(z(t))dt \le \int_0^{t_1} p(y(t))dt + 2\varepsilon.$$

Finally, combining the above reasonings, we have

$$K_{x_0} \ge \int_0^\infty p(x(t))dt - \varepsilon \ge \int_0^{t_1} p(x(t))dt - \varepsilon$$
$$\ge \int_0^{t_1} p(y(t))dt - \varepsilon - \omega\delta \ge \int_0^\infty p(y(t))dt - 3\varepsilon - \omega\delta$$
$$\ge J_{x_0} - 3\varepsilon - \omega\delta.$$

As both ε and δ can be chosen to be arbitrarily small, this implies that $J(x_0) \le K_{x_0}$. This leads directly to the theorem. $\square$

Corollary 6.15. *Under the condition of Theorem 6.14, we have $K_* = J_*$, and K_{x_0} is a continuous function of x_0 in $\mathbf{R}^n$.*

Corollary 6.16. *Under the condition of Theorem 6.14, any optimal state trajectory of the switched linear system is an optimal process of the relaxed differential inclusion.*

Corollary 6.16 indicates a method to find the optimal state trajectory (and hence the optimal switching signal) of the switched linear system. Indeed, in the literature, there have already been many works on finding the optimal process of the differential inclusion. If an optimal process $x(\cdot)$ is a boundary process, *i.e.*,

$$\dot{x}(t) \in \{A_1 x(t), \cdots, A_m x(t)\}$$

for almost all $t \in [0, \infty)$, then, the process is also an optimal state trajectory of the switched linear system. If the relaxed differential inclusion does not have any boundary optimal process, then, the switched linear system has no optimal state trajectory. In particular, if the relaxed differential inclusion does not have any optimal process, then, the switched linear system does not have any optimal state trajectory, either.

6.3.4 Discrete-time Case

For the discrete-time unforced switched linear system

$$x(k + 1) = A_{\sigma(k)} x(k) \tag{6.30}$$

several results presented in Sections 6.3.1 and 6.3.2 can be obtained in a parallel manner. In this subsection, we present the main results and outline the key points of the proofs.

Suppose that $\{p_i\}_{i \in M}$ is a set of functions defined on $\mathbf{R}^n$. Assume that each function is continuous and bounded by positive definite polynomials of state. That is, there are two sets of polynomials $\{\gamma_i(\cdot)\}_M$ and $\{\gamma^i(\cdot)\}_M$ with $\gamma_i(0) = 0$ and $\gamma^i(0) = 0$, such that

$$\gamma_i(x) \leq p_i(x) \leq \gamma^i(x) \quad \forall \ i \in M \ \ x \in \mathbf{R}^n.$$

The problem of optimal switching is to find, for a given x_0, a switching path θ_{x_0} that minimizes the cost function

$$J(x_0, \theta_{x_0}) = \min_{\theta \in \mathcal{S}} J(x_0, \theta)$$

where

$$J(x_0, \theta) = \sum_{k=0}^{\infty} p_{\theta(k)}(x(k)) \tag{6.31}$$

and $\mathcal{S}$ is the set of switching paths defined on $\mathbf{N}_+$.

The optimal switching path, the optimal switching law, and the optimal state trajectory can be defined in the same manner as in continuous time.

Definition 6.17. *The* optimal cost *associated with the cost function at x_0 is*

$$J(x_0) = \inf_{\theta \in \mathcal{S}} J(x_0, \theta).$$

The worst case optimal cost *(in the unit ball) associated with the cost function is*

$$J_* = \sup_{\|x_0\| \leq 1} J(x_0).$$

The following technical lemma follows directly from the fact that $\{\gamma_i(\cdot)\}_M$ and $\{\gamma^i(\cdot)\}_M$ are positive definite polynomials.

Lemma 6.18. *Define two functions on $\mathbf{R}^n$ by*

$$\underline{\gamma}(x) = \min_{i \in M}\{\gamma_i(x)\} \ and \ \bar{\gamma}(x) = \max_{i \in M}\{\gamma_i(x)\}.$$

Then, for any $s \geq 0$, there are two positive real numbers ρ_s and ρ^s such that

$$\rho_s\|x\| \leq \underline{\gamma}(x) \leq \bar{\gamma}(x) \leq \rho^s\|x\| \quad \forall \ x \in \mathbf{B}_s.$$

Without loss of generality, we assume that ρ^s is increasing and ρ_s is decreasing as s increases.

Theorem 6.19. *The following statements are equivalent*:

(i) *the optimal cost $J(x_0)$ is finite for any given $x_0 \in \mathbf{R}^n$;*
(ii) *the worst case optimal cost is finite; and*
(iii) *the switched system is asymptotically stabilizable.*

Proof. We establish the equivalence between (i) and (iii). The equivalence between (ii) and (iii) can be proven in the same manner.

To prove that $(i) \implies (iii)$, suppose that the optimal cost $J(x_0)$ is finite for any given x_0. This means that for each $x_0 \in \mathbf{R}^n$, there is a switching path $\theta_{[0,\infty)}$ such that

$$\sum_{k=0}^{\infty} p_{\theta(k)}(\phi(k;0,x_0,\theta)) < \infty.$$

It can be seen that

$$p_{\theta(k)}(\phi(k;0,x_0,\theta)) \geq \underline{\gamma}(\phi(k;0,x_0,\theta))$$

where function $\underline{\gamma}$ is defined in Lemma 6.18. Accordingly, we have

$$\sum_{k=0}^{\infty} \underline{\gamma}(\phi(k;0,x_0,\theta)) < \infty.$$

This implies that

$$\lim_{k \to \infty} \underline{\gamma}(\phi(k;0,x_0,\theta)) = 0$$

which in turn implies that trajectory $\phi(t;0,x_0,\theta)$ converges. By Theorem 3.53, the switched system is asymptotically stabilizable.

To show that $(iii) \implies (i)$, suppose that the switched system is asymptotically stabilizable. Then, by Theorem 3.53, the system is exponentially stabilizable. That is, there exist a switching signal σ, and positive real numbers α and β, such that

$$\|\phi(t;0,x_0,\sigma)\| \leq \beta e^{-\alpha t}\|x_0\| \quad \forall \, x_0 \in \mathbf{R}^n \ t \in \mathbf{N}_+. \tag{6.32}$$

It can be seen that

$$J(x_0) \leq J(x_0,\sigma) \leq \sum_{t=0}^{\infty} \bar{\gamma}(\phi(t;0,x_0,\sigma))$$

where function $\bar{\gamma}$ is defined in Lemma 6.18. It follows from Lemma 6.18 that

$$\bar{\gamma}(x) \leq \rho^{\beta\|x_0\|}\|x\| \quad x \in \mathbf{B}_{\beta\|x_0\|}.$$

Therefore, we have

$$J(x_0) \leq \sum_{k=0}^{\infty} \rho^{\beta \|x_0\|} \beta e^{-\alpha k} \|x_0\| = \frac{\beta e^{\alpha}}{\alpha(e^{\alpha} - 1)} \rho^{\beta \|x_0\|} \|x_0\| \qquad (6.33)$$

which sets an upper bound for the optimal cost. $\square$

Given an $s \geq 0$, define

$$J_s = \sup\{J(x_0) : \|x_0\| \leq s\}.$$

It follows from (6.33) that

$$J_s \leq \frac{\beta s e^{\alpha}}{\alpha(e^{\alpha} - 1)} \rho^{\beta s} \qquad (6.34)$$

which provides a useful upper bound for J_s.

Theorem 6.20. *Suppose that the switched system is asymptotically stabilizable. Then, any optimal state trajectory is exponentially convergent.*
Proof. Suppose that $x(\cdot)$ is an optimal state trajectory starting from $x_0 \neq 0$. Let $\theta(\cdot)$ be the associated switching path, that is,

$$x(k + 1) = A_{\theta(k)} x(k) \quad \forall \, k \in \mathbf{N}_+.$$

It is clear that

$$J(x_0) = J(x_0, \theta) = \sum_{t=0}^{\infty} p_{\theta(t)}(x(t)) < \infty. \qquad (6.35)$$

From the proof of Theorem 6.19, it can be seen that

$$\sum_{k=0}^{\infty} p_{\theta(k)}(x(k)) \geq \sum_{k=0}^{\infty} \underline{\gamma}(x(k)). \qquad (6.36)$$

Let $r_1 = \frac{\|x_0\|}{2}$.

As pointed out in the proof of Theorem 6.19, the state trajectory must be convergent. Accordingly, the state trajectory is bounded, *i.e.*, $\|x(k)\| \leq \nu$ for some ν and all $k \in N_+$. Define

$$k_1 = \min\{k \geq 0 : x(k) \in \mathbf{B}_{r_1}\}.$$

It is clear that $x(k) \notin \mathbf{B}_{r_1}$ for $k \leq k_1$. Therefore, we have

$$\sum_{k=0}^{k_1} \underline{\gamma}(x(k)) \geq \sum_{k=0}^{k_1} \rho_{\nu} \frac{\|x_0\|}{2} \geq \rho_{\nu} \frac{\|x_0\|}{2} k_1.$$

This, together with inequality (6.36), implies that

$$k_1 \leq \frac{2J(x_0)}{\rho_{\nu} \|x_0\|}.$$

Furthermore, it follows from (6.34) that

$$k_1 \le \frac{2\beta e^\alpha}{\rho_\nu \alpha (e^\alpha - 1)} \rho^{\beta \|x_0\|} \overset{def}{=} \delta.$$

This means that for any optimal state trajectory $x(\cdot)$, there is a time instant $k \le \delta$ such that $x(k) \in \mathbf{B}_{\frac{\|x(0)\|}{2}}$. On the other hand, let

$$y(k) = x(k + k_1) \quad k \in \mathbf{N}_+.$$

By the optimality of $x(\cdot)$, the state trajectory $y(\cdot)$ is also an optimal state trajectory starting from $y(0) = x(k_1)$. Accordingly, there is a time $s_2 \le \delta$ such that $y(s_2) \in \mathbf{B}_{\frac{\|y(0)\|}{2}}$, which implies that

$$x(k_2) \in \mathbf{B}_{\frac{\|x(k_1)\|}{2}}$$

where $k_2 = k_1 + s_2$. Continuing with this process, we can find a monotone time sequence $0 = k_0 < k_1 < k_2 < \cdots$ such that

(i) $k_{i+1} - k_i \le \delta \quad \forall\, i \in \mathbf{N}_+$;
(ii) $x(k_{i+1}) \in \mathbf{B}_{\frac{\|x(k_i)\|}{2}} \quad \forall\, i \in \mathbf{N}_+$; and
(iii) $\|x(k)\| \le e^{\eta \delta} \|x(k_i)\| \quad \forall\, t \in [t_i, t_{i+1}] \quad i \in \mathbf{N}_+$, where $\eta = \max_{i \in M} \{\|A_i\|\}$.

These facts suffice to conclude that the state trajectory is exponentially convergent. □

Theorem 6.21. *Suppose that the switched system is asymptotically stabilizable. Then, the optimal cost $J(x_0)$ is a continuous function of x_0.*
Proof. The theorem can be proven in the same manner as the proof of Theorem 6.13, and we hence omit the details. □

6.4 Mixed Optimal Switching and Control

In this section, we address the optimization problem for the switched linear system where both the control input and the switching signal are design variables. For such a problem, we need to find the optimal switching signal and optimal control input simultaneously. To address this problem, a key issue is to understand the interaction between the switching signal and control input, which is very difficult and challenging. Here, we focus on the simplest case where both the time horizon and the switching number are finite.

6.4.1 A Two-stage Optimization Approach

Switched Linear Quadratic Optimal Control Problem   Given a switched linear control system

$$\dot{x}(t) = A_\sigma x(t) + B_\sigma u(t) \quad x(t_0) = x_0 \tag{6.37}$$

a fixed end-time $t_f < \infty$, and an upper bound l of the number of allowed switches, find (if possible) a piecewise continuous input $u \colon [t_0, t_f] \mapsto \mathbf{R}^n$, and a switching signal σ which switches at most l times in $[t_0, t_f]$, such that the quadratic cost function

$$J(x_0, \sigma, u) = \frac{1}{2} x^T(t_f) Q_f x(t_f) + \int_{t_0}^{t_f} (\frac{1}{2} x^T Q x + \frac{1}{2} u^T R u) dt \tag{6.38}$$

is minimized, where $Q_f \geq 0$, $Q \geq 0$ and $R > 0$.

In the above formulation, the assumption of the fixed end time is not crucial as we can easily convert a free end-time problem into a fixed end-time one by introducing an additional state variable. However, the assumption that the number of switches is bounded by a fixed number is crucial in the approach to be developed. In many practical applications, the high frequency switching is not desired and a positive dwell time is applied. In this case, an upper bound for the switching number is the length of horizon $t_f - t_0$ divided by the dwell time.

In the problem, we need to find an optimal control solution (σ^*, u^*) such that

$$J(x_0, \sigma^*, u^*) = \min_{\sigma \in \mathcal{S}, u \in \mathcal{U}} J(x_0, \sigma, u) \tag{6.39}$$

where $\mathcal{S}$ is the set of well-defined switching paths over $[t_0, t_f]$, $\mathcal{U}$ is the set of piecewise continuous vector functions over $[t_0, t_f]$.

As the number of switches is bounded by l, the problem can be equivalently formulated as:

Find a nonnegative integer $k \leq l$, an index sequence $i_0, \cdots, i_k$ in M, a time sequence $t_1, \cdots, t_k$ in $[t_0, t_f]$, and a control input $u \in \mathcal{U}$, such that the cost function

$$J(x_0, t_1, \cdots, t_k, i_0, \cdots, i_k, u) = J(x_0, \theta, u)$$

is minimized, where θ is the switching path with the switching sequence

$$\{(t_0, i_0), (t_1, i_1), \cdots, (t_k, i_k)\}.$$

If we fix the switching signal, the problem is reduced to a conventional optimal control problem for linear time-varying systems. This simple observation leads us towards considering the problem as a two-stage optimization problem. That is, we decompose the problem into two subproblems. The following lemma provides a support to this two-stage decomposition.

Lemma 6.22. *For a given $x_0 \in \mathbf{R}^n$, suppose that*

(i) an optimal solution (σ^*, u^*) exists; and

(ii)for any fixed switching index sequence $i_0, \cdots, i_k$, there exist a time sequence $t_1, \cdots, t_k$, and a control input u, such that the cost function

$$J(x_0, t_1, \cdots, t_k, i_0, \cdots, i_k, u)$$

is minimized.

Then, we have

$$J(x_0, \sigma^*, u^*) = \min_{i_0, \cdots, i_k} \min_{t_1, \cdots, t_k, u} J(x_0, t_1, \cdots, t_k, i_0, \cdots, i_k, u). \quad (6.40)$$

Proof. The proof is straightforward and is hence omitted. □

Based on the lemma, we propose the following two-stage optimization approach.

Stage 1. Fixing a switching index sequence, solve the optimal control problem for the corresponding time-varying system. That is, fixing a nonnegative integer $j \leq l$, and a switching index sequence $i_0, i_1, \cdots, i_j$, find (if possible) a switching time sequence $t_1, \cdots, t_j$ in $[t_0, t_f]$, and piecewise continuous input $u: [t_0, t_f] \mapsto \mathbf{R}^p$, such that the cost function $J(x_0, t_1, \cdots, t_k, i_0, \cdots, i_k, u)$ is minimized; and

Stage 2. Regarding the optimal cost for each switching index sequence $i_0, i_1, \cdots, i_j$ as a function

$$J_1 = J_1(i_0, i_1, \cdots, i_j) = \min_{t_1, \cdots, t_k, u} J(x_0, t_1, \cdots, t_k, i_0, \cdots, i_k, u)$$

minimize J_1 w.r.t. the switching index sequence.

As l is known, the number of possible switching index sequences can be computed to be $k_m^l \stackrel{def}{=} \sum_{j=0}^{l} m(m-1)^j$, where m is the number of subsystems. In particular, when $m = 2$ the number is $2(l+1)$ which is linear w.r.t. l. Accordingly, in the second stage, we just need to compare the k_m^l number of costs:

$$J^* = \min\{J_1(i_0, \cdots, i_j) : j \leq l, i_s \in M \quad s = 1, \cdots, j\}.$$

As a consequence, the second stage is conventional but the first is the core and difficult step for solving the problem. Indeed, in the first stage, we need to determine the optimal switching time sequence, which is the most challenging issue.

To illustrate the approach, consider the simplest case where the switched system has two subsystems and there is only one switch with fixed index sequence.

Single Switch Optimal Control Problem. For a switched system

$$\dot{x} = A_1 x + B_1 u \quad t_0 \leq t < t_1$$
$$\dot{x} = A_2 x + B_2 u \quad t_1 \leq t \leq t_f \quad (6.41)$$

find an optimal switching instant t_1 and an optimal input u to minimize the cost function (6.38).

By introducing a state variable x_{n+1} corresponding to the switching instant t_1, and a new scaled time variable τ with

$$t = t_0 + (x_{n+1} - t_0)\tau \quad \text{for } 0 \leq \tau \leq 1$$
$$t = x_{n+1} + (t_f - x_{n+1})(\tau - 1) \quad \text{for } 1 \leq \tau \leq 2$$

the problem is converted to finding an optimal x_{n+1} and an optimal control u for the system

$$\begin{aligned} \frac{dx(\tau)}{d\tau} &= (x_{n+1} - t_0)(A_1 x + B_1 u) \\ \frac{dx_{n+1}}{d\tau} &= 0 \end{aligned} \quad 0 \leq \tau < 1$$

$$\begin{aligned} \frac{dx(\tau)}{d\tau} &= (t_f - x_{n+1})(A_2 x + B_2 u) \\ \frac{dx_{n+1}}{d\tau} &= 0 \end{aligned} \quad 1 \leq \tau \leq 2$$

with the quadratic cost function

$$J = \frac{1}{2} x^T(2) Q_f x(2) + \int_0^1 (x_{n+1} - t_0) L(x, u) d\tau$$
$$+ \int_1^2 (t_f - x_{n+1}) L(x, u) d\tau \tag{6.42}$$

where

$$L(x, u) = \frac{1}{2} x^T Q x + \frac{1}{2} u^T R u.$$

First, assuming that we are given a fixed x_{n+1}, we can apply the principle of optimality as follows. Suppose that the optimal cost function is

$$V^*(x, \tau, x_{n+1}) = \frac{1}{2} x^T P(\tau, x_{n+1}) x \tag{6.43}$$

where $P^T(\tau, x_{n+1}) = P(\tau, x_{n+1})$. The Hamilton-Jacobi-Bellman (HJB) equation is

$$-\frac{\partial V^*}{\partial \tau}(x, \tau, x_{n+1}) = \min_u \{(x_{n+1} - t_0)[L(x, u)$$
$$+ \frac{\partial V^*}{\partial x}(x, \tau, x_{n+1})(A_1 x + B_1 u)]\} \tag{6.44}$$

in the interval $\tau \in [0, 1)$ and

$$-\frac{\partial V^*}{\partial \tau}(x, \tau, x_{n+1}) = \min_u \{(t_f - x_{n+1})[L(x, u)$$
$$+ \frac{\partial V^*}{\partial x}(x, \tau, x_{n+1})(A_2 x + B_2 u)]\} \tag{6.45}$$

in the interval $\tau \in [1, 2]$.

Using the idea for solving the conventional optimal control problems, we can prove that the solution to (6.44) is

$$u(x, \tau, x_{n+1}) = -K(\tau, x_{n+1})x(\tau, x_{n+1}) \tag{6.46}$$

where

$$K(\tau, x_{n+1}) = R^{-1}B_1^T P(\tau, x_{n+1})$$

and $P(\tau, x_{n+1})$ satisfies the following parametrized Riccati equation

$$-\frac{\partial P}{\partial \tau} = (x_{n+1} - t_0)(Q + PA_1 + A_1^T P - PB_1 R^{-1}B_1^T P). \tag{6.47}$$

Similarly, the solution to (6.45) is also (6.46) with

$$K(\tau, x_{n+1}) = R^{-1}B_2^T P(\tau, x_{n+1})$$

and P satisfies the following parametrized Riccati equation

$$-\frac{\partial P}{\partial \tau} = (t_f - x_{n+1})(Q + PA_2 + A_2^T P - PB_2 R^{-1}B_2^T P). \tag{6.48}$$

The boundary condition at $\tau = 2$ is given by

$$P(2, x_{n+1}) = Q_f. \tag{6.49}$$

As a result, any optimal trajectory satisfies the following constraints:

$$-\frac{\partial P}{\partial \tau} = (x_{n+1} - t_0)(Q + PA_1 + A_1^T P - PB_1 R^{-1}B_1^T P) \quad \tau \in [0, 1)$$

$$-\frac{\partial P}{\partial \tau} = (t_f - x_{n+1})(Q + PA_2 + A_2^T P - PB_2 R^{-1}B_2^T P) \quad \tau \in [1, 2]$$

$$P(2, x_{n+1}) = Q_f. \tag{6.50}$$

The parametrized optimal cost at $\tau = 0$ is

$$J_1(t_1) = J_1(x_{n+1}) = V^*(x_0, 0, x_{n+1}) = \frac{1}{2}x_0^T P(0, x_{n+1})x_0. \tag{6.51}$$

Next, if $J_1(x_{n+1})$ is obtained for each $x_{n+1} \in [t_0, t_f]$, then, the optimal cost can be calculated by seeking the minimal cost

$$J^* = \min_{x_{n+1} \in [t_0, t_f]} J_1(x_{n+1}) = \min_{x_{n+1} \in [t_0, t_f]} \frac{1}{2}x_0^T P(0, x_{n+1})x_0.$$

This means that the optimal switching instant x_{n+1} should be either the extreme point (t_0 or t_f), which means that no switch occurs, or an inner point with the derivative equal to zero, that is

$$\frac{dJ_1}{dx_{n+1}}(x_{n+1}) = 0.$$

To determine $\frac{dJ_1}{dx_{n+1}}(x_{n+1})$, differentiate the cost function $w.r.t.$ x_{n+1}

$$\frac{dJ_1}{dx_{n+1}}(x_{n+1}) = \frac{1}{2}x_0^T \frac{\partial P}{\partial x_{n+1}}(0, x_{n+1})x_0. \tag{6.52}$$

To obtain this cost, we need to know $\frac{\partial P}{\partial x_{n+1}}$, which can be computed from (6.47) and (6.48) as

$$-\frac{\partial}{\partial \tau}(\frac{\partial P}{\partial x_{n+1}}) = (Q + PA_1 + A_1^T P - PB_1R^{-1}B_1^T P)$$

$$+(x_{n+1} - t_0)(\frac{\partial P}{\partial x_{n+1}}A_1 + A_1^T \frac{\partial P}{\partial x_{n+1}}$$

$$-\frac{\partial P}{\partial x_{n+1}}B_1R^{-1}B_1^T P - PB_1R^{-1}B_1^T \frac{\partial P}{\partial x_{n+1}}) \tag{6.53}$$

in the interval $\tau \in [0, 1)$ and

$$-\frac{\partial}{\partial \tau}(\frac{\partial P}{\partial x_{n+1}}) = (Q + PA_2 + A_2^T P - PB_2R^{-1}B_2^T P)$$

$$+(t_f - x_{n+1})(\frac{\partial P}{\partial x_{n+1}}A_2 + A_2^T \frac{\partial P}{\partial x_{n+1}}$$

$$-\frac{\partial P}{\partial x_{n+1}}B_2R^{-1}B_2^T P - PB_2R^{-1}B_2^T \frac{\partial P}{\partial x_{n+1}}) \tag{6.54}$$

in the interval $\tau \in [1, 2]$. The boundary condition is

$$\frac{\partial P}{\partial x_{n+1}}(2, x_{n+1}) = 0. \tag{6.55}$$

Accordingly, the solution of $\frac{\partial P}{\partial x_{n+1}}(0, x_{n+1})$ can be obtained and the optimal x_{n+1} can be determined.

As the analytic expression for the optimal solution is usually not available, numerical solutions have to be sought. To this end, we propose an efficient numerical procedure to compute the optimal switching time.

Searching Algorithm for Optimal Switching Time

1. Set the iteration index $k = 0$. Choose an initial time $s_k \in [t_0, t_f]$.
2. Compute $J_1(s_k)$ and $\frac{dJ_1}{dx_{n+1}}(s_k)$.
3. Using the gradient projection method, update s_k to be $s_{k+1} = s_k + a_k dt_k$ (the step size a_k can be chosen using the Armijo's rule [13]). Set the iteration step $k = k + 1$.
4. Repeat Steps 2 and 3, until $\frac{dJ_1}{dx_{n+1}}(s_k)$ is within a pre-specified neighborhood of the origin.

Remark 6.23. It can be seen that there is no difficulty in applying the above method to the optimal control problem with several subsystems and more than one switch. For this, we can convert the problem to an equivalent problem in $\tau \in [0, k + 1]$ if there are k switches. It is then straightforward to differentiate the Riccati equations, which are parameterized by $x_{n+1}, \cdots, x_{n+k}$, so as to obtain additional differential equations for $\frac{\partial J}{\partial x_{n+i}}$'s. Along with the corresponding boundary condition at $\tau = k + 1$, we can solve the resultant initial value ordinary differential equations backwards in τ to find the values at $\tau = 0$. By doing this, we obtain the accurate values of $\frac{\partial J}{\partial x_{n+i}}$'s. What remains is to search for an optimal switching time sequence, and this can be done by means of standard numerical algorithms in optimization.

Example 6.24. Consider the switched linear system $\Sigma(A_i, B_i)_{\bar{2}}$ with

$$A_1 = \begin{bmatrix} 0 & 0 \\ 0 & 0.5 \end{bmatrix} \quad b_1 = \begin{bmatrix} 1 \\ 0 \end{bmatrix} \quad A_2 = \begin{bmatrix} 0.1 & 1 \\ -1 & 0.1 \end{bmatrix} \quad b_2 = \begin{bmatrix} 0 \\ 0 \end{bmatrix}. \quad (6.56)$$

Let $x_0 = [1, -1]^T$ and $t_f = 5$. We are to minimize the cost function

$$J(x_0, u, \sigma) = x^T(t_f)x(t_f) + \int_0^{t_f} (x^T(t)x(t) + u^2(t))dt$$

where the switching signal is allowed to switch at most once. According to the two-state optimization approach, we first classify the four switching index sequences as

$$\{1\} \quad \{2\} \quad \{1, 2\} \quad \{2, 1\}.$$

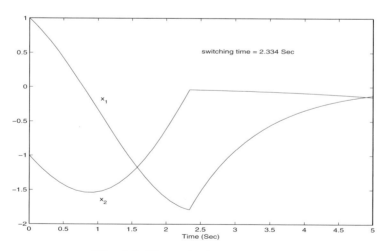

Fig. 6.3. The optimal state trajectory

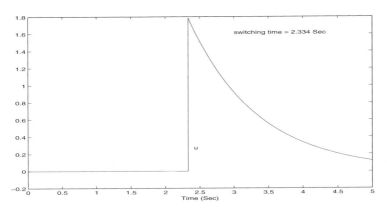

Fig. 6.4. The optimal input trajectory

Then, for the latter two switching sequences, determine the switching times that minimize the cost function. The optimal cost can be obtained by comparing the costs associated with the switching sequences. The optimal switching sequence is computed to be

$$\{(0, 2), (2.334, 1)\}$$

and the optimal cost is

$$J(x_0) \approx 9.1808.$$

The optimal state and input trajectories are shown in Figures 6.3 and 6.4, respectively. It is clear that the state trajectory approaches the origin although both subsystems are individually unstable.

6.4.2 Piecewise Constant Feedback Suboptimal Control

In the finite horizon optimization problem, even for a linear time-invariant system, the optimal control input is a linear time-varying feedback of the state. The time-varying nature of the state feedback is not desired in many practical situations. In this subsection, we seek a piecewise time-invariant linear state feedback control input to optimize the cost index. This scheme has been extensively addressed in the literature for linear time-invariant systems. Here, we re-visit and extend the scheme to switched linear systems.

For a switched linear system

$$\dot{x}(t) = A_\sigma x(t) + B_\sigma u(t) \quad x(t_0) = x_0 \tag{6.57}$$

if the switching signal is known, then, the switched system is given by

$$\dot{x}(t) = A(t)x(t) + B(t)u(t) \quad x(t_0) = x_0 \tag{6.58}$$

where $A(t) = A_{\sigma(t)}$ and $B(t) = B_{\sigma(t)}$, for $t \geq t_0$.

The optimal linear regulator problem is to determine the control u on $[t_0, t_f]$ which minimizes the quadratic cost function:

$$J(x_0, t_0, t_f, u) = x^T(t_f)Q_f x(t_f) + \int_{t_0}^{t_f} \left(x^T(t)Qx(t) + u^T(t)Ru(t) \right) dt$$

where the end time t_f is fixed, the terminal state $x(t_f)$ is unconstrained, Q_f and Q are positive semi-definite matrices, while matrix R is positive definite.

It is well known that the optimal control u^* is generated by the linear feedback law

$$u^*(t) = -R^{-1}B^T(t)K(t)x(t) = -L^*(t)x(t)$$

where $K(t)$ is the unique (positive semi-definite) solution of the matrix Riccati differential equation

$$\dot{K}(t) = -A^T(t)K(t) - K(t)A(t) - Q + K(t)B(t)R^{-1}B^T(t)K(t)$$

with the boundary condition $K(t_f) = Q_f$. Furthermore, the matrix $K(t)$ has the property that, for arbitrary $t \in [t_0, t_f)$ and $x \in \mathbf{R}^n$, we have

$$J(x, t, t_f, u)|_{u=-L^*(t)x(t)} = \min_u J(x, t, t_f, u) = x^T K(t)x. \tag{6.59}$$

In particular, substituting x_0 and t_0 for x and t, respectively, we have

$$\min_u J(x_0, t_0, t_f, u) = x_0^T K(t_0)x_0.$$

Consider the control law $u_L(t) = -L(t)x(t)$ with $L(t) \neq L^*(t)$. Clearly, u_L is not optimal. Therefore, by comparing the cost associated with the use of $L(t)$ as opposed to $L * (t)$, it is possible to obtain bounds on the matrix $K(t)$. This is illustrated by the following lemma which was reported in [80].

Lemma 6.25. *Let $L(t)$ be an arbitrary $p \times n$ time-varying matrix defined for $t \in [t_0, t_f]$. Let $V_L(t)$ denote the (unique positive semi-definite) solution of the linear matrix differential equation*

$$\dot{V}_L(t) = -V_L(t)\left(A(t) - B(t)L(t)\right) - \left(A(t) - B(t)L(t)\right)^T V_L(t)$$
$$-Q - L^T(t)RL(t)$$

satisfying the boundary condition $V_L(t_f) = Q_f$. Then

$$K(t) \leq V_L(t) \quad \forall \ t \in [t_0, t_f].$$

Next, we turn to the suboptimal control of the switched linear system. We are seeking an optimal switching path θ with switching sequence

$\{(t_0, i_0), \cdots, (t_j, i_j)\}$, where the number of switches j is less than or equal to a given number l, and optimal linear state feedback gains $F_0, \cdots, F_j$, such that the quadratic cost index

$$J(x_0, \theta, u) = x^T(t_f)Q_f x(t_f) + \int_{t_0}^{t_f} \left(x^T(t)Qx(t) + u^T(t)Ru(t) \right) dt$$

$$= x^T(t_f)Q_f x(t_f) + \Sigma_{k=0}^{j} \int_{t_k}^{t_{k+1}} \frac{1}{2} x^T(t)(Q + F_k^T RF_k)x(t)dt \quad (6.60)$$

is minimized, where $t_{j+1} \stackrel{def}{=} t_f$, and $u(t) = F_k x(t)$ for $t \in [t_k, t_{k+1})$.

By the two-state optimization approach introduced in the previous subsection, we first fix the switching index sequence and seek the optimal switching times and the optimal gain matrices, then solve the optimization problem by comparing the costs with respect to the set of switched index sequences.

We focus on the first stage and suppose that the switching index sequence $i_0, \cdots, i_j$ is given. Let $\{t_k\}_{k=1}^{j}$ be a set of switching times such that

$$t_0 < t_1 < \cdots < t_j < t_{j+1} = t_f.$$

Let $\{L_k\}_{k=0}^{j}$ be a set of gain matrices. Define the time-varying gain matrix $L(\cdot)$ by

$$L(t) = \begin{cases} L_0 & t \in [t_0, t_1) \\ \vdots \\ L_j & t \in [t_j, t_f]. \end{cases}$$

In addition, let θ be the switching path corresponding to the switching sequence $\{(t_0, i_0), \cdots, (t_j, i_j)\}$, u_L be $u_L(t) = -L(t)x(t)$ for $t \in [t_0, t_f]$, $A(\cdot)$ and $B(\cdot)$ be $A(t) = A_{\theta(t)}$ and $B(t) = B_{\theta(t)}$ for $t \in [t_0, t_f]$, and $\Phi_L(s_1, s_2)$ be the transition matrix corresponding to the matrix $A(t) - B(t)L(t)$. Then, we have

$$J(x_0, \theta, u_L) = x_0^T V_L(t_0)x_0 \quad (6.61)$$

where

$$V_L(x_0) = (\Phi_L(t_f, t_0))^T Q_f \Phi_L(t_f, t_0)$$
$$+ \int_{t_0}^{t_f} (\Phi_L(t, t_0))^T \left(Q + L^T(t)RL(t) \right) \Phi_L(t, t_0)dt.$$

For a fixed $k \in \bar{j}$, this equation can be re-written as

$$V_L(x_0) = (\Phi_L(t_1, t_0))^T \cdots (\Phi_L(t_f, t_j))^T Q_f \Phi_L(t_f, t_j) \cdots \Phi_L(t_1, t_0)$$
$$+ \int_{t_0}^{t_k} (\Phi_L(t, t_0))^T \left(Q + L^T(t)RL(t) \right) \Phi_L(t, t_0)dt + (\Phi_L(t_k, t_0))^T$$
$$\times \int_{t_k}^{t_f} (\Phi_L(t, t_k))^T \left(Q + L^T(t)RL(t) \right) \Phi_L(t, t_k)dt\Phi_L(t_k, t_0).$$

By differentiating $J(x_0, \theta, u_L)$ w.r.t. L_k (c.f. [80]), we have

$$\frac{\partial}{\partial L_k} J(x_0, \theta, u_L) = P_k x_0 x_0^T \tag{6.62}$$

where

$$P_k = \int_{t_k}^{t_{k+1}} \left(B_{i_k}^T V_L(t) - RL_k \right) \Phi_L(t, t_0) (\Phi_L(t, t_0))^T dt. \tag{6.63}$$

Similarly, by differentiating $J(x_0, \theta, u_L)$ w.r.t. t_k, and using the relations

$$\frac{d}{dt} \Phi_L(t, s) = (A(t) - B(t)L(t)) \Phi_L(t, s)$$

$$\frac{d}{ds} \Phi_L(t, s) = -\Phi_L(t, s)(A(s) - B(s)L(s))$$

and

$$L(t_k-) = L_k \qquad L(t_k+) = L_{k+1}$$

we obtain

$$\frac{\partial}{\partial t_k} J(x_0, \theta, u_L) = x_0^T (\Phi_L(t_k, t_0))^T S_k \Phi_L(t_k, t_0) x_0 \tag{6.64}$$

where

$$\begin{aligned} S_k = {} & V_L(t_k) \left(A_{i_{k+1}} + B_{i_{k+1}} L_{k+1} - A_{i_k} - B_{i_k} L_k \right) \\ & + \left(A_{i_{k+1}} + B_{i_{k+1}} L_{k+1} - A_{i_k} - B_{i_k} L_k \right)^T V_L(t_k) \\ & + (L_{k+1}^T R L_{k+1} - L_k^T R L_k) \quad k = 0, \cdots, j. \end{aligned} \tag{6.65}$$

Suppose that $\{L_k^*\}_{k=0}^j$ and $\{t_k^*\}_{k=0}^j$ minimize the cost index. Then, we have

$$\frac{\partial}{\partial L_k^*} J(x_0, \theta, u_{L^*}) = 0 \text{ and } \frac{\partial}{\partial t_k^*} J(x_0, \theta, u_{L^*}) = 0 \quad k = 0, \cdots, j.$$

By (6.62) and (6.64), this is equivalent to

$$P_k x_0 x_0^T = 0 \text{ and } x_0^T (\Phi_{L^*}(t_k^*, t_0))^T S_k \Phi_{L^*}(t_k, t_0) x_0 = 0 \tag{6.66}$$

for $k = 0, \cdots, j$, which gives a necessary condition for the optimal solution.

As it is virtually impossible to obtain explicit analytic expressions for $\{L_k^*\}_{k=0}^j$ and $\{t_k^*\}_{k=0}^j$, we need to develop an iterative scheme to solve Equations (6.62) and (6.64). That is, given a set of gain matrices $\{L_k^1\}_{k=1}^j$ and a set of switching times $\{t_k^1\}_{k=1}^j$, we need to find new sets $\{t_k^2\}_{k=1}^j$ and $\{L_k^2\}_{k=1}^j$ such that the resultant cost is reduced. To be more specific, there exist nonnegative real numbers $\epsilon_0, \cdots, \epsilon_j$ such that

$$t_k^2 = t_k^1 - \epsilon_k \frac{\partial}{\partial t_k^1} J(x_0, \theta^1, u_{L^1}) \quad k = 0, \cdots, j$$

$$t_0 < t_1^2 < \cdots < t_j^2 < t_f \quad J(x_0, \theta^2, u_{L^1}) \leq J(x_0, \theta^1, u_{L^1}) \qquad (6.67)$$

where θ^2 is the switching path corresponding to the new switching times $\{t_k^2\}_{k=0}^j$, and u_{L^1} corresponds to $\{L_k^1\}_{k=0}^j$ and $\{t_k^1\}_{k=0}^j$. The numbers $\epsilon_0, \cdots, \epsilon_j$ are chosen according to the well-known gradient or steepest descent schemes. Similarly, there exist nonnegative real numbers $\varepsilon_0, \cdots, \varepsilon_j$ such that

$$L_k^2 = L_k^1 - \varepsilon_k \frac{\partial}{\partial L_k^1} J(x_0, \theta^2, u_{L^1}) \quad k = 0, \cdots, j$$

$$J(x_0, \theta^2, u_{L^2}) \leq J(x_0, \theta^2, u_{L^1}) \qquad (6.68)$$

where u_{L^2} corresponds to $\{L_k^2\}_{k=0}^j$ and $\{t_k^2\}_{k=0}^j$. The numbers $\varepsilon_0, \cdots, \varepsilon_j$ are chosen according to one of the well-known gradient or steepest descent schemes.

The procedure is summarized as follows:

1) choose initial iterate sets $\{L_k^1\}_{k=0}^j$ and $\{t_k^1\}_{k=0}^j$ with corresponding cost $J(x_0, \theta^1, u_{L^1})$;

2) calculate $\{t_k^2\}_{k=0}^j$ using (6.64) and (6.65), and the gradient scheme in (6.67) so that $J(x_0, \theta^2, u_{L^1}) \leq J(x_0, \theta^1, u_{L^1})$;

3) calculate $\{L_k^2\}_{k=0}^j$ using (6.62) and (6.63), and the gradient scheme in (6.68) so that $J(x_0, \theta^2, u_{L^2}) \leq J(x_0, \theta^2, u_{L^1})$; and

4) re-set $\{L_k^1\}_{k=0}^j$ by $\{L_k^2\}_{k=0}^j$ and $\{t_k^1\}_{k=0}^j$ by $\{t_k^2\}_{k=0}^j$, and repeat the procedure successively.

Since the cost function decreases at each iteration, the convergence to a local minimum is guaranteed.

Remark 6.26. In this scheme, we allow the feedback gains to vary according to the active subsystem index, and the switching times. That is, $i_{k_1} = i_{k_2}$ does not necessarily mean that $L_{k_1} = L_{k_2}$. However, the scheme can be modified so as to be applicable to the piecewise linear feedback framework where each subsystem is assigned a unique feedback gain.

Remark 6.27. The optimal choice of switching times and feedback gains generally depend on the initial state x_0. In some practical situations, the initial state is not known *a priori*. In this case, we can minimize a cost function independent of the initial state. A nice choice for this purpose is the trace of $V_L(x_0)$, which is n times the average of the cost in (6.61) as x_0 varies over the unit sphere. Such an initial-independent optimal problem can be addressed using the approach presented in this subsection.

6.5 Notes and References

Optimization is a mature topic in mathematics and has been extensively addressed in various branches and fields of application. Optimization for switched and hybrid systems has also attracted much attention in recent years. In particular, general versions of the maximum principle have been developed, and they are applicable to switched systems [146]. Various aspects of the optimization problems have been explored in the literature [18, 116, 37, 114, 55, 94, 10, 16, 28, 11] too.

The optimal convergence rate introduced in Section 6.2 was adopted from [143]; see also [121] for related work on simultaneous triangularization. The discussions on the relationship between the asymptotic stabilizability and the infinite horizon optimal switching problem, presented in Section 6.3 for continuous-time systems and for discrete-time systems, were adopted from the recent work [136].

In Section 6.4, a couple of mixed (sub)optimal switching/control problems were introduced. The two-stage optimization method in Subsection 6.4.1 was mainly adopted from [167]. This method actually provides a basic framework for approaching the optimization problems of switched and hybrid systems; see also [165, 166, 9] for similar treatments. The suboptimal scheme presented in Subsection 6.4.2 was extended from the early works [80, 81, 12] for suboptimal control of linear time-varying systems.

7

Conclusions and Perspectives

7.1 Summary of the Book

In this book, we focused on the control and design issues of switched linear systems where the subsystems are linear time-invariant, while the switching signal and the control input are design variables. The topics include the design of switching signals to robustly stabilize autonomous switched systems; the joint design of the switching signal and the control input to achieve controllability, observability, and stability; and the search for the optimal switching signal and the optimal control input with respect to a performance cost. While most of the topics are conventional in the system and control literature, the involvement of the switching mechanism brings new insights as well as new challenges to the topics.

In Chapter 1, we briefly described the system formulations and discussed the related background and motivations. In Section 1.3, we presented several general concepts and fundamental observations, which provide a sound basis for the book.

Chapter 2 gathered the mathematical preliminaries and tools needed for the development of the book. Most of the material is elementary and can be found in the classical textbooks. A few exceptions include the concept and construction of multiple controllable subspaces in Section 2.4, and the Generalized Chow's Theorem in Section 2.10.

In Chapter 3, the switching signal design methodology was investigated for the stability and robustness of the unforced switched linear system. Unlike in conventional control theory, where the control input is the design variable, here the only design variable is the switching signal whose role is far from being well understood in the literature. The main results presented in this chapter include:

- the equivalence among switched convergence, asymptotic stabilizability, and exponential stabilizability (Theorems 3.9 and 3.53);

- the equivalence between the consistent stabilizability and the periodic stabilizability (Theorems 3.12 and 3.54);
- the combined switching strategy (3.45) and the establishment of its the robustness properties (Theorem 3.36);
- the observer design for robust stabilization (Theorem 3.38).

These results, together with others, provide insights into the capability and limitations of the switching scheme, and pave the way for further investigation.

Controllability, observability, and system decomposition issues have been addressed in Chapter 4. We proved that, for continuous-time systems, the controllable set and reachable set always coincide with each other, and the set is exactly the multiple controllable subspace which can be directly calculated from the system matrices (Theorems 4.17 and 4.18). This complete geometric criterion greatly facilitates the development of the constructive approach for the stabilization of switched linear systems. The observability and reconstructibility were addressed by the principle of duality (Theorems 4.26 and 4.27). For discrete-time systems, the above criteria still hold for a fairly large class of systems (Theorems 4.36 and 4.37), while examples were presented to show that the controllable and reachable sets are not subspaces in general (Examples 4.34 and 4.35).

Based on the controllability/observability criteria, a switched linear system can be transformed into the canonical form with a clear system structure (Theorem 4.46). In addition, we proved that, a controllable multi-input switched system can be reduced to a controllable single-input system by means of nonregular state feedback (Theorem 4.51), hence each controllable system admits a single-input controllable normal form (Theorem 4.49). These pave the way for addressing the feedback stabilization problem in Chapter 5. We also showed that the controllability is preserved by means of equidistance sampling under almost any sampling rate (Theorem 4.56). This provides a bridge between a continuous-time switched system and its discrete-time counterpart, thus enabling us to address the problems of digital control and regular switching in a unified framework. Finally, several further issues related to controllability were discussed and the results provide complementary insights from various points of view.

In Chapter 5, we examined the problem of stabilization by means of switching/input design. We adopted the approach that combines the stabilizing switching design in Chapter 3 and the piecewise linear state/output feedback control design based on the normal forms presented in Chapter 4. First, for a single process controlled/measured by multiple controllers/sensors, we designed separately the piecewise state feedback controller and the piecewise state estimator (observer) based on the system canonical decomposition. By incorporating the estimator into the feedback loop and establishing the separation property of the overall system, the problem of dynamic output feedback stabilization was solved in a thorough and elegant manner (Theorem 5.15). Second, if the summation of the controllability subspaces of the individual subsystems is the total state space, then, a stabilizing state-feedback control scheme with a dwell time was developed (Theorem 5.19). Third, for switched systems in the controllable normal forms, a sufficient condition was obtained for the piecewise linear quadratic stabilization problem (Theorem 5.24), and was applied to the second and third order systems by the detailed classification. In addition, the general stabilization problem was briefly discussed and a sufficient condition was obtained (Theorem 5.31). In the discrete-time case, deadbeat feedback controllers were designed for third-order systems (Theorem 5.41) as well as other special classes of systems (Theorems 5.39 and 5.40).

Finally, several optimization problems have been addressed in Chapter 6. In general, the optimization over the switching signals is highly non-convex and non-smooth in nature, hence the problems are usually very difficult to address. We classified the problems into optimal switching problems where the switching signal is the only optimization variable, and mixed problems where both the optimal switching signal and optimal control input are to be sought. For the optimal switching problems, we formulated the optimal convergence rate for the systems which are simultaneously triangularisable (Theorem 6.7); set up the connections between the finiteness of the infinite horizon optimal cost and the asymptotic stabilizability (Theorems 6.10-6.13, and 6.19-6.21); and established the equalization of the optimal costs for the switched linear system and the relaxed differential inclusion (Theorem 6.14). For the mixed problems, a two-stage optimization methodology was proposed for determining the optimal and suboptimal switching signal and control input.

7.2 Concluding Remarks

To conclude, we have the following general comments.

1. The switched linear systems are essentially nonlinear systems.

    Due to the involvement of the switching signal, the switched linear system is essentially a nonlinear system. Different switching signals may lead to different complex system behavior such as multiple limit cycles [118], and chaos [20], which are very different from linear behavior. In the book, we also presented several essentially nonlinear characteristics of the switched linear system, including the examples showing that stabilizability does not imply linear feedback stabilizability (Example 5.23), and that the controllable set is the union of an infinite number of maximal components (Example 4.35). The nonlinear nature of the switched linear system brings challenges to theoretical analysis as well as great value to practical applications. Indeed, it is the nonlinear features that make switched linear system widely represented in industrial and engineering practice. From the viewpoint of hybrid systems, the switched system includes both the continuous dynamics and discrete event (switching). This is a critical factor of switched systems which has been attracting increasing attention in the literature.

2. The switched linear system possesses many linear characteristics.

As made clear in Section 1.3.5, switched linear systems have several nice features that are valuable in analysis and design. These include:

- the time invariance property (Proposition 1.6);
- the radially linear property (Proposition 1.7);
- the analytic solution is readily obtained in terms of the system parameters.

By means of these nice properties, many linear and multi-linear tools are applicable or extensible to the analysis and design of the switched linear system. These tools include Wonham's geometric approach (especially the multiple invariant subspace scheme), and linear algebra (*e.g.*, the technical lammas in Section 4.2.4).

3. The bottleneck for understanding the switched linear system is the switching mechanism.

While both the switching signal and the control input play important roles in determining the behavior of the switched linear system, the bottleneck for understanding the switched system is the switching mechanism. This is quite natural because the role of the control input is relatively well understood in the literature. In this sense, the study of the unforced switched system is most needed to reveal the intrinsic features of switched systems.

4. The field of switched linear systems is largely unexploited.

In the field of switched linear systems, there have already been many good results and more are emerging. However, most fundamental issues are still in need of further exploration. Generally speaking, the switched linear system theory is still in an early stage of development. Perhaps the only exception is the continuous-time controllability/observability theory where both verifiable criteria and constructive path planning algorithms are available. Nevertheless, as more and more powerful tools are being introduced or developed to cope with the systems, it seems reasonable to expect a comprehensive switched linear system theory which extends the standard linear theory on one hand and applies to more real world problems on the other hand.

7.3 Perspectives and Open Problems

As the switched linear control theory is far from being well established, there are in fact numerous important problems which are still not well understood and open to further investigation. In the previous chapters, we mentioned explicitly a few open problems that naturally arise from the topics under study, while more are implicitly behind the words. Here, we highlight some open problems with detailed discussion whenever possible. The problems are chosen mainly based on the following two criteria. First, to the authors' knowledge, these problems are still open and have not been solved in the literature. Second, in the authors' opinion, these problems have, or potentially have, important impacts on the development of the core theory of switched linear systems. These criteria, of course, are highly subjective and mainly reflect the authors' personal experience and points of view. Nevertheless, for ease of reference, we try to describe the problems in a self-contained way, and discuss briefly the possible tools and ideas towards the further understanding of the problems. The discussions, however, usually are made in an elementary and intuitive manner.

Problem 7.1. Finite representation of stabilizing switching paths

For a pointwise asymptotically stabilizable switched system

$$\dot{x}(t) = A_\sigma x(t)$$

the problem is to find a set of well-defined switching paths

$$\mathcal{S}_J = \{\sigma_j\}_{j \in J} \subseteq \mathcal{S}_{[t_0,\infty)}$$

with a minimum number of elements (*i.e.*, cardinal of J$\to$ min), such that $\forall\, x_0 \in \mathbf{R}^n$, there exists a $k \in J$, such that the solution of equation

$$\dot{x}(t) = A_{\sigma_k(t)} x(t) \quad x(t_0) = x_0$$

is asymptotically convergent (to the origin).

Roughly speaking, the question is: How many switching paths do we need to stabilize a stabilizable switched system? In other words, is it possible to stabilize the system by means of a finite number time-driven switching paths? This problem is crucial in many practical situations. Indeed, if we can find a finite number of such switching paths, then, we can stabilize the system in a 'multi-controller' manner: we choose, according to the initial condition, a switching path from a number of candidates that makes the switched system stable. In this case, the switched system is 'reduced' to a finite number of time-varying systems and hence the linear time-varying system theory is applicable.

To further understand the problem, let us recall some known facts (*c.f.* Section 3.2.2) related to this problem. Owing to the Finite Covering Theorem, we can partition the unit sphere into a finite number of open (in the relative topology) and path-connected regions

$$\mathbf{S}_1 = \cup_{i=1}^l \Omega_i$$

such that each region Ω_i can be assigned to a switching path $\theta_i \colon [0,T] \mapsto M$ satisfying

$$\phi(T; 0, x_0, \theta_i) \in \mathbf{B}_{\frac{1}{2}} \quad \forall\, x_0 \in \Omega_i$$

where T is a finite real number. Based on this observation, we propose a stabilizing strategy as follows. For any initial state $x(0) = x_0 \neq 0$, choose an index i_0 such that $\frac{x_0}{\|x_0\|} \in \Omega_{i_0}$. Let the i_0 subsystem be activated during the period $[0, T)$, and we have $x(T) = \phi(T; 0, x_0, \theta_{i_0})$. Then, choose an index i_1 such that $\frac{x(T)}{\|x(T)\|} \in \Omega_{i_1}$, and let the i_1 subsystem be activated during the period $[T, 2T)$. Continuing with this process, we have the sequence $i_0, i_1, \cdots$. It can be seen that, the switching path

$$\sigma_{x_0}(t) = \begin{cases} \theta_{i_0}(t) & t \in [0, T) \\ \theta_{i_1}(t - T) & t \in [T, 2T) \\ \vdots \end{cases}$$

makes the state exponentially convergent. Note that the switching path σ_{x_0} is in fact a concatenation of the paths in the finite set $\{\theta_i\}_{i=1}^l$. This means that we can associate each x_0 with an infinite sequence $\Lambda(x_0) = \{i_0, i_1, \cdots\}$ with $i_j \in L \overset{def}{=} \{1, \cdots, l\}$. For notational convenience, we say that the sequence $\Lambda(x_0)$ stabilizes x_0. Let

$$\Upsilon = L^\infty = \{(i_0, i_1, \cdots): i_j \in L \quad j = 0, 1, \cdots\}.$$

It is clear that each state can be stabilized by a switching path in Υ. This means that Υ provides a universal set of switching paths for the stabilizability of the switched system.

Unfortunately, the set Υ still contains an (uncountable) infinite number of elements. However, if each state permits an eventually cyclic stabilizing sequence, then we have a universal set with a countably infinite number of elements. Furthermore, if each state permits a cyclic stabilizing sequence with a period bounded by a known number, then we have a universal set with a finite number of elements.

Though we do not know the general answer to the question, the above analysis is useful in leading us to the following practically interesting conclusion.

Proposition 7.2. *Suppose that Ω_1 is a bounded set in $\mathbf{R}^n$, and Ω_2 a neighborhood of the origin. Then, there exists a set of switching paths, $\mathcal{S}_J$, with a finite cardinality, such that each state in Ω_1 can be steered into Ω_2 in a finite time by at least one switching path in $\mathcal{S}_J$.*

We do not know how far the above analysis can go towards the solution of the problem. This remains a subject for further study.

Problem 7.3. The existence of a time-invariant smooth converse Lyapunov function

For the unforced switched system, a real-valued function $V: \mathbf{R}^n \mapsto \mathbf{R}$ is said to be a Lyapunov function of the system, if the function is positive definite, and if there exists a switching signal σ_0, such that the function decreases along each non-trivial state trajectory of the switched system under σ_0. The question is, suppose that the switched system is pointwise asymptotically stabilizable, does a Lyapunov function for the switched system exist? If yes, does a smooth one exist?

It is well recognized that converse Lyapunov theorems play an important role in stability analysis and robustness design of control systems. For switched systems, such a converse Lyapunov theorem will be useful in many situations.

Now, we present a partial solution to the problem. Suppose that there is a pure-state feedback switching signal

$$\sigma(t) = \varphi(x(t)) \quad \forall \, t \geq t_0$$

which makes the switched system exponentially stabilizable. Define the function $W : \mathbf{R} \times \mathbf{R}^n \mapsto \mathbf{R}_+$ as

$$W(t_0, x_0) = \int_{t_0}^{\infty} \|\phi(t; t_0, x_0, \sigma)\|^2 dt \quad \forall \, t_0 \in \mathbf{R} \;\; x_0 \in \mathbf{R}^n.$$

The exponential stability guarantees the well-definedness of the function. By the time-invariance of the switching signal, it can be seen that $W(t_1, x) = W(t_2, x)$ for all t_1, t_2 and x. That is, the function is time-invariant.

Next, define function $V : \mathbf{R}^n \mapsto \mathbf{R}_+$ by $V(x) = W(0, x)$, $\forall \, x \in \mathbf{R}^n$. It is clear that $V(\cdot)$ is positive definite and strictly decreasing along each state trajectory $\phi(\cdot; t_0, x_0, \sigma)$.

The above derivation lead to the following proposition.

Proposition 7.4. *Suppose that the switched system is exponentially stabilizable by a switching signal in the pure-state-feedback form. Then, the system admits a (not necessarily smooth) Lyapunov function.*

In this proposition, the pure-state-feedback requirement of the switching signal is crucial. In fact, if the stabilizing switching signal is time-driven or mixed time/event-driven, then the resultant system is time-varying in nature and hence function $W(t, x)$ may not be time-invariant.

Example 7.5. Consider the switched system with $n = 2$, $m = 2$, and

$$A_1 = \begin{bmatrix} 2 & 0 \\ 0 & -1 \end{bmatrix} \text{ and } A_2 = \begin{bmatrix} 0 & 1 \\ -1 & 0 \end{bmatrix}.$$

This switched system is exponentially stabilizable and a stabilizing switching law can be formulated as follows. For any initial state $x(t_0) = x_0$, first check whether x_0 is in the x_2-axis, *i.e.*, $x_0(1) = 0$. If not, then activate the second subsystem until the state reaches the x_2-axis. This is always possible because the second subsystem can bring a state rotating to any direction without changing its 2-norm. Once the state reaches the x_2-axis, let the first subsystem be activated for good. Note that the the x_2-axis is a stable invariant subspace of the first subsystem. Accordingly, the above switching strategy always makes the switched system exponentially stable.

It can be seen that the switching signal is in the pure-state-feedback form. Applying Proposition 7.4, the Lyapunov function can be computed to be

$$V(x) = \left(s(x) + \frac{1}{2}\right)(x_1^2 + x_2^2)$$

where $s(x)$ is the time needed to bring the state from x to the x_2-axis. Let $\theta(x)$ denote the angle (in $[0, 2\pi)$) between vector x and the x_1-axis, $i.e.$,

$$x = [\|x\| \cos \theta(x), \|x\| \sin \theta(x)]^T.$$

By the rotative characteristic of the second subsystem, we have

$$s(x) = \begin{cases} \theta(x) + \frac{\pi}{2} & \theta(x) \in [0, \pi/2) \\ \theta(x) - \frac{\pi}{2} & \theta(x) \in [\pi/2, 3\pi/2) \\ \theta(x) + \frac{3}{2}\pi & \theta(x) \in [3\pi/2, 2\pi). \end{cases}$$

Note that $s(x)$ does not depend on x continuously when x is on the x_2-axis. For example, let $y_1 = [0, 1]^T$ and $y_2 = [1, 0]^T$, then, we have

$$\lim_{r \downarrow 0} V(y_1 + ry_2) = V(y_1) = \frac{1}{2} \text{ and } \lim_{r \uparrow 0} V(y_1 + ry_2) = \pi + \frac{1}{2}.$$

This means that function V is not continuous at point y_1.

Of course, the example itself does not exclude the possibility that there exists a stabilizing switching law that results in a smooth Lyapunov function.

Problem 7.6. Maximal and minimal dwell times for stability and stabilizability

Given an unforced switched linear system $\Sigma(A_i)_M$ and a switching signal σ with dwell time τ, if the switching signal asymptotically stabilizes the system, then, we say that the system admits a stabilizing dwell time τ. The maximal dwell time for stabilizability is the largest possible dwell time τ_{sup} the switched system can possess, that is,

$$\tau_{sup} = \sup\{\tau : \mathcal{S}_\tau \cap \mathcal{S}^s \neq \emptyset\}$$

where $\mathcal{S}_\tau$ is the set of switching signals with dwell time τ, and $\mathcal{S}^s$ is the set of stabilizing switching signals. Similarly, the minimal dwell time for stability is the smallest possible dwell time τ_{inf} among the stabilizing signals, that is,

$$\tau_{inf} = \inf\{\tau : \mathcal{S}_\tau \subseteq \mathcal{S}^s\}.$$

The problem is to determine the minimum and maximum dwell times for stability and stabilizability, respectively.

It is clear that the minimal and maximal dwell times are defined only when the switched system is stabilizable. For a switched unstable system where each of the subsystems is unstable, it follows from the proofs of Theorems 3.9 and 3.11 that $\tau_{sup} > 0$. In addition, in Section 3.5, we proposed a combined switching strategy with relatively large dwell time. However, the determination of the exact value of τ_{sup} seems to be very difficult and no general result is available so far.

On the other hand, for switched stable systems where all the subsystems are stable, it is obvious that $\tau_{sup} = \infty$, that is, any switching path with sufficiently large dwell time makes the switched system stable. If the subsystems share a common Lyapunov function, then, $\tau_{inf} = 0$, which means that each switching path stabilizes the switched system.

In general, to find an upper bound for τ_{inf}, the most direct way is to find, for each matrix A_i, real constants α_i and β_i such that

$$\|e^{A_i t}\| \leq \beta_i e^{\alpha_i t} \quad \forall \, t \geq 0.$$

Note that for each $\alpha_i \in (\bar{\alpha}_i, 0)$, such a β_i always exists, where $\bar{\alpha}_i$ is the maximum real part of the eigenvalues of A_i. It is clear that

$$\tau_{inf} \leq \max\{-\frac{\ln \beta_i}{\alpha_i} : i \in M\}.$$

Though this upper bound is easily calculated, it is probably over conservative. A less conservative method is to verify the norm pairwise as follows.

Consider the case that the system contains only two subsystems, A_1 and A_2. For each matrix A_i and a nonnegative real number h, define

$$\beta_h(A_i) = \max_{t \geq h}\{\|e^{A_i t}\| e^{-\bar{\alpha}_i(t-h)}\}.$$

Take a nonsingular complex matrix Q such that $J_1 \overset{def}{=} Q^{-1} A_1 Q$ is in the Jordan form. Let $J_2 = Q^{-1} A_2 Q$. Suppose that $\{(1, h_0), (2, h_1), \cdots\}$ is a switching duration sequence. It is clear that

$$e^{A_1 h_0} e^{A_2 h_1} e^{A_1 h_2} \cdots = Q^{-1} e^{J_1 h_0} e^{J_2 h_1} e^{J_1 h_2} \cdots Q.$$

Let

$$\tau_1 = \min\{h \geq 0 : \beta_h(J_2) \geq e^{-\bar{\alpha}_1 h}\}.$$

Then, this τ_1 is an upper bound for τ_{inf}. Accordingly, we have

$$\tau_{inf} \leq \min_Q\{h \geq 0 : \beta_h(Q^{-1} A_2 Q) \geq e^{-\bar{\alpha}_1 h}\}$$

$$\tau_{inf} \leq \min_R\{h \geq 0 : \beta_h(R^{-1} A_1 R) \geq e^{-\bar{\alpha}_2 h}\}$$

where the minimum is taken over Q and R, which turn $Q^{-1} A_1 Q$ and $R^{-1} A_2 R$ into the Jordan form, respectively.

The exact calculation of the minimal dwell time is still an open problem.

Problem 7.7. Well-definedness of pure-observer-driven stabilizing switching signal

For the unforced switched linear system

$$\dot{x}(t) = A_\sigma x(t)$$
$$y(t) = C_\sigma x(t)$$

find, if possible, a state observer

$$\dot{\bar{x}} = \bar{A}_\sigma \bar{x} + \bar{B}_\sigma y$$

and an observer-driven switching law

$$\sigma(t+) = \varphi(\sigma(t), y(t), \bar{x}(t))$$

such that the overall system

$$\dot{x}(t) = A_\sigma x$$
$$\dot{\bar{x}}(t) = \bar{A}_\sigma \bar{x}(t) + \bar{B}_\sigma y(t)$$
$$y(t) = C_\sigma x(t)$$
$$\sigma(t+) = \varphi(\sigma(t), y(t), \bar{x}(t))$$

is well-posed and asymptotically stable.

We have designed several observer-based switching laws in Sections 3.4.3, 3.5.3, and 5.2, respectively. However, the switching laws are either for practical stabilization, or time-driven, or mixed time/event driven. In other words, none of the switching laws are purely observer-driven. In many practical situations, pure observer-driven switching laws are desirable when it is not appropriate to explicitly incorporate the time factor.

The problem of pure observer-driven stabilization has been addressed in [44] where a state-space-partition-based pure observer switching law was proposed. While the stability of the overall system has been proven by the Lyapunov approach, the well-posedness of the system was not established. In fact, the well-posedness is exactly the core issue of the problem. As pointed out in Section 3.4.1, even an exponentially convergent perturbation may make a well-posed nominal system ill-posed. Accordingly, when an observer is incorporated into the system, the well-posedness of the overall system has to be addressed. This brings about a challenging issue which deserves further investigation.

Problem 7.8. Structural stability

By structural stability, we mean a 'perturbation' of a nominal stabilizing switching signal and its influence in the system stability. The problem of structural stability is to find a stabilizing switching signal such that the system is still stable if the switching signal undergoes small perturbations.

Intuitively, for a robust stabilizing switching signal, a small perturbed switching signal should also make the switched system stable. A question thus arises naturally: How should we characterize the distance between two switching signals?

As the state transition matrix is of the form

$$e^{A_{i_k}(t-t_k)} \cdots e^{A_{i_1}(t_2-t_1)} e^{A_{i_0}(t_1-t_0)} \quad k = 0, 1, \cdots$$

which we refer to as the transition (matrix) chain. If the above matrix chain is convergent (to the zero matrix), then, the corresponding switching path makes the state convergent for all initial states, and vice-versa. On the other hand, a variation of a switching path means variations of the switching time sequence $t_0, t_1, \cdots$ and switching index sequence $i_0, i_1, \cdots$, which can be seen as variations of the transition chain.

Suppose that there is a time delay ϵ in each switching instance caused by the switching device. It is obvious that the delayed switching path is with switching sequence

$$\{(t_0 + \epsilon, i_0), (t_1 + \epsilon, i_1), (t_2 + \epsilon, i_2), \cdots\}.$$

It can be seen that the two paths have nearly the same transition sequence but they are not equal at time intervals of an infinite length. This excludes the reasonableness of formulating the distance between two switching paths p_1 and p_2 by

$$\text{meas}\{t \colon p_1(t) \neq p_2(t)\}$$

where 'meas' denotes the Lebesgue measure, though this seems to be the most straightforward way.

A possible way to characterize the distance is by the concatenation of switching paths. Suppose that we have two sequences of pairs $S_1 = \{(i_1, h_1), (i_2, h_2), \cdots\}$ and $S_2 = \{(j_1, \tau_1), (j_2, \tau_2), \cdots\}$, a concatenation (in the generalized sense) of them, denoted by $S_3 = S_1 \wedge S_2$, is a new sequence $S_3 = \{(k_1, \mu_1), (k_2, \mu_2), \cdots\}$ with the property that, we can split the set $\mathbf{N}^+$ into two (monotone) sequences $\upsilon_1, \upsilon_2, \cdots$ and $\nu_1, \nu_2, \cdots$, such that

$$(i_s, h_s) = (k_{\upsilon_s}, \mu_{\upsilon_s}) \text{ and } (j_s, \tau_s) = (k_{\nu_s}, \mu_{\nu_s}) \quad s = 1, 2, \cdots.$$

We say that S_3 is a common generalized sequence of S_1 and S_2, if there exist sequences S_4 and S_5 such that

$$S_3 = S_1 \wedge S_4 = S_2 \wedge S_5.$$

The distance between S_1 and S_2 is defined by

$$\inf\{|S_4| + |S_5| \colon S_1 \wedge S_4 = S_2 \wedge S_5\}$$

where $|S| = \sum_k h_k$ for $S = \{(i_1, h_1), (i_2, h_2), \cdots\}$. The distance between two switching paths can be defined to be the distance between their switching duration sequences. In this way, the distance between any two switching path is defined. It can be verified that the distance satisfies the positiveness, the symmetry, and the triangular inequality. However, we still do not know if such a definition can lead to reasonable structural stability analysis.

Problem 7.9. Minimum switching for controllability/reachability

Let N_n^m be the minimum number of switches for controllability of switched systems, where n is the order of the systems and m is the number of subsystems. This means that, for each nth order switched linear system $\Sigma(A_i, B_i)_M$ with m subsystems, an arbitrary given state in the controllable subspace $\mathcal{C}(A_i, B_i)_M$ can be transferred to the origin within N_n^m times of switching. The problem is to formulate N_n^m in terms of n and m.

Note that N_n^m is defined over a class of systems rather than over an individual system. It can be readily seen that the following simple properties hold for N_n^m:

(i) $N_n^1 = 0$;
(ii) $N_n^m = N_n^n$, for $m \geq n$;
(iii) $N_{n_1}^{m_1} \geq N_{n_2}^{m_2}$ for $n_1 \geq n_2$ and $m_1 \geq m_2$; and
(iv) $N_n^m \leq \sum_{k=0}^{n-1} m(mn)^k - 1$.

The last upper bound is taken from the proof of Theorem 4.17.

By (ii), we assume without loss of generality that $m \leq n$.

A lower bound can be given by

$$N_n^m \geq \frac{m(2n - m + 1)}{2} - 1.$$

This can be verified by the following example.

Example 7.10. Consider the switched system given by

$$A_1 = 0 \quad B_1 = e_1 \quad A_i = e_i e_{i-1}^T \quad B_i = 0 \quad i = 2, \cdots, n$$

where e_i is the ith unit column.

Extensive calculation shows that a controllable switching path with minimum number of switches is with the switching index sequence

$$1, 2, 1, 3, 2, 1, \cdots, n, n - 1, \cdots, 1.$$

Let $k_i = \frac{i(i+1)}{2}$, $i = 1, 2, \cdots$. Routine calculation gives

$$\mathcal{C}(0) = \mathrm{span}\{e_1\}$$
$$\mathcal{C}(1) = \mathrm{span}\{e_1 + h_1 e_2 : h_1 > 0\}$$
$$\mathcal{C}(2) = \mathrm{span}\{e_1, e_2\}$$
$$\vdots$$
$$\mathcal{C}(k_i) = \mathrm{span}\{e_1, \cdots, e_{i-1}, e_i + h_{k_i} e_{i+1} : h_{k_i} > 0\}$$
$$\vdots$$
$$\mathcal{C}(k_i + i) = \mathrm{span}\{e_1, \cdots, e_{i+1}\}$$

where $\mathcal{C}(k)$ is the controllable set along the switching path within k times of switches.

It is interesting to notice that, the first new direction (e_2) appears in 2 switches, the second in another 3 switches, and the $n-1$th in another n switches. Accordingly, the minimal number of switches needed for complete controllability is

$$2 + 3 + \cdots + n = \frac{n(n+1)}{2} - 1.$$

For the case where $m \leq n$, by properly adding $A_{m+1}, \cdots, A_n$ into $A_1, \cdots, A_m$, it can be verified that the minimal number of switches is

$$n + (n-1) + \cdots + (n-m+1) - 1 = \frac{m(2n-m+1)}{2} - 1.$$

We conjecture that this is exactly the minimal number needed for controllability, that is

$$N_n^m = \frac{m(2n-m+1)}{2} - 1 \quad n \geq m \geq 2.$$

However, we are still not in a position to prove (or disprove) the conjecture.

For discrete-time systems, the minimum switching problem was addressed in [26]. Like in the continuous time case, the general problem remains a subject for further investigation.

Problem 7.11. Path planning for local reachability

The problem of path planning for local reachability is to find a switching path to steer the unforced switched system from a given initial state x_0 to a given target state in $R(x_0)$ in a finite time, where $R(x_0)$ denotes the reachable set from x_0.

According to Theorem 4.79, for the unforced switched linear system $\Sigma(A_i)_M$ and a pathwise connected open set Ω in $\mathbf{R}^n$, if

$$\dim\{A_1 x, \cdots, A_m x\}_{LA}(x) = n \quad \forall\, x \in \Omega$$

then, the system is locally weakly controllable in Ω. Furthermore, if the system is symmetric, then the system is locally reachable in Ω. However, Theorem 4.79 does not provide any constructive information on the path planning. Besides, the general properties of the reachable set are largely unknown. These are interesting problems deserving further study.

Problem 7.12. Decidability of discrete-time reachability/controllability

As proven in Theorem 4.31, a discrete-time switched linear system $\Sigma(A_i, B_i)_M$ is reachable if and only if there exist an integer $k < \infty$, and an index sequence $i_0, \cdots, i_k$, such that

$$A_{i_k} \cdots A_{i_1} \operatorname{Im} B_{i_0} + \cdots + A_{i_k} \operatorname{Im} B_{i_{k-1}} + \operatorname{Im} B_{i_k} = \mathbf{R}^n. \tag{7.1}$$

If we know the upper bound of such a k, then, the reachability property is verifiable in a finite time by exhaustively searching among the candidate switching sequences. Accordingly, the reachability is decidable, if for a given n, there is a number k_n, such that for any reachable switched system $\Sigma(A_i, B_i)_M$ of dimension n, (7.1) holds for some $k \leq k_n$ and $i_0, \cdots, i_k \in M$. The question is: Does such a k_n exist for each n?

It can be seen that the reachability is decidable if either $n = 2$, or $\operatorname{Im} A_i + \operatorname{Im} B_i = \mathbf{R}^n$ for all $i \in M$. In the latter case, there exist gain matrices F_i such that $A_i + B_i F_i$ is nonsingular for all $i \in M$, and hence the switched system can be transformed into a reversible system by means of a state feedback controller. Note that the property of controllability is invariant under feedback transformations, and the controllability of a reversible switched linear system is decidable. This means that the controllability of the original non-reversible system is also decidable.

It was conjectured that the controllability/reachability is not decidable [56]. However, no proof has been reported in the literature to the authors' knowledge.

Problem 7.13. Nonlinear feedback design for the feedback stabilization problem

As shown in Example 5.23, piecewise linear state feedback control laws are not sufficient for the stabilization of switched linear control systems, even if the systems are controllable. This intrinsic phenomenon necessitates the design of (piecewise) nonlinear state feedback laws for the stabilization problem. However, by using nonlinear state feedback controllers, a switched linear system turns into a switched nonlinear system, for which the stability analysis is usually very challenging which remains a subject for further study.

Problem 7.14. Flexibility of the stabilizing switching/control laws

The design of a stabilizing strategy for a general switched linear control system can be divided into the design of the controllable part and the design of the uncontrollable part. However, the two design schemes are not independent as we have to use the same stabilizing switching signal for both parts. This poses a challenging problem. To cope with the problem, we can first design a stabilizing switching signal for the uncontrollable part. With this switching signal being the potential switching candidate for the whole system, we then design a feedback control input law to steer the controllable part stable.

Note that in the proposed scheme, there is some flexibility that we can make use of. First, suppose that we have found a stabilizing switching signal for the uncontrollable part, then, a switching signal nearby is probably also a stabilizing switching signal. That is, we have flexibility in choosing the switching signal for stabilizing the uncontrollable part. Second, for the controllable part, we have much flexibility in choosing the feedback controllers, such as high-gain feedback laws, when necessary. The exploitation of flexibility in the design of stabilizing switching/control laws is an important subject for further investigation.

Problem 7.15. The existence of an optimal solution

For the optimal switching/control of switched linear systems, an optimal solution is the switching signal/control input that minimizes the cost function. The existence of an optimal solution is one of the most important issues in optimization problems, yet it also seems to be one of the most challenging issues.

In Chapter 6, the only existence result presented is for the optimal convergence rate of simultaneously triangularizable switched linear systems. In this special case, there exists a set of optimal switching signals which are periodic. In the literature, several existence results have been reported for optimal control problems of special classes of switched systems [119, 18, 10].

For the simple quadratic cost function given by

$$J(x_0, u, \sigma) = \int_{t_0}^{t_f} x^T(t) P x(t) + u^T(t) Q u(t) dt$$

where $P \geq 0$ and $Q > 0$, and $t_f \leq \infty$, if the switching signal is given, then, by conventional optimal control theory, the optimal control problem admits a unique solution. Hence, the question is reduced to whether or not there exists an optimal switching signal, a core issue worthy of further study.

Problem 7.16. The essential features of the optimal switching

For an infinite horizon optimal switching problem of the unforced switched linear system, suppose that there exists an optimal switching signal. The number of switches may be zero (no switch at all), or finite (no switch occurs eventually) or infinite. In particular, if each subsystem is unstable, then, for most initial states, the corresponding optimal switching paths have an infinite number of switches. The question is: Do the optimal switching paths share any common essential features? More concretely, are the optimal switching index sequences eventually cyclic and/or the optimal switching time sequence eventually synchronous or multi-rate in nature?

Problem 7.17. Various suboptimal switching/control problems

As it is usually difficult to find the optimal solution, we have to be consent with finding a suboptimal solution. Here, sub-optimality may mean that the cost is within a prescribed neighborhood of the optimal cost, or mean that the solution is optimal only for a specific class of feasible design variables. For the switched linear control system, as both the switching signal and the control input are design variables, there are several types of suboptimal solutions which are of interest in various theoretical and/or practical situations. Several examples are listed below.

(i) Optimal switching time sequence

In many practical applications, the switching index sequences are governed by given logic-based devices. In this case, the optimal switching is to seek the switching times that optimize a cost function. This problem is intrinsically discontinuous and the solution may reveal the essential nature of switched systems.

(ii) Optimal switching index sequence

Suppose that the switching times are given in advance, then, the optimal switching problem is reduced to seeking a switching index sequence that optimizes the cost function. A typical example is when the switched system is controlled by a digital device, or when the system itself is of discrete time in nature. As discussed in Section 6.4.1, if the time horizon is finite, the number of feasible switching index sequence is also finite, hence, an exhaustive search is possible, at least theoretically. However, for a large time horizon, effective pruning algorithms must be sought to prevent computational exploration. For the infinite horizon, new schemes have to be sought to address the problem.

(iii) Optimal periodic switching signal

For an infinite horizon optimization problem, a possible scheme for suboptimality is to seek an appropriate periodic switching signal that optimizes the cost function. Referring to the discussions in Section 3.2.3, such a suboptimal cost is finite if the switched system is consistently asymptotically stabilizable. The problem is to obtain an optimal period, and an optimal switching sequence within a period.

(iv) Optimal closed-form switching/control

A switching signal is said to be in closed form if it is event-driven, and a control input is said to be in closed form if it is piecewise (time-invariant) state/output feedback. The problem of suboptimal closed-form switching/control is of particular interest from both the theoretical and practical points of view. An important question is: How large is the difference between the (open-form) optimal cost and the (closed-form) suboptimal cost? When do they coincide with each other?

A key point here is whether or not the set of closed-form switching signals and the set of closed-form control inputs are rich enough to approximate each of the open-form switching signals/control inputs. Indeed, if the set of closed-form switching signals is a dense subset of the set of switching signals with respect to an appropriate functional topology, then each switching signal can be approached by a sequence of closed-form switching signals. A similar point holds for the control input. Alternatively, let Ω and Ω_0 be the set of state trajectories corresponding to the open-form and closed-form switching/input, respectively. The core question is whether or not each optimal state trajectory in Ω can be sufficiently approximated by a state trajectory in Ω_0 with respect to the cost index.

Problem 7.18. Efficient computational algorithms for optimization

It is well recognized that the computation of the optimal solution is not easy for most optimization problems. Hence, the reduction of computational burdens becomes an important technology which motivates the rich and vast optimization methods and techniques in the literature. For the optimization of switched linear systems, despite the fact that the subsystem dynamics are very simple, the involvement of the switching signals makes exhaustively searching impractical due to combinational explosion. In Section 6.4.2, numerical algorithms were presented to search for the optimal or suboptimal strategies. Unfortunately, the numerical algorithms either lead to local solutions or run in exponential times. Efficient computation is still widely open for further investigation.

References

1. Agrachev AA, Liberzon D. Lie-algebraic stability criteria for switched systems. SIAM J Contr Optimiz 2001; 40(1): 253–269.
2. Antsaklis PJ (ed.). Special Issue on Hybrid Systems. In: Proc the IEEE 2000; 88(7): 879–1123.
3. Antsaklis PJ, Nerode A (eds.). Special Issue on Hybrid Control Systems. In: IEEE Trans Automat Contr 1998; 43(4): 457–579.
4. Astolfi A. Discontinuous control of nonholonomic systems. Syst Contr Lett 1996; 27(1): 37–45.
5. Babaali M, Egerstedt M. Pathwise observability and controllability are decidable. In: Proc 42nd IEEE CDC 2003; 5771–5776.
6. Barabanov NE. Lyapunov indicator of discrete inclusions I. Avtomatika i Telemekhanika 1988; 4: 40–46. (Translation from Russian)
7. Bacciotti A. Stabilization by means of state space depending switching rules. Syst Contr Lett 2004; 53(3-4): 195–201.
8. Barker GP, Conner LT Jr, Stanford DP. Complete controllability and contracibility in multimodel systems. Linear Algeb Appl 1988; 110: 55–74.
9. Bemporad A, Giua A, Seatzu C. Synthesis of state-feedback optimal controllers for continuous-time switched linear systems. In: Proc 41st IEEE CDC 2002; 3182–3187.
10. Bengea SC, DeCarlo RA. Optimal and suboptimal control of switching systems. In: Proc 42nd IEEE CDC 2003; 5295–5300.
11. Bengea SC, DeCarlo RA. Optimal control of switching systems. Automatica 2005; 41(1): 11–27.
12. Bertsekas D. Note on the design of linear systems with piecewise constant feedback gains. IEEE Trans Automat Contr 1970; 15(2): 262–263.
13. Bertsekas D. Nonlinear Programming (2nd ed.) 1999. Athena Scientific.
14. Blanchini F. Set invariance in control. Automatica 1999; 35(11): 1747–1767.
15. Blanchini F. The gain scheduling and the robust state feedback stabilization problems. IEEE Trans Automat Contr 2000; 45(11): 2061–2070.
16. Borrelli F. Constrained Optimal Control of Linear and Hybrid Systems. New York: Springer-Verlag, 2003.
17. Branicky MS. Multiple Lyapunov functions and other analysis tools for switched and hybrid systems. IEEE Trans Automat Contr 1998; 43(4): 475–482.

18. Branicky MS, Borkarand VS, Mitter SK. A unified framework for hybrid control: Model and optimal control theory. IEEE Trans Automat Contr 1998; 43(1): 31–45.

19. Brockett RW, Wood JR. Electrical networks containing controlled switches. In: Applications of Lie Groups Theory to Nonlinear Networks Problems, Supplement to IEEE Inter Symp Circuit 1974; 1–11.

20. Chase C, Serrano J, Ramadge PJ. Periodicity and chaos from switched flow systems: Contrasting examples of discretely controlled continuous systems. IEEE Trans Automat Contr 1993; 38(1): 70–83.

21. Chen CT. Linear System Theory and Design (3rd ed.). New York: Oxford Univ Press, 1999.

22. Cheng DZ, Chen HF. Accessibility of switched linear systems. In: Proc 42nd IEEE CDC 2003; 5759–5764.

23. Cheng DZ, Guo L, Huang J. On quadratic Lyapunov functions. IEEE Trans Automat Contr 2003; 48(5): 885–890.

24. Chizeck HJ, Willsky AS, Castanon D. Discrete-time markovian jump linear quadratic optimal control. Int J Contr 1986; 43(1): 213–231.

25. Cohen AL, Rodman L, Stanford DP. Pointwise and uniformly convergent sets of matrices. SIAM J Matrix Anal Appl 1999; 21(1): 93–105.

26. Conner LT Jr, Stanford DP. State deadbeat response and obsevability in multimodal systems. SIAM J Contr Optimiz 1984; 22(4): 630–644.

27. Conner LT Jr, Stanford DP. The structure of the controllable set for multimodal systems. Linear Algebra Applic 1987; 95: 171–180.

28. Corona D, Giua A, Seatzu C. Optimal control of hybrid automata: design of a semiactive suspension. Contr Engin Practice 2004; 12(10): 1305–1318.

29. Costa OLV, Tuesta EF. Finite horizon quadratic optimal control and a separation principle for Markovian jump linear systems. IEEE Trans Automat Contr 2003; 48(10): 1836–1842.

30. Dayawansa WP, Martin CF. A converse Lyapunov theorem for a class of dynamical systems which undergo switching. IEEE Trans Automat Contr 1999; 44 (4): 751–760.

31. DeCarlo RA, Branicky MS, Pettersson S, Lennartson B. Perspective and results on the stability and stabilizability of hybrid systems. Proc IEEE 2000; 88(7): 1069–1082.

32. De Dona JA, Goodwin GC, Moheimani SOR. Combining switching, oversaturation and scaling to optimise control performance in the resence of model uncertainty and input saturation. Automatica 2002; 38(7): 1153–1162.

33. De Koning WL. Digital optimal reduced-order control of pulse-width-modulated switched linear systems. Automatica 2003; 39(11): 1997–2003.

34. Di Bernardo M, Chungand HS, Tse CK (eds.). Special Issue on Switching and Systems. In: IEEE Trans Circ Syst I 2003; 50(8): 983–1120.

35. Dion J-M, Commault C, Van Der Woude J. Generic properties and control of linear structured systems: a survey. Automatica 2003; 39(7): 1125–1144.

36. Drager LD, Foote RL, Martin CF, Wolfer J. Controllability of linear systems, differential geometry curves in Grassmannians and generalized Grassmannians, and Riccati equations. Acta Applicandae Mathematicae 1989; 16(3): 281–317.

37. Egerstedt M, Ogren P, Shakernia O, Lygeros J. Toward optimal control of switched linear systems. In: Proc 39th IEEE CDC 2000; 587–592.

38. Ezzine J, Haddad AH. On the controllability and observability of hybrid systems. In: Proc 1988 Amer Contr Conf 1988; 41–46.

39. Ezzine J, Haddad AH. On the stabilization of two-form hybrid systems via averaging. In: Proc 22nd Annual Conf Inform Sci Syst 1988; 579–584.

40. Ezzine J, Haddad AH. Controllability and observability of hybrid systems. Int J Contr 1989; 49(6): 2045-2055.

41. Ezzine J, Haddad AH. Error bounds in the averaging of hybrid systems. IEEE Trans Automat Contr 1989; 34(11): 1188–1192.

42. Fang Y, Loparo KA. Stabilization of continuous-time jump linear systems. IEEE Trans Automat Contr 2002; 47(10): 1590–1603.

43. Feng G, Ma J. Quadratic stabilization of uncertain discrete-time fuzzy dynamic systems. IEEE Trans Circ Syst I 2001; 48(11): 1337–1344.

44. Feron E. Quadratic stabilizability of switched systems via state and output feedback. Center for Intelligent Control Systems, Massachusetts Institute of Technology, Tech Rep CICS-P-468, 1996.

45. Feuer A, Goodwin GC, Salgado M. Potential benefits of hybrid control for linear time invariant plants. In: American Contr Conf 1997; 2790–2794.

46. Fliess M. Reversible linear and nonlinear discrete-time dynamics. IEEE Trans Automat Contr 1992; 37(8): 1144–1153.

47. Fryszkowski A, Rzezuchowski T. Continuous version of Filippov-Wazewski relaxation theorem. J Diff Equat 1991; 94(2): 254–265.

48. Fu M, Barmish B. Adaptive stabilization of linear systems via switching control. IEEE Trans Automat Contr 1986; 31(12): 1097–1103.

49. Gaines FJ, Thompson RC. Sets of nearly triangular matrices. Duke Math J 1968; 35(3): 441–454.

50. Gantmacher FR. The Theory of Matrices. New York: Chelsea, 1959.

51. Ge SS, Hang CC, Lee TH, Zhang T. Stable Adaptive Neural Network Control. Boston: Kluwer, 2001.

52. Ge SS, Sun Z, Lee TH. Reachability and controllability of switched linear discrete-time systems. IEEE Trans Automat Contr 2001; 46(9): 1437–1441.

53. Ge SS, Sun Z, Lee TH. Nonregular feedback linearization for a class of second-order systems. Automatica 2001; 37(11): 1819–1824.

54. Ge SS, Wang ZP, Lee TH. Adaptive stabilization of uncertain nonholonomic systems by state and output feedback. Automatica 2003; 39(8): 1451–1460.

55. Giua A, Seatzu C, Van Der Mee C. Optimal control of autonomous linear systems switched with a pre-assigned finite sequence. In: Proc IEEE Int Symp Intell Contr 2001; 145–149.

56. Gurvits L. Stabilities and controllabilities of switched systems (with applications to the quantum systems). In: Proc 15th Int Symp Math Theory Network Syst 2002.

57. Hahn W. Stability of Motion. Berlin: Springer-Verlag, 1967.

58. Harary F. Graph Theory. Reading: Addison-Wesley, 1969.

59. Hautus M. A simple proof of Heymann's lemma. IEEE Trans Automat Contr 1977; 22(5): 885–886.

60. Hermann R. Diffirential Geometry and the Calculas of Variations. New York: Academic Press, 1968.

61. Hespanha JP. Uniform stability of switched linear systems: Extensions of LaSalle's invariance principle. IEEE Trans Automat Contr 2004; 49(4): 470–482.

62. Hespanha JP, Liberzon D, Morse AS. Logic-based switching control of a non-holonomic system with parametric modeling uncertainty. Syst Contr Lett 1999; 38(3): 167–177.

63. Hespanha JP, Liberzon D, Morse AS. Overcoming the limitations of adaptive control by means of logic-based switching. Syst Contr Lett 2003; 49(1): 49–65.
64. Hespanha JP, Morse AS. Stability of switched systems with average dwell-time. In: Proc 38th IEEE CDC 1999; 2655-2660.
65. Heymann M. Comments 'On pole assignment in multi–input controllable linear systems'. IEEE Trans Automat Contr 1968; 13(6): 748–749.
66. Horn RO, Johnson CR. Matrix Analysis. New York: Combridge University Press, 1985.
67. Hu B, Xu XP, Antsaklis PJ, et al. Robust stabilizing control laws for a class of second-order switched systems. Syst Contr Lett 1999; 38(3): 197–207.
68. Hu B, Zhai GS, Michel AN. Common quadratic Lyapunov-like functions with associated switching regions for two unstable second-order LTI systems. Int J Contr 2002; 75(14): 1127–1135.
69. Hu B, Zhai GS, Michel AN. Hybrid static output feedback stabilization of second-order linear time-invariant systems. Linear Algebra Applic 2002; 351-352: 475–485.
70. Hybrid System I-V. Lecture Notes in Computer Science. Berlin: Springer-Verlag, 1993, 1995-97, 1999.
71. Hybrid Systems: Computation and Control. Lecture Notes in Computer Science. Berlin: Springer-Verlag, 1998-2002.
72. Ingalls B, Sontag ED, Wang Y. An infinite-time relaxation theorem for differential inclusions. Proc Amer Math Soc 2003; 131: 487–499.
73. Ishii H, Francis BA. Stabilizing a linear system by switching control with dwell time. IEEE Trans Automat Contr 2002; 47(12): 1962–1973.
74. Isidori A. Nonlinear Control Systems (3rd ed.). London: Springer-Verlag, 1995.
75. Johnson CR, Kerr MK, Stanford DP. Semipositivity of matrices. Linear Multilinear Algebra 1994; 27: 265–271.
76. Johnson TL. Synchronous switching linear systems. In: Proc of 24th IEEE CDC 1985; 1699-1700.
77. Kailath T. Linear Systems. Englewood Cliffs: Prentice Hall, 1980.
78. Kaplan W. Introduction to Analytic Functions. Reading: Addison-Wesley, 1966.
79. Khalil HK. Nonlinear Systems (2nd ed.). Upper Saddle River: Prentice Hall, 1996.
80. Kleinman D, Athans M. The design of suboptimal linear time-varying systems. IEEE Trans Automat Contr 1968; 13(2): 150–159.
81. Kleinman D, Fortmann T, Athans M. On the design of linear systems with piecewise-constant feedback gains. IEEE Trans Automat Contr 1968; 13(4): 354–361.
82. Kolmanovsky I, McClamroch NH. Developments in nonholonomic control problems. IEEE Contr Syst Mag 1995; 15(6): 20–36.
83. Kolmanovsky I, McClamroch NH. Hybrid feedback laws for a class of cascade nonlinear control systems. IEEE Trans Automat Contr 1996; 41(9): 1271–1282.
84. Kozin F. A survey of stability of stochastic systems. Automatica 1969; 5(1): 95–112.
85. Krastanov MI, Veliov VM. On the controllability of switching linear systems. Automatica 2005, to appear.
86. Laffey TJ. Simultaneous triangularisation of matrices–low rank case and the nonderogatory case. Linear Multilinear Algebra 1978; 6(1): 269–305.

87. Leith D, Shorten R, Leithead W, Mason O, Curran P. Issues in the design of switched linear systems: A benchmark study. Int J Adapt Contr 2003; 17(2): 103–118.

88. Leonessa A, Haddad WM, Chellaboina V. Nonlinear system stabilization via hierarchical switching control. IEEE Trans Automat Contr 2001; 46(1): 17–28.

89. Li ZG, Wen CY, Soh YC. Observer-based stabilization of switching linear systems. Automatica 2003; 39(3): 517–524.

90. Liberzon D. Switching in Systems and Control. Boston: Birkhauser, 2003.

91. Liberzon D, Hespanha J, Morse AS. Stability of switched linear systems: a Lie-algebraic condition. Technical report, CVC-9801, Yale University, 1998.

92. Liberzon D, Morse AS. Basic problems in stability and design of switched systems. IEEE Contr Syst Mag 1999; 19(5): 59–70.

93. Lin Y, Sontag ED, Wang Y. A smooth converse Lyapunov theorem for robust stability. SIAM J Contr Optimiz 1996; 34(1): 124–160.

94. Lincoln B, Bernhardsson B. LQR optimization of linear system switching. IEEE Trans Automat Contr 2002; 47(10): 1701–1705.

95. Linz P. An Introduction to Formal Languages and Automata (3rd ed.). Sudbury: Jones and Bartlett, 2001.

96. Loparo KA, Aslanis JT, Hajek O. Analysis of switching linear systems in the plane, Part 2, Global behavior of trajectories, controllability and attainability. J Optim Theory Appl 1987; 52(3): 395–427.

97. Mariton M. Jump Linear Systems in Automatic Control. New York: Marcel Dekker, 1990.

98. McClamroch NH, Kolmanovsky I. Performance benifits of hybrid control design for linear and nonlinear systems. Proc IEEE 2000; 88(7): 1083–1096.

99. Michel AN. Recent trends in the stability analysis of hybrid dynamical systems. IEEE Trans Circ Syst I 1999; 46(1): 120–134.

100. Michel AN, Hu B. Towards a stability theory of general hybrid dynamical systems. Automatica 1999; 35(3): 371–384.

101. Molchanov AP, Pyatnitskiy YeS. Criteria of asymptotic stability of differential and difference inclusions encountered in control theory. Syst Contr Lett 1989; 13(1): 59–64.

102. Morari M, Baotic M, Borrelli F. Hybrid systems modelling and control. Euro J Contr 2003; 9(2–3): 177–189.

103. Morgan B Jr. The synthesis of linear multivariable systems by state-variable feedback. IEEE Trans Automat Contr 1964; 9(4): 405–411.

104. Mori Y, Mori T, Kuroe Y. A solution to the common Lyapunov function problem for continuous-time systems. In: Proc. IEEE Conf. Decision Contr. 1997; 3530–3531.

105. Morse AS. Supervisory control of families of linear set-point controllers- Part 1: Exact matching. IEEE Trans Automat Contr 1996; 41(10): 1413–1431.

106. Morse AS, Pantelides CC, Sastry SS, Schumacher JM (eds.). Special Issue on Hybrid Systems. Automatica 1999; 35(3): 347–535.

107. Nair GN, Evans RJ. Exponential stabilisability of finite-dimensional linear systems with limited data rates. Automatica 2003; 39(4): 585–593.

108. Narendra KS, Balakrishnan J. A common Lyapunov function for stable LTI systems with commuting A-matrices. IEEE Trans Automat Contr 1994; 39(12): 2469–2471.

109. Narendra KS, Balakrishnan J. Adaptive control using multiple models. IEEE Trans Automat Contr 1997; 42(2): 171–187.

110. Narendra KS, Xiang C. Adaptive control of discrete-time systems using multiple models. IEEE Trans Automat Contr 2000; 45(9): 1669–1686.
111. Nijmeijer H, Van der Schaft AV. Nonlinear Dynamical Control Systems. New York: Springer-Verlag, 1990.
112. Pettersson S, Lennartson B. Exponential stability of hybrid systems using piecewise quadratic Lyapunov functions resulting in an LMI problem. In: Proc 14th IFAC World Congress 1999.
113. Radjavi H, Rosenthal P. Simultaneous Triangularization. New York: Springer-Verlag, 1999.
114. Rantzer A, Johansson M. Piecewise linear quadratic optimal control. IEEE Trans Automat Contr, 2003; 45(4): 629–637.
115. Ravindranathan M, Leitch R. Model switching in intelligent control systems. Artif Intellig Engin 1999; 13(2): 175–187.
116. Riedinger P, Kratz F, Iungand C, Zanne C. Linear quadratic optimization for hybrid systems. In: Proc 38th IEEE CDC 1999; 3059–3064.
117. Savkin AV, Evans RJ. Hybrid Dynamical Systems: Controller and Sensor Switching Problems. Boston: Birkhauser, 2002.
118. Savkin AV, Skafidas E, Evans RJ. Robust output feedback stabilizability via controller switching. Automatica 1999; 35(1): 69-74.
119. Seidman TI. Optimal control for switching systems. In: Proc 21st Annual Conf Inform Sci Syst 1987; 485–489.
120. Shamma JS, Xiong DP. Set-valued methods for linear parameter varying systems. Automatica 1999; 35(6): 1081–1089.
121. Shorten RN, Narendra KS. On the existence of a common quadratic Lyapunov functions for linear stable switching systems. In: Proc 10th Yale Workshop Adapt Learn Syst 1998.
122. Shorten RN, Narendra KS. On common quadratic Lyapunov functions for pairs of stable LTI systems whose system matrices are in companion form. IEEE Trans Automat Contr 2003; 48(4): 618–621.
123. Shorten RN, Narendra KS, Mason O. A result on common quadratic Lyapunov functions. IEEE Trans Automat Contr 2003; 48(1): 110–113.
124. Skafidas E, Evans RJ, Savkin AV, Petersen IR. Stability results for switched controller systems. Automatica 1999; 35(4): 553–564.
125. Smirnov GV. Introduction to the Theory of Differential Inclusions. Providence: AMS, 2002.
126. Stanford DP. Stability for a multi-rate sampled-data system. SIAM J Contr Optimiz 1979; 17(3): 390–399.
127. Stanford DP, Conner LT Jr. Controllability and stabilizability in multi-pair systems. SIAM J Contr Optimiz 1980; 18(5): 488–497.
128. Stanford DP, Conner LT Jr. Addendum: Controllability and stabilizability in multi-pair systems. SIAM J Contr Optimiz 1981; 19(5): 708–709.
129. Stanford DP, Urbano JM. Some convergence properties of matrix sets. SIAM J Matrix Anal Appl 1994; 14(4): 1132–1140.
130. Sun Z. Stabilizability and insensitiveness of switched systems. IEEE Trans Automat Contr 2004; 49(7): 1133-1137.
131. Sun Z. A modified stabilizing law for switched linear systems. Int J Contr 2004; 77(4): 389–398.
132. Sun Z. Combined stabilizing strategies for switched linear systems. Submitted to IEEE Trans Automat Contr, provisionally accepted.

133. Sun Z. Sampling and control of switched linear systems. J Franklin Inst 2004; 341: 657–674.

134. Sun Z. Canonical forms of switched linear control systems. In: Proc Amer Contr Conf 2004; 5182–5187.

135. Sun Z. Reachability analysis of switched linear systems with switching/input constraints. IFAC World Congress 2005, accepted.

136. Sun Z. Stabilization and optimal switching of switched linear systems, submitted to Automatica, 2004.

137. Sun Z. Feedback equivalence and canonical forms of switched linear systems. Preprint, May, 2003.

138. Sun Z, Ge SS. Sampling and control of switched linear systems. In: Proc 41st IEEE CDC 2002; 4413–4418.

139. Sun Z, Ge SS. Dynamic output feedback stabilization of a class of switched linear systems. IEEE Trans Circ Syst I 2003; 50(8): 1111–1115.

140. Sun Z, Ge SS. Nonregular feedback linearization: A nonsmooth approach. IEEE Trans Automat Contr 2003; 48(10): 1772–1776.

141. Sun Z, Ge SS. Analysis and synthesis of switched linear control systems. Automatica 2005; 41(2): 181–195.

142. Sun Z, Ge SS, Lee TH. Reachability and controllability criteria for switched linear systems. Automatica 2002; 38(5): 775–786.

143. Sun Z, Shorten R. On convergence rates of switched linear systems. In: Proc 42nd IEEE CDC 2003; 4800–4805.

144. Sun Z, Xia X. On nonregular feedback linearization. Automatica 1997; 33(7): 1339–1344.

145. Sun Z, Zheng D. On reachability and stabilization of switched linear control systems. IEEE Trans Automat Contr 2001; 46(2): 291–295.

146. Sussmann HJ. New theories of set-valued differentials and new versions of the maximum principle of optimal control theory. In: Isidori A, Lamnabhi-Lagarrigue F, Respondek W (eds.) Nonlinear Control in the Year 2000, Vol 2. London: Springer-Verlag, 2001; 487–526.

147. Sworder DD. On the control of stochastic systems: I. Int J Contr 1967; 6(2): 179–188.

148. Sworder DD. On the control of stochastic systems: II. Int J Contr 1969; 10(3): 271–277.

149. Sworder DD. Control of systems subject to sudden changes. Proc IEEE 1976; 64(8): 1219–1225.

150. Szigeti F. A differential-algebraic condition for controllability and observability of time varying linear systems. In: Proc 31st IEEE CDC 1992; 3088–3090.

151. Tanaka K, Sano M. A robust stabilization problem of fuzzy control systems and its application to backing up control of a truck-trailer. IEEE Trans Fuzzy Syst 1994; 2(2): 119-134.

152. Tokarzewski J. Stability of periodically switched linear systems and the switching frequency. Int J Syst Sci 1987; 18(4): 697–726.

153. Troch I. Comments on 'Driving a linear constant system by piecewise constant control'. Int J Syst Sci 1990; 21(11): 2379–2386.

154. Tsinias J, Kalouptsidis N. Transforming a controllable multiinput nonlinear system to a single input controllable system by feedback. Syst Contr Lett 1981; 1(3): 173–178.

155. Tsinias J, Kalouptsidis N. Controllable cascade connections of nonlinear systems. Nonlinear Anal Theory Method Appl 1987; 11(11): 1229–1244.

156. Van der Schaft AJ, Schumacher, JM. An Introduction to Hybrid Dynamical Systems. New York: Springer-Verlag, 2000.

157. Vidyasagar M. Nonlinear Systems Analysis (2nd ed.). Eaglewood Cliffs: Prentice Hall, 1993.

158. Wicks MA, Peleties V, DeCarlo RA. Construction of piecewise Lyapunov functions for stabilizing switched systems. In: Proc 33rd IEEE CDC 1994; 3492–3497.

159. Wicks MA, Peleties P, DeCarlo RA. Switched controller synthesis for the quadratic stabilization of a pair of unstable linear systems. Euro J Contr 1998; 4(2): 140–147.

160. Wonham WM. Linear Multivariable Control – A Geometric Approach. Berlin: Spinger-Verlag, 1979.

161. Xie GM, Wang L. Controllability and stabilizability of switched linear-systems. Syst Contr Lett 2003; 48(2): 135–155.

162. Xie GM, Zheng DZ, Wang L. Controllability of switched linear systems. IEEE Trans Automat Contr 2002; 47(8): 1401-1405.

163. Xu X, Antsaklis PJ. On the reachability of a class of second-order switched systems. Technical report ISIS-99-003, University of Notre Dame, 1999.

164. Xu XP, Antsaklis PJ. Stabilization of second-order LTI switched systems. Int J Contr 2000; 73(14): 1261–1279.

165. Xu X, Antsaklis PJ. Optimal control of switched systems: new results and open problems. In: Proc Amer Contr Conf 2000; 2683–2687.

166. Xu X, Antsaklis PJ. Optimal control of switched systems via nonlinear optimization based on direct differentiations of value functions. Int J Contr 2002; 75(16-17): 1406–1426.

167. Xu X, Antsaklis PJ. Optimal control of switched systems based on parameterization of the switching instants. IEEE Trans Automat Contr 2004; 49(1): 2–16.

168. Yang Z. An algebraic approach towards the controllability of controlled switching linear hybrid systems. Automatica 2002; 38(7): 1221–1228.

169. Zaanen AC. Linear Analysis. Amerterdam: North Holland, 1960.

170. Zhai GS, Hu B, Yasuda K, Michel AN. Disturbance attenuation properties of time-controlled switched systems. J Franklin Inst 2001; 338(7): 765–779.

171. Zhai GS, Lin H, Antsaklis PJ. Quadratic stabilizability of switched linear systems with polytopic uncertainties. Int J Contr 2003; 76(7): 747–753.

172. Zhai GS, Takai S, Michel AN, Xu X. Improving closed-loop stability of second-order lti systems by hybrid static output feedback. Int J Hybrid Syst 2003; 3(2-3): 237–250.

173. Zhao J. Hybrid control for global stabilization of a class of systems. In: Advanced Topics in Nonlinear Control Systems 2001; Singapore: World Scientific; 129–160.

174. Zhao J, Dimirovski GM. Quadratic stability of a class of switched nonlinear systems. IEEE Trans Automat Contr 2004; 49(4): 574–578.

175. Zhao J, Spong MW. Hybrid control for global stabilization of the cart-pendulum system. Automatica 2001; 37(12): 1941–1951.

176. Zhao Q, Zheng D. Stable and real-time scheduling of a class of hybrid dynamic systems. J DEDS 1999; 9(1): 45–61.

Index